PART II

MAINTENANCE AND

OPERATIONS OF

BUILDINGS AND

GROUNDS

Coordinator:
Gary L. Reynolds
The Colorado College

01/07

APPA: The Association for Higher Education Facilities Officers, 1643 Prince Street, Alexandria, Virginia 22314-2818

Printed in the United States of America
Library of Congress Catalog Card Number: 97-73729
International Standard Book Number: 0-913359-99-8

Editor: Steve Glazner

Editorial Production and Typography: Wolf Publications, Inc.
Printing: Port City Press, Inc.

Printed on acid-free paper

CONTENTS

Part II

SECTION II-A

PLANT MANAGEMENT

Editor:
Paul F. Tabolt
University of Colorado at Boulder

INTRODUCTION

Plant Management

This section ventures from planning for the expected to planning for the unexpected. The chapters portray a level of sophistication in the facilities profession that helps clarify where a facilities organization lies on its path toward excellence. An organization in the early stages of its evolution can read these chapters to chart a course of action for the future. A mature organization can visit and revisit them to navigate a more immediate course of action. A declining organization, or an organization that has surpassed its peak performance, can read them to realign its path with commonly accepted facilities management practices.

If I were to advise you of the sequence you should follow to read these chapters, I would begin with Chapter 21, "Work Management Systems." After all, facilities officers can expect immediate demand for the management of incoming requests for service, repair, and alterations from the moment a building is opened for business. The proactive facilities professional will anticipate this demand and begin to lay the foundations for the preventive maintenance program so aptly defined by chapter author Jay Klingel. The underpinnings and system foundation Klingel describes can be put into place to plan, estimate, and deliver services to provide cost-effective and timely customer service. These systems are a necessity for the facilities professional to move to Chapter 19, "Capital Renewal and Deferred Maintenance Programs," and Chapter 20, "Facilities Condition Assessments."

We should be thankful for Harvey Kaiser's ongoing wisdom as he once again outlines the purpose and process of conducting a facilities assessment. Kaiser captures one of the most important management responsibilities for those who elect to perform a facilities assessment program—marketing the dilemma. Gaining support for the audit can be comparable to convincing the cobbler to make shoes for his or her children. Kaiser has helped many of us to establish a commitment to stewardship from the decision makers who must be aware of the risks and liabilities that arise from inadequate attention to facilities or neglect. He also diligently demonstrates the role the facilities professional can play in developing a comprehensive strategy for capital asset management and renewal.

Logically, Chapter 22, "Emergency Management," is the final chapter of this section. A facilities professional who does not read the other chapters in this section can expect this chapter to be required reading. Certainly not all emergencies are of one's own making, but the absence of proactive facilities management can lead to crisis. The program Richard Fowler describes in this chapter can save lives and property and can reduce the time it takes for recovery from an emergency or disaster. Planning in advance for the emergency or disaster is easier for all of us when we realize that it is not a matter of whether disaster will strike our campus; rather, it is a question of when.

—Paul F. Tabolt

CHAPTER 19

Capital Renewal and Deferred Maintenance Programs

Harvey H. Kaiser
Harvey H. Kaiser & Associates

19.1 INTRODUCTION

Facilities management responsibilities include programs for acquiring new capital assets; programs governing operations, maintenance, and repairs; capital renewal programs; and deferred maintenance programs. Organizational structures for managing these programs vary. Larger institutions may divide the responsibilities between operations and maintenance management and facilities planning departments; at smaller institutions a facilities department may manage both responsibilities. Because the programs emerge from two concepts—accounting and plant operations—there is a built-in confusion about their meanings. The issue of deferred maintenance further complicates clear delineation of terms and operational applications to achieve the basic goal: extending the life of existing facilities or replacing them with new facilities.

This chapter discusses the overall topic of capital renewal and deferred maintenance programs within the context of integrated financial planning for capital assets.

19.2 HISTORICAL PERSPECTIVE

Deferred Maintenance

Today's buildings, grounds, infrastructure, and equipment—amassed to house and support academic programs—are the legacy of the dramatic growth of new and existing campuses. More than half of the current campus facilities organizations were developed after World War II, when enrollment grew 600 percent, from 2.3 million students in 1950 to more than 14 million students by 1995. During the same period, the total number of institutions grew from 1,800 to 3,768. New institutions, primarily publicly supported two-year colleges, opened at a rate of one every two weeks from 1955 to 1974.

Facilities also expanded dramatically during this period. Of the current total of more than 4 billion gross sq. ft. of space, it is estimated that 2.5 billion was built between 1950 and 1990. Buildings were opened annually, sometimes several in the same year, and construction cranes on the skyline and mud-covered walkways were common campus experiences. Driven by financing demands for capital construction and tight delivery schedules, many of the new facilities evidence a lack of durability and adaptability for alternative uses.

The major share of construction was for new facilities, with little reinvestment allocated for existing facilities. The glamour of planning new facilities and the excitement of ribbon-cutting ceremonies overshadowed the obsolescence and decay of earlier campus buildings. "Old Main" was slowly fraying at the edges while newer facilities entered their own cycle of deterioration. Currently, many institutions not only are facing the challenge of an increasing debt burden due to an aging facilities organization, but are also planning to replace or renew their facilities using debt.

The burdensome problems of major maintenance and capital renewal/ replacement have troubled higher education since the 1970s. The term *deferred maintenance* emerged in the early 1970s as college and university administrators began to recognize the serious nature of plant problems on their campuses. The deteriorated plant conditions produced by ignoring older facilities during higher education's post–World War II expansion were compounded by the following:

- Poor designs for institutional durability
- Cost cutting that rapidly produced space with inferior construction techniques, and innovative materials that showed early failures
- Soaring utility costs
- Inflation-reduced operations and maintenance budget reductions
- Inadequate funding for capital renewal and major maintenance
- Increased government regulations resulting in reallocation of resources and further deferral of maintenance

A Shift in Focus

The early alerts to the deferred maintenance problem were heralded in news magazine stories in the late 1970s. The graphic portrayals of scaffold-protected exterior walls, students huddled in overcoats in unheated classrooms, and laboratories with "Closed for Repair" signs introduced an awareness of the issues. Coincidentally, a national debate on the infrastructure crisis illustrated by crumbling bridges, collapsing utilities, and disasters involving loss of life reinforced higher education's problems.

The 1980s saw campuses, through a variety of initiatives by legislators, governing board members, campus presidents, business officers, and facilities directors, begin surveys of conditions. The availability of a tool for inspecting facilities conditions was jointly sponsored by APPA: The Association of Higher Education Facilities Officers and the National Association of College and University Business Officers (NACUBO) in 1982, with publication of the *Facilities Audit Workbook*. This simple format, building on work by the Tennessee Board of Higher Education and military agencies, described a process that produced comparative ratings of campus facilities conditions. The Association of Governing Boards of Colleges and Universities's (AGB) publication of *Crumbling Academe* (1984), NACUBO and APPA's *The Decaying American Campus*[1] (1988), NACUBO's *Managing the Facilities Portfolio* (1990), and APPA's *A Foundation to Uphold* (1996) were supported by seminars and journal articles charting the way to proceed. APPA's *The Facilities Audit* (1993) provided a cost-deficiency technique to measure the extent of deferred maintenance and guidelines for capital renewal.

Results of actions varied, with some campuses responding with effective deferred maintenance reduction programs, others moving the problem onto priority agendas to seek funding, and still others struggling with overwhelming costs and stalled on directly addressing the issue. Where there were successes, they can be attributed to a determination to break the cycle of facilities deterioration. Those institutional leaders who sponsored unpopular positions of reallocating resources saw ways to develop multiple funding sources.

Perceptive facilities and financial managers, seeing that the phrase *deferred maintenance reduction* presented semantic barriers and resistance to acceptance, sought other descriptive terms. Sometimes wrought with pessimism, frustration, and cynicism by business officers and facilities managers, *deferred maintenance* was a pejorative term. It spoke of failures in management, judgment, and stewardship, as though deferred maintenance was a trait of inept management. What followed was a shift in focus to programs of comprehensive financial and facilities planning, adopting the terms *capital renewal, facilities equilibrium,* and *capital asset management.* They all offer opportunities for programs consistent with institutional strategic planning.

19.3   DEFINITIONS

Concurrent Facilities Maintenance Programs

An institution may be simultaneously conducting programs for major maintenance, capital renewal, and deferred maintenance reduction, along with operations and maintenance activities. The decision to create separate programs is not unusual and usually evolves from different sources of funding and/or different components of facilities management designated to supervise the programs.

Defining terms in their general usage will help in the process of selecting priorities for work and budgeting practices. Consistency in terms also assists in separating different budget components to aggregate overall needs required for comprehensive financial planning for facilities. Terms used in this chapter are defined as follows:

- *Operations*: Activities related to normal performance of the functions for which a building is used (i.e., utilities, janitorial services, waste treatment, etc.).
- *Maintenance*: Work necessary to realize the originally anticipated life of a fixed asset, including buildings, fixed equipment, and infrastructure.
- *Repairs*: Work to restore damaged or worn-out facilities to normal operating condition. Repairs are curative, whereas maintenance is preventative.
- *Replacements*: An exchange of one fixed asset for another that has the same capacity to perform the same function. In contrast to repair, replacement generally involves a complete identifiable item of reinvestment (i.e., a major building component or subsystem).
- *Alterations*: Work performed to change the interior arrangements or other physical characteristics of an existing facility or fixed equipment so that it can be used more effectively for its current designated purpose or adapted to a new use.

Major Maintenance

The Classification of Accounts jointly developed by APPA and NACUBO provides the following definition for *major repairs and renovations:* ". . . expenditures for those major jobs or projects that must be accomplished but are not funded by normal maintenance resources received in the annual operating budget cycle" (see Section 6 of Appendix 15-A, "Major Repairs and Renovations"). The distinction between major repairs and minor repairs should be defined by the institution. The key ingredients in this definition are the source of funds and the institutionally set cost limits for the lowest value of major maintenance and highest value for minor repairs.

Major maintenance is sometimes included as a routine part of current fund operations and maintenance. However, an institutional limit for the cost of maintenance work can shift the designation to the category of a capitalized project. Thus, an accounting decision can distinguish capital renewal from major maintenance. Rules are not fixed on the distinctions between the two categories, which can lead to confusion in allocating projects for selection of funding priorities.

Capital Renewal and Replacement

Capital renewal and replacement is defined as a systematic management process to plan and budget for known cyclic repair and replacement requirements that extend the life and retain usable condition of facilities and systems and are not normally contained in the annual operating budget. Capital renewal is a planned investment program that ensures that facilities will function at levels commensurate with the academic priorities and missions of an institution. Included are major building and infrastructure systems and components that have a maintenance cycle in excess of one year.

Renewal and replacement is an accounting term used to distinguish a subgroup of plant fund assets from capitalized plant additions and improvements. However, institutional accounting practices vary; decisions are sometimes made to capitalize portions of major maintenance and renewal and replacement. Replacements in the form of new construction are routinely designated as capitalized and are grouped together with renewals as capital renewal and replacement programs. As a form of capitalized construction, replacements are interchangeable with new construction, whether they are actually replacing an existing facility or are an addition to plant. Linking capital renewals with replacements is a more accurate way to describe a program for renewal of existing plant assets as distinguished from totally new additions to plant assets.

The scope, complexity, cost, and duration of a project can dictate whether major maintenance should be supervised by maintenance management or by a separate design and construction department. As an alternative to using in-house maintenance and design staff, a major maintenance project requiring plans, specifications, and competitive bidding can be designed by consultants and constructed by contractors. Capital renewal and replacement usually requires external assistance in design and construction administration to avoid dedicating facilities management staff to lengthy, time-consuming projects. Regardless of the choice made, major maintenance and capital renewal and replacement require supervision by facilities management staff to coordinate campus conditions (i.e., access during construction, interim relocations, utilities) and assure project delivery in conformance with specifications, budgets, and schedules.

Deferred Maintenance

Deferred maintenance is defined as maintenance work that has been deferred on a planned or unplanned basis to a future budget cycle or postponed until funds are available. Roof replacements, major building component repairs, mechanical equipment, underground utilities, and roads and walkways are projects that are often deferred to the next annual funding cycle.

This definition could serve just as well for major maintenance and offers a temptation to bypass the use of annual operating budgets and fund major maintenance through a deferred maintenance reduction program. The difference is that a deferred maintenance program is a comprehensive, one-time approach, often extended over several years, to control a massive backlog of maintenance work.

Deferred maintenance reduction programs result from a campus policy to group deferred major maintenance projects, and sometimes other plant needs, into a program funded separately from major maintenance or capital renewal and replacement.

Major maintenance and deferred maintenance are expenditure programs designed to accommodate the deterioration process of facilities; both programs cope with facilities renewal. As a strategy to achieve funding to eliminate problems of facilities deterioration, deferred maintenance reduction programs can be expanded to include life safety, code compliance requirements, and provisions for accessibility. In contrast, major maintenance is a planned activity of facilities renewal funded by the annual operating budget. Failure to perform needed repair, maintenance, and renewal as part of normal maintenance management creates deferred maintenance.

Functional Improvements

Space modifications to accommodate program needs, sometimes called *program improvements* or *alterations,* are often overlooked in budgeting for facilities needs. It is common to use maintenance funds as the only available source for improvements. Thus, an erosion in facilities maintenance results from this practice. A preferred practice is to set up a specific budget line item for functional improvements and to attempt to coordinate major maintenance projects into planning. For example, a revision to a suite of laboratories could include replacements for heating, ventilation, and air conditioning (HVAC); electrical systems; and plumbing systems, creating a project with a larger scope than the space modification project.

19.4   BUDGETING FOR CAPITAL ASSETS

Magnitude of the Capital Renewal/Replacement and Deferred Maintenance Problem

Questions repeatedly asked by federal and state agencies, legislators, governing boards, foundations, and corporate donors are, "How large is the problem of facilities deterioration?" and "How much is needed to remedy conditions?"

As reported by APPA in *A Foundation to Uphold,*[3] the total accumulated deferred maintenance for all 3,768 higher education institutions in

1994 was $26 billion, with $5.7 billion defined as "urgent" needs. The 1994 data provided by a comprehensive survey identifies deferred maintenance backlogs by Carnegie classification. Changes in conditions after a baseline year of 1988, the publication date of *The Decaying American Campus*,[2] vary among institutions. For example, more than half the institutions reported increases in accumulated deferred maintenance, approximately one-fourth saw declines, and the remainder either stayed the same or did not report the status of condition.

Benchmarking data highlights comparisons for conditions at similar institutions. The inability to fund deferred maintenance from current operating budgets exists for many institutions, and it is evident that external funding is necessary to assist in remedying unsatisfactory facilities conditions.

Faculty, staff, students, and alumni will testify to the needs reported in *A Foundation to Uphold*.[3] Prospective students and donors also convey the message by admission applications and contributions. The importance of facilities appearance to student recruitment is underscored by the 1987 study of the Carnegie Foundation for the Advancement of Teaching on how students choose a college. For 62 percent of the students surveyed, "the appearance of the buildings and grounds was the most influential factor during a campus visit."[3]

Federal and state legislators, campus administrators, and governing boards want more than a general estimate of need. Hard data is demanded to justify the claims for campus capital renewal and replacement and deferred maintenance needs. At the campus level, a response can be prepared by conducting detailed facilities audits and condition assessments. Rigorously prepared and creatively presented, often with persistence and persuasiveness, these surveys and assessments of conditions have proven effective in securing funding.

Liabilities of Deferred Maintenance

Renewal and replacement needs vary by region, building type, the extent of facilities use and abuse, and quality of original construction and maintenance management. Levels of current operating budgets and special appropriations for capital renewal and deferred maintenance also affect required funding levels. However, inevitably, building systems and components deteriorate and need replacement. Plumbing wears out, roofing breaks down and leaks, window frames warp, patched-up electrical wiring becomes dangerous, HVAC systems fail to heat or cool, and equipment can no longer be replaced.

Underfunding of major maintenance and capital renewal and replacement inevitably results in backlogs of deferred maintenance. Unsafe buildings and unreliable infrastructure create hazardous conditions.

Failing HVAC, electrical, and plumbing systems jeopardize the usability of spaces necessary for academic, student, and administrative activities. Unattractive building interiors deter enrollment strategies essential to tuition-dependent institutions. All of these factors add up to liabilities not shown on a college or university balance sheet.

Viewing campus facilities as liabilities rather than assets should change financial perspectives and encourage strategic plans for new construction, maintenance, and repairs budgeting and surveys of existing conditions to develop deferred maintenance reduction programs. Recognition that a significant portion of a campus's facilities are in unsatisfactory condition and are a substantial liability should spur action.

Funding Sources

Funds flow into facilities improvements from two funding sources— current and plant funds—for preventing deterioration and renewing or replacing facilities (Figure 19-1). *Current funds* routinely provide for major maintenance, and *plant funds* provide for capital improvements. Deferred maintenance is also supported by plant funds. An additional funding source, an annual renewal allowance, is budgeted from current funds.

An important reminder is that annual funding for facilities operations and maintenance is expected to accommodate major maintenance and thereby compensate for the aging process of facilities and equipment. Separate funding for functional improvements in operating budgets is necessary to protect funding of capital renewal. Major maintenance is typically treated as a residual category after budgeting for plant administration, building and equipment maintenance, custodial services, utilities, and grounds maintenance. The residual treatment—often leaving major maintenance and functional improvements unfunded—has proven to be inadequate to meet plant needs and is how most campuses reached their current levels of deferred maintenance. The preferred approach is to establish an appropriate level of funding for major maintenance and capital renewal in the operating budget to prevent continued obsolescence of facilities and equipment.

Annual allocations for facilities renewal can be made either in the major maintenance component of annual budgets for operations and maintenance or as a special line item with identified specific projects. The choice is made according to the strategy most acceptable to campus budgeting practices. The important principle for policy makers to remember is that a one-time elimination of deferred maintenance priorities does not solve the problem of facilities renewal. Campus facilities continue to deteriorate and become obsolete. An annual allocation for facilities renewal is required to prevent future accumulation of deferred maintenance. An appropriate level of funding established at the beginning of a comprehensive facilities funding

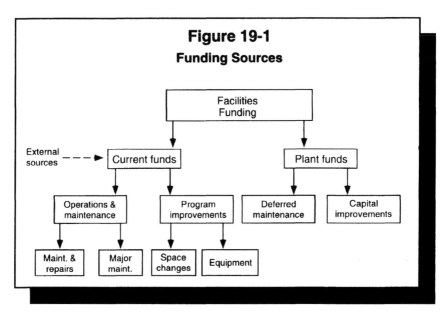

Figure 19-1

Funding Sources

program may have to include catch-up costs. As needs are reduced to manageable proportions, the operating budget can accommodate priorities as they are identified. The end result is a program that maintains campus facilities in good repair, functionally adequate for teaching research, campus life, and public service.

Integrating Maintenance and Repair Financial Planning

The goal of integrated financial planning is to ensure that new capital assets are acquired based on well-defined needs and are cost-effective additions to plant. This means that utilization of existing space is fully examined prior to commitments for planning and funding new construction. In addition, operations and maintenance, along with repairs, replacements, and renovating, must be fully funded to prevent accumulation of backlogs. Maintenance backlogs, when present, should be addressed through facilities audits and assessment of priorities to define funding needs. The integration of the components of facilities management responsibilities, discussed in this chapter as a model program, requires strategic planning and regular adjustments to fit changing conditions (Figure 19-2).

Budgeting vs. Financial Planning

The overall cost of maintaining an institution's facilities organization has received varied treatment in the past. Primary attention has been devoted to *budgeting* techniques for operations and maintenance, with an evident lack of guidelines that can be applied to the wide-ranging conditions and needs of higher education. Some attention has been given to allocations for annual capital renewal to forecast appropriate funds to offset deterioration, similar to the concept of an annual depreciation applied in corporate balance sheet entries for declining plant value. Omitted from these approaches are concepts of *financial planning* for capital asset productivity of facilities organizations that integrate acquisition of new assets, operations and maintenance, capital renewal, and deferred maintenance.

Planning budget requirements for major maintenance and capital renewal and replacement derives from principles of capital asset depreciation. It is an economic fact of life that facilities have a limited productive or useful life. The life cycle concept defines the useful life of a facility as the aggregate of the durability of individual building components and systems. In common terms, a building does not wear out all at once but fails gradually, by individual systems and components. A planned program of major maintenance and capital renewal by systems and components is necessary to restore deterioration and extend a facility's life.

A common budget management pitfall is the use of major maintenance and capital renewal funding for renovations. The first category deals with deteriorated facilities conditions and protection of capital assets; the second category addresses modifications for functional inadequacies, obsolescence, and new academic or other program improvements. As final

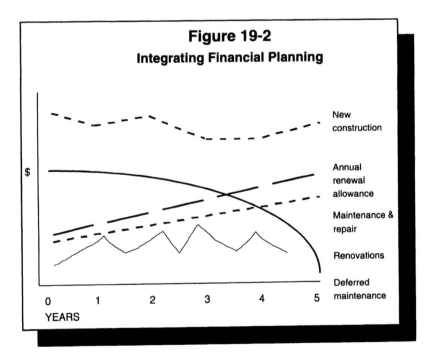

Figure 19-2
Integrating Financial Planning

budget decisions are reached, pressure is often applied to shift major maintenance and renewal funding to meet program needs. This is an element of "budget politics" that compromises protection of plant assets.

Major Maintenance and Capital Renewal and Replacement Strategy

The key components of a major maintenance and capital renewal and replacement strategy are as follows:

- Planning new construction and/or acquisition of capital assets based on utilization of existing assets and evaluation of alternatives for adaptation of existing space
- Funding maintenance and repairs at adequate levels to avoid accumulation of backlogs
- Conducting facilities audits and assessments of conditions
- Prioritizing critical deferred maintenance needs
- Developing multiple funding sources for capital renewal
- Adjusting operating and capital budgeting practices to emphasize maintenance and capital renewal

A capital renewal strategy results in a plan with the following characteristics:

- A continually adjusted process
- A short-term need for deferred maintenance and long-term needs for life cycle renewal
- A coordinated facilities improvement program, funding plan, and monitoring method

19.5   ESTIMATING MAJOR MAINTENANCE AND CAPITAL RENEWAL/REPLACEMENT FUNDING REQUIREMENTS

Funding Requirements

A campus's awareness that deteriorating facilities conditions have reached the point of significant liability immediately opens the question: How much is needed to correct the problem?

An effective capital renewal and deferred maintenance reduction program requires reliable estimates of funding requirements and thorough planning. A successful program should estimate funding needs in the following categories:

- Long-term capital renewal needs
- Estimates of short-term programs to reduce deferred maintenance backlogs to acceptable levels

Long-term and short-term needs should be identified concurrently for an institution to achieve desired goals for capital renewal and deferred maintenance reduction programs. This approach recognizes that 1) facilities conditions continually deteriorate over time and require ongoing investments to maintain functional and financial value and 2) historical facilities underfunding problems must be addressed through a short-term remedial program of deferred maintenance reduction.

Concerns for the condition of the nation's infrastructure resulted in a study by the Building Research Board of the National Research Council. In *Committing to the Cost of Ownership: Maintenance and Repair of Public Buildings*,[4] the Committee on Advanced Maintenance Concepts for Buildings examined issues of financial planning for facilities. The committee's study addressed an array of aspects of the costs of acquiring, maintaining, and replacing facilities to guide financial planning for integrating maintenance and repairs and the backlog reduction of deferred maintenance.

The Building Research Council's conclusions and recommendations are based on the finding that *underfunding of maintenance and repair is a widespread and persistent problem*. To overcome this problem, maintenance and repair budgets should be structured to explicitly identify the expenditures associated with routine maintenance and repair and activities to reduce the backlog of deferred maintenance. The council concluded that an appropriate total budget allocation for routine maintenance *and* capital renewal is in the range of 2 to 4 percent of the aggregate current replacement value of those facilities (excluding major infrastructure). When a backlog of deferred maintenance has been allowed to accumulate, spending must exceed this minimum level until the backlog has been eliminated.

The specific percentage for a facility depends on a wide range of factors, and the relationship between maintenance and repair requirements and current replacement value may vary widely, for any one building may be outside the proposed range (Figure 19-3). The 2 to 4 percent range is most valid as a budget guide for a large inventory of buildings and over time periods of several years. However, even with small inventories, the 2 to 4 percent rule of thumb may be applied over a longer period of time, such as 5 to 10 years. An important and often misunderstood point is that this range *does not include* "one-time" funding to reduce deferred maintenance backlogs.

In addition to the council's conclusions regarding overall routine maintenance and capital renewal annual funding, the results of empirical studies of the life cycles of individual components provide general parameters for annual capital renewal allowances, separate from maintenance. Acknowledging variances for ages and types of facilities, a recommended range for the annual capital renewal component of total is 1.5 to 3 percent of the total replacement value of plant. Some evaluations of plant conditions and needs recommend higher ranges. For example, a research-intensive institution will have a high rate of obsolescence and deterioration owing to changing technologies and usage of facilities. Institutions that have implemented a deferred maintenance reduction program will see benefits in lower capital renewal and replacement needs.

In summary, the range of 2 to 4 percent for total funding *includes* components of 0.5 to 2.5 percent for maintenance and repairs and 1.5 to 3 percent for capital renewal. These ranges heighten the importance of an accurate forecast for annual capital renewal allowance and accurate condition assessments to determine additional needs for deferred maintenance.

Selecting the Appropriate Method for Estimating Capital Funding Needs

Selecting the appropriate method for estimating an annual renewal allowance forecast of capital renewal funding needs (Figure 19-4) requires an understanding of an organization's fiscal planning needs and available resources for estimating.

The goal is to provide an adequate, realistic budget reserve for allocating funds for specific projects. As projects are identified, funds are allocated and expended as required. An allowance provides flexibility in determining which projects will be funded in any given year and gives facilities managers confidence that funds will be available to meet capital renewal needs.

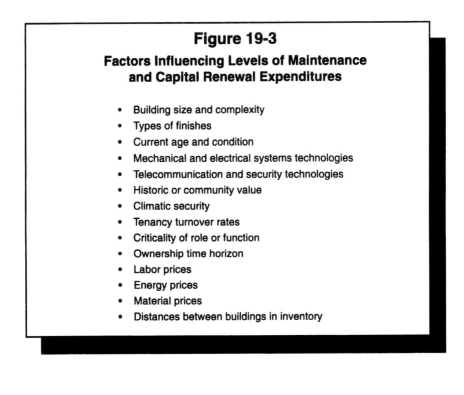

Figure 19-3

Factors Influencing Levels of Maintenance and Capital Renewal Expenditures

- Building size and complexity
- Types of finishes
- Current age and condition
- Mechanical and electrical systems technologies
- Telecommunication and security technologies
- Historic or community value
- Climatic security
- Tenancy turnover rates
- Criticality of role or function
- Ownership time horizon
- Labor prices
- Energy prices
- Material prices
- Distances between buildings in inventory

Several methods for forecasting annual renewal allowances have been outlined by Sherman and Dergis in *A Funding Model for Building Renewal*, Harvey H. Kaiser in *Crumbling Academe*, Cushing Phillips Jr., in *Facilities Renewal: The Formula Approach*, and NACUBO's *Managing the Facilities Portfolio*. Estimating methods are characterized by the amount of detail and level of refinement needed for an annual renewal allowance for deferred maintenance project cost estimates.

Methods for estimating *annual renewal allowance* to offset facilities deterioration are based on data developed from building and infrastructure characteristics. Calculations that determine either a dollar amount for annual funding or percentages of total plant replacement value are derived from life cycles of building and infrastructure systems and components. This approach provides an overall estimate of needs and can be done with minimal field inspections of conditions. It is also cost-effective because of the reduced effort involved in gathering data and generating calculations.

The Sherman and Dergis method is a simple but effective set of calculations based on building age. Data requirements are minimal and forecasts are provided for individual buildings and total campus needs.

Another method for estimating an annual renewal forecast is based on calculations of building and infrastructure component renewal cycles. The underlying concept is that building components age at different rates and require different renewal cycles. Facilities renewal forecasts for buildings and infrastructure are developed with the following procedures:

1. *Building attributes.* For each building, the following attributes are identified:

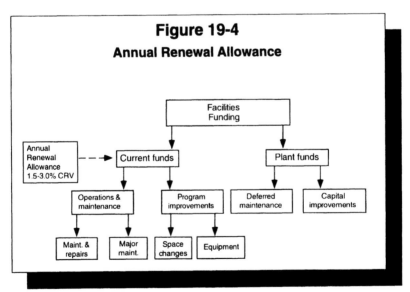

Figure 19-4

Annual Renewal Allowance

- Type of construction (reinforced concrete, membrane roofing, etc.)
- Building use (library, laboratory, residence hall, etc.)
- Gross area (e.g., 48,000 gross sq. ft.)
- Current replacement cost

2. *Building component depreciation.* A component depreciation table is prepared for each different building type (Figure 19-5):

- Percentage cost of gross building value
- Life cycle (e.g., membrane roofing—20 years)
- Renewal percentage at the end of the life cycle
- Individual component renewal profiles

3. *Present plant condition.* Each major building system (component) is assessed to account for the present condition of each building. The condition is expressed as a percentage used of the component relative to its life cycle. For example, if the HVAC system has a 20-year useful life but was completely renovated 2 years ago, its percentage used is 10 percent.

4. *Life cycle cost calculation.* The anticipated replacement year(s) for each component is calculated using values for life cycle and the percentage used. The cost of replacement is calculated by using the building replacement cost; the percentage of each of each component in each building; and the percentage of the component replaced. This cost is assigned to the year(s) of anticipated replacement.

The following example illustrates the calculations:

Building	Sample hall
Building system	Roof
Life cycle	20 years
Renewal percentage at the end of life cycle	95 percent
Percentage used	50 percent
Sample Hall current replacement value	$3,540,000
Percentage of current replacement value of roof	4.3 percent
Projected life cycle period	50 years

Given that, in 1994, the roof was 50 percent through its lifetime of 20 years, its first replacement will occur in 10 years (2004), and subsequent replacements will occur in 20-year intervals after that (2024, 2044). The replacement cost is 95 percent of the value of the roof, which in turn is 4.3 percent of the replacement cost of the building ($3,540,000). Therefore, the replacement cost is:

$$95\% \times 4.3\% \times \$3,540,000 = \$144,609$$

The cost of $144,609 is assigned to years 2004, 2024, and 2044 for roof replacement on Sample Hall.

Figure 19-5
Sample Component Depreciation Table

BUILDING NAME Square foot replacement cost = $132

Building System	Renewal Cycle (years)	Percentage of Total Building Replaced	Percentage of System
Foundations	100	10.52	5
Vertical support	100	0.00	5
Floor	75	0.00	5
Roof	75	18.39	10
Roofing	20	5.95	100
Exterior walls	75	8.03	65
Windows	30	2.78	80
Exterior doors	30	0.54	85
Partitions	50	6.97	65
Flooring	7	13.12	95
Ceiling	20	1.02	55
Furnishings	25	1.73	85
Plumbing	40	7.54	65
Fire protection	50	0.00	0
HVAC	20	13.21	75
Controls	20	0.15	85
Elec. service	30	0.84	5
Distribution	30	3.68	25
Lighting	30	3.68	95
Emergency	30	1.85	80

After the calculations for each component in each building are completed, the costs assigned to each year are added together into the total renewal cost for that year. The total renewal cost can be calculated for one building, a group of buildings, or the entire campus and can be charted into a renewal expense curve. Annual costs can be monitored to manage special conditions, emergencies, or the need to reallocate a pool of annual funds as required.

Facilities Audits

Routine questions asked of the facilities manager are:

- What are the conditions of our facilities?
- Do we face a significant deferred maintenance problem?

- How much will be needed to correct the problem?
- How do we plan a capital renewal program?

The facilities audit, described in detail in Chapter 20, identifies the existing physical condition and functional performance of buildings and infrastructure, as well as maintenance deficiencies. From the information gathered during an audit, capital renewal and replacement requirements can be estimated for prioritizing and phasing projects. The facilities audit provides a basis for decision-making on routine maintenance, capital renewal/deferred maintenance, functional improvements, replacements, and disposal of an institution's facilities.

19.6 SETTING PRIORITIES

Capital renewal and replacement and deferred maintenance programs require clear guidelines and procedures for setting priorities among potential projects. The desire to balance competition for scarce resources can be satisfied by the funding source's determination of categories of projects and selection criteria.

Priorities by Categories of Work

Setting priorities requires consistent treatment of deficiencies and functional improvement funding requests. Typically, categorizing involves data collection, estimating project requests, and then summarizing project requests for a five- or six-year budgeting cycle. Selection of priorities is based on a systematic categorization established by an institution or governing board.

Suggested groupings in broad descriptive categories for selecting priorities are:

- *Liability proposals:* special matters requiring early attention to remove jeopardy through life safety, property damage, regulatory, or court-ordered actions.
- *Program and operational purposes:* actions necessary to support an organization's mission and meet operational requirements.
- *Economy and efficiency measures:* projects that also support program and operational objectives but deserve special attention because they will result in immediate or eventual cost savings.

A suggested outline for ranking priorities by point values is shown in Figure 19-6. Three priority levels for a five-year program are described that can be interpreted from the component rating forms. Categories and subcategories can change annually. Careful judgment is necessary in choosing the priorities to fit with strategic planning and other policy considerations. Selection guidelines and project categories should be reexamined annually to ensure compatibility with institutional goals.

Separate Priority Lists

The setting of priorities and selection criteria for capital renewal should be reviewed annually, or on a schedule compatible with submission of capital requests. A triage process to assign projects to the most appropriate funding source can discover opportunities that were overlooked in the past. For example, an initial inspection of project requests should segregate potential funding sources, including operating budgets, quasi-endowment funds, reserves, and grants. In reviewing all projects, opportunities should be analyzed to "package" several projects for economies of scale. For example, roofing repairs and replacements for several buildings are commonly grouped together into a single project to allow for lower unit pricing. Similar operations, such as erecting scaffolding or suspending use of portions of buildings, also lend themselves to cost efficiencies and minimize building use inconveniences.

Intangible Factors

Other factors do not readily lend themselves to categories but should be considered when making funding decisions. Faculty and staff morale make a positive contribution to overall productivity and can be influenced by sufficient space and properly functioning, well-furnished and well-equipped, attractive, and well-maintained facilities. Faculty and staff recruitment and retention are similarly affected by the physical appearance of facilities and the architectural qualities of buildings and site aesthetics.

Historic preservation is an important aspect of a campus's traditions and image. Facilities that are in marginal condition and being considered for replacement may be more valuable, because of their historical importance or as a focal point for a community, if they are retained and improved. Organizing these categories and intangible factors into a specific set of selection guidelines enables decisions based on technical evaluations and an institution's requirements.

Management Philosophy

Two concepts influencing final priority decisions are need and risk. For example, are projects for improving quality of the environment selected before life safety or operating economy projects?

In the final analysis, selection of priorities by management is the relative weight given to the protection of plant assets, possible fiscal instability caused by postponing deferred maintenance or energy conservation measures, the visual image of the institution, and the risk of erosion of function and quality of environment. Although these matters may seem relatively intangible, they can be as debilitating as the more

Figure 19-6
Project Priority Levels

Priority Level	Point Value

PRIORITY LEVEL I (Year One)

I-1 Life Safety and Legal Compliance — 10
 a. Hazardous life safety building or site conditions that jeopardize people, programs, equipment; unless corrected will cause suspension of facilities use
 b. Repairs, renovation, and improvements required for immediate compliance with local, state, and federal agencies

I-2 Damage or Deterioration to Facilities — 9
 a. Repairs, renovations, and improvements to facilities that unless corrected will lead to a loss of a facility

I-3 Cost-effective Measures — 8
 a. Repairs, renovations, and improvements required to prevent serious facilities deterioration and significantly higher labor costs if not immediately corrected
 b. Energy conservation to reduce consumption with a rapid return on investment

PRIORITY LEVEL II (Year Two)

II-1 Mission Support — 7
 Actions required to support functional activities

II-2 Delayed Priority I — 6
 Repairs and renovations less compelling than Priority I

II-3 Deferred Maintenance — 5
 Deferral of repairs or renovations that will lead to major damage to a facility and loss of use, hamper program activities, or affect economies of operation

PRIORITY LEVEL III (Years Three through Five)

III-1 Project Completion — 4
 Building or site improvements uncompleted because of inadequate funding or other reasons. Improvements are necessary for proper functioning, economic maintenance, and suitable appearance of new construction.

III-2 Delayed Deferred Maintenance — 3
 Repairs, renovations, and improvements that can be postponed

III-3 Anticipating Actions — 2
 Actions carried out in anticipation of longer range development including land acquisition, infrastructure elements, and advance planning for capital projects

III-4 Reduction in scope — 1
 Modify scope to a smaller scale or consolidate with other project

obvious physical consequences of deferring high-priority building and site repairs.

Selection of priorities for major maintenance and capital renewal and replacement includes temptations similar to those that occur in the development of annual operating budgets. The annual operating budget, which is established in a political environment of satisfying short-term needs that conflict with institutional goals, becomes a compromise between alternatives. Overcoming this traditional approach to distribution of resources requires clear policy guidelines in selecting priority projects and a strong partnership between the facilities manager and senior administration. The following principles should guide the priority selection process:

- Major maintenance and repairs in the annual operating budget should be reserved only for projects offsetting facilities deterioration and extending the life of plant assets.
- New construction or major renovations for program improvements should be funded separately from major maintenance and repairs.
- Deferred maintenance or renovations should be funded as special appropriations on a project-by-project basis.

19.7   CAPITAL BUDGET REVIEW PROCESS

The vicissitudes of the budget process can be minimized by a capital budget review committee that includes the campus chief executive officer, chief academic and business officers, and facilities administrator. Monitoring of progress on major maintenance and capital renewal is done by the committee on a frequent basis, preferably quarterly. This permits a routine opportunity for introducing emergencies and new priorities and enhances management of institutional cash flow. Reports provided by the facilities administrator summarize the status of authorized and funded projects and lists proposed projects for the following budget percent. An additional report summarizing anticipated projects for a "rolling" five-year period assists in overall campus long-range budgeting.

An annual meeting of the committee provides the formal approval for projects to be incorporated into the budget cycle for the following year. Use of this process integrates major maintenance, capital renewal/replacement, and capital improvements. Deferred maintenance reduction programs, where established, are included in the review committee's responsibilities of overall institutional fiscal management.

The process of selecting projects for funding is supported by priority guidelines. A preliminary evaluation is made, classifying projects by funding source. Self-amortizing projects are evaluated on their own merits. Next, capital construction is defined and set aside from the priority selection process under criteria different from major maintenance and program improvements. Capital projects for new construction of major additions are usually self-defined by program scope, complexity, and costs. Distinguishing between maintenance and capitalized construction eases the task of prioritizing major maintenance and program improvements to be funded from annual operating budgets. Classification of a project as capital construction is usually guided by minimum cost ranges (e.g., $50,000 to $500,000) and should be based on an institution's budgeting practices of funding facilities improvements.

Invariably, the funding needs of major maintenance and program improvements exceed available budgets. Special allocations for unfunded projects from operating budgets can be aided by following the guidelines for priority selection. The process of selecting project priorities consists of systematic categorization to arrive at funding decisions. Occasionally, first priorities for available funds are bypassed, and lower priorities are advanced when improvement projects are selected before repair and renovation projects. For these reasons, it is essential that an institution use the facilities audit as the basis for developing facilities improvement policy to meet the funding needs for observed conditions.

19.8   CAPITAL RENEWAL PLANNING

Capital Renewal Planning Process

Capital renewal planning is a continuous process, beginning with a preliminary plan covering several years and evolving into a final plan adjusted on an annual basis. A preliminary plan for capital renewal defines overall goals for short-term needs for deferred maintenance and long-term needs for life cycle renewal of facilities components. Such a plan for guiding the shifting in the level of facilities condition from marginal to desirable will be influenced by institutional mission and strategic plan. These factors have become increasingly important as declining resources have affected restructuring, resulting in downsizing, shifts in emphasis from research to undergraduate teaching, and demands to improve the quality of campus residential life. Thus, a resource allocation model for capital renewal is an integral part of an overall strategic plan.

The process of developing a preliminary plan includes the following:

- Project prioritization
- Determination of the rate of annual capital reinvestment
- Determination of the duration of a deferred maintenance reduction program

A final capital renewal plan matches the rate of capital reinvestment over a period of time with the desired duration of a deferred maintenance reduction program. The result is a coordinated approach for capital renewal and maintenance that is designed to protect capital assets, is based on a funding plan, and includes monitoring of the program.

Estimating Short-Term and Long-Term Renewal Needs

Short-term and long-term needs must be addressed concurrently if an institution is to begin the capital renewal process within a reasonable period of time. Deferred maintenance, which typically results from insufficient operating budgets for maintenance, is the focus of a short-term renewal program. However, unless long-term needs for adequate maintenance and renewal are also addressed, backlogs of maintenance work will continue to accrue. Historical facilities underfunding must be addressed through a short-term program of deferred maintenance reduction, and a facilities renewal component must simultaneously be added to the operating budget to offset ongoing facilities deterioration.

Short-term renewal needs are estimated by a facilities assessment program that forms the basis of a deferred maintenance reduction plan. The facilities audit is the first step in estimating needs, followed by a tabulation and prioritization of deficiencies. The final steps are to determine available funding for deferred maintenance, compile a preliminary plan, and document individual projects.

Long-term facilities renewal needs are developed using the techniques for facilities renewal forecasting described earlier.

Facilities Condition Index

A method for measuring the relative condition of a single facility or group of facilities is useful in setting annual funding targets and the duration of deferred maintenance reduction. The facilities condition index (FCI) serves this purpose. The FCI is the ratio of the cost of remedying facilities deficiencies to the current replacement value.

$$\text{Facilities Condition Index (FCI)} = \frac{\text{Deficiencies}}{\text{Current Replacement Value}}$$

The following definitions are used in calculating an FCI:

- *Deficiencies*: The total dollar amount of existing major maintenance repairs and replacements, identified by a comprehensive facilities audit of buildings, grounds, fixed equipment, and infrastructure. The amount does not include projected maintenance and replacements or other types of work, such as program improvements or new construction. Those items should be treated as separate capital needs.

- *Current replacement value*: The estimated cost of constructing a new facility containing an equal amount of space that is designed and equipped for the same use as the original building, meets the current commonly accepted standards of construction, and also complies with environmental and regulatory requirements.

The FCI provides a readily available and valid indication of the relative condition of a single facility or group of facilities. It also enables the comparison of conditions with other facilities or groups of facilities. The higher the FCI, the worse the conditions. For example, after conducting an inspection of buildings and infrastructure, a campus with 3.5 million gross sq. ft. finds it has $60 million in deferred maintenance. Thus, using an example current replacement value of $100 per square foot ($350,000,000), the FCI is 0.171, an indication of poor conditions. Similar calculations for individual buildings can provide comparisons of relative conditions.

Suggested ratings for comparative purposes based on results of comprehensive facilities audits at a number of higher education institutions are assigned FCI ranges as follows:

FCI Range	Condition Rating
Under 0.05	Good
Between 0.05 and 0.10	Fair
Over 0.10	Poor

Costs for correcting facilities deficiencies obtained from an audit and a calculation of the current replacement value allows modeling of the variables for annual and total funding needs and the rate of backlog reduction. For example, if only 1 percent of the current replacement value is available, the change in the FCI can be calculated, or a determination to achieve an FCI of 5 percent in 10 years can produce a calculation of annual capital renewal needs.

A rule of thumb for the annual reinvestment rate is 1.5 to 3.0 percent of current replacement value. However, experience is showing a preference for the upper end of the range (2.0 to 3.0 percent) to prevent further accumulation of a deferred maintenance backlog. As noted earlier, this is separate from funding required to eliminate immediate critical needs of deferred maintenance. A capital renewal plan must include funding for two factors: deferred maintenance backlog reduction and component renewal. This concept is fundamental to capital renewal funding planning. Adequate analysis of various funding options for facilities conditions is done with backlog and funding projection models.

Backlog and Funding Projection Models

Short-term capital renewal plans ideally reduce deferred maintenance backlogs to acceptable levels. Targets established by the FCI offer choices in determining backlog and funding levels.

How much to spend on capital renewal is guided by the results of an audit and the total cost of prioritized projects. Resource allocation questions are:

1. What are the effects of different amounts of annual expenditures for capital reinvestment on total backlog reduction?
2. What is a desirable rate of annual expenditures for reducing marginal facilities conditions?

Restated, this could be posed as: How will x dollars be spent for some number of years to reduce the backlog of deficiencies at the end of the period? Or, how much must be spent over a certain number of years to reach a desired level of conditions for all campus facilities?

Factors to be considered in developing backlog and funding models[5] are as follows:

- *The rate of inflation.* A deficiency will cost more to repair next year than it will cost this year because of increases in labor and material rates.
- *The rate of overall plant deterioration.* Facilities are in a constant state of deterioration. While the identified deficiencies are being corrected, other deficiencies are being created.
- *The rate of backlog deterioration.* A component with an existing deficiency will usually deteriorate at a somewhat faster rate than a component that is in good condition.
- *The rate of plant growth.* As the gross square footage of the institution increases, so does the potential for maintenance and repair deficiencies.

To apply the backlog and funding projection models, the following terms are used:

- *Current replacement value (V)*: The cost of replacing all assets. The initial value (Vo) can be determined as of a given date and then projected for future years, based on inflation and facility growth.
- *Projected plant growth rate (G):* The planned percentage increase or decrease in gross square feet per year from new construction, acquisition, or disposal. The plant growth rate is expressed as a percentage of current replacement value per year. It can be assumed to be constant or it can vary from year to year.
- *Maintenance and repair backlog (B)*: The deferred maintenance backlogs as of a given point in time. The initial level of the repair and maintenance backlogs (Bo) is the cost, as of a given date, to eliminate all facilities deficiencies identified as deferred maintenance in

the facilities audit. The calculation of future maintenance and repair backlogs takes into account the time elapsed, the rates of plant and backlog deterioration, and annual funding levels. The FCI can provide guidance in establishing backlog targets.

- *Assumed annual funding levels (F)*: Projected to determine the level of maintenance and repair backlog at the end of a certain number of years. The funding will be applied to all maintenance and repair.
- *Projected annual inflation rate (I)*: Expressed as a percentage per year. It can be assumed constant or it can vary from year to year.
- *Estimated annual rates of backlog deterioration (D)*: Expressed as a percentage per year. The rates can be assumed to be constant or vary from year to year. A component with an existing deficiency will usually deteriorate at an annual rate that is somewhat greater than that for a component that is in good condition. On this basis, values for average annual rates of backlog deterioration range from approximately 2 to 10 percent.
- *Estimated annual rates of overall plant deterioration (P)*: Includes all maintenance and repair as described earlier, except the reduction of backlog. The items primarily include minor maintenance and repair, equipment maintenance, component renewal, and other maintenance and repair. The rate of overall plant deterioration is expressed as a percentage of current replacement value per year. It can be assumed to be constant, or it can vary from year to year.

Both the backlog projection and the funding projection models demonstrate the impact of various funding levels on overall facilities condition levels. These models can be powerful management support tools. When used properly with facilities data, the models can assist decision makers in managing capital renewal.

Backlog Projection Model   The backlog projection model projects the level of deferred maintenance backlog that will be obtained when assuming a certain funding level. It helps answers the question: If annual funding for maintenance and repair is provided at a given level for *n* years, what will the backlog level be at the end of the period? To apply the model, the current replacement value and backlog level must be determined; the annual rates of inflation, backlog deterioration, plant deterioration, and plant growth must be assumed; and the theoretical annual funding level must be applied.

The formula to project backlog is as follows:

$$Bn = (Bn - 1) (1 + In + Dn) + (Vn) (Pn) - Fn,$$

where:

Bn = Backlog at end of year
Vn = Current replacement value at end of year n,
$Vn = (Vn - 1)(1 + In + G_n)$
In = Inflation rate in year n
Dn = Backlog deterioration rate in year n
Pn = Plant deterioration rate in year n
Gn = Average plant growth rate in year n
Fn = Planned funding in year n

A backlog reduction projection model is illustrated in Figure 19-7. This illustration assumes a backlog of $20 million and a current replacement value of $400 million. The chart shows the relationship between various levels of annual repair and maintenance funding and the resulting levels of backlog at the end of a five-year period. A reinvestment rate in this example of less than 2.5 percent of initial replacement value causes overall facilities conditions to deteriorate from the current 5.0 percent FCI.

Funding Projection Model   The funding projection model projects the level of annual maintenance and repair funding required to produce a certain backlog level. It helps answer the question: What annual funding levels will be required to meet a given backlog target in t years? To use the model, the current replacement value and current backlog level must be determined; the annual rates of inflation backlog deterioration, plant deterioration, and plant growth must be assumed; and the target backlog must be applied.

The formula to project funding is as follows:

$$Fn = (Bn\text{-}1)(1 + In + Dn) + (Vn)(Pn) - Bn$$

where:

Fn = Projected annual funding
Bn = Backlog at end of year n
Vn = Current replacement value at end of year n,
$Vn = (Vn\text{-}1)(1 + In + G_n)$
In = Inflation rate in year n
Dn = Backlog deterioration rate in year n
Pn = Plant deterioration rate in year n
Gn = Average plant growth rate in year n

A funding backlog projection model is illustrated in Figure 19-8. The relationship is shown between different levels of target backlogs at the end of five years and the average annual funding required over the five-year period to achieve different target levels. Calculations are performed for different target backlogs using the FCI as a measure.

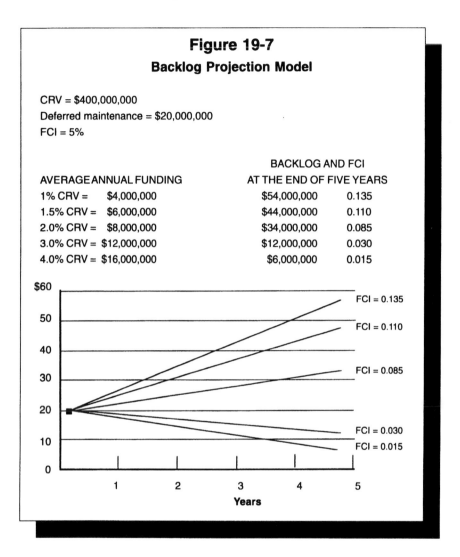

Figure 19-7

Backlog Projection Model

CRV = $400,000,000
Deferred maintenance = $20,000,000
FCI = 5%

AVERAGE ANNUAL FUNDING		BACKLOG AND FCI AT THE END OF FIVE YEARS	
1% CRV =	$4,000,000	$54,000,000	0.135
1.5% CRV =	$6,000,000	$44,000,000	0.110
2.0% CRV =	$8,000,000	$34,000,000	0.085
3.0% CRV =	$12,000,000	$12,000,000	0.030
4.0% CRV =	$16,000,000	$6,000,000	0.015

19.9   PROTECTING CAPITAL ASSETS

Funding Capital Renewal

Seeking funds for capital renewal on the scale required to reduce deferred maintenance backlogs is a challenging venture for higher education. For some colleges and universities, the traditional method of funding capital

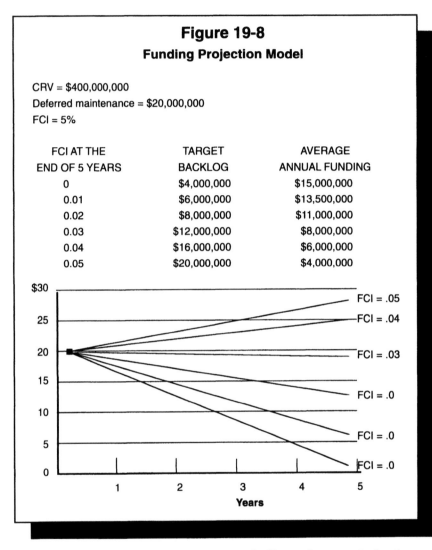

Figure 19-8
Funding Projection Model

CRV = $400,000,000
Deferred maintenance = $20,000,000
FCI = 5%

FCI AT THE END OF 5 YEARS	TARGET BACKLOG	AVERAGE ANNUAL FUNDING
0	$4,000,000	$15,000,000
0.01	$6,000,000	$13,500,000
0.02	$8,000,000	$11,000,000
0.03	$12,000,000	$8,000,000
0.04	$16,000,000	$6,000,000
0.05	$20,000,000	$4,000,000

improvements from the traditional sources of gifts and grants is inadequate for the tasks faced by those universities. Successful examples show that multiple funding sources are necessary, providing a stream of funding that meets capital and component renewal project priorities. The spreading out of projects allows pooling of multiple sources to meet annual needs. This principle enables funding planning that can incorporate some of the following experiences of public systems and independent institutions.

Bond Issues Borrowing for capital projects is a routine practice for public systems of higher education and is used occasionally by independent institutions. The urgent need for capital renewal has made acceptable the issuing of general obligation bonds, revenue bonds, or other options for new construction to use for reinvesting in existing facilities. The Commonwealth of Virginia and the states of Georgia, California, Mississippi, and others have initiated this practice. Vanderbilt University borrowed $150 million to finance renovation and deferred maintenance projects.

Operating Budgets   Some institutions have begun supplementing annual operating budgets with additional funds for capital renewal. Even in difficult financial times, states and independent institutions have both begun to reallocate financial priorities by establishing an amount in the operating budget specifically for deferred maintenance.

The College of Wooster initiated a funding model in 1977 using a "capital charge" budgeting concept to develop a reliable source of capital renewal funding for five-year planning cycles. An amount was incorporated annually into the operating budget and defined as a charge to create a reserve for funding capital renewal and debt reduction. An unrestricted gift was allocated to a reserve fund to initiate the concept. Each year's charge to the annual operating budget includes an average of projects budgeted in the current year and estimates of projects to be done in each of the next four years. The pool of reserve funds is drawn on as required by priority projects.

The Commonwealth of Virginia developed a maintenance reserve appropriation in 1982, distributing funds to public institutions using a formula developed by Douglas R. Sherman and William A. Dergis (*A Funding Model for Building Renewal,* 1981). Each institution is required to prepare a maintenance reserve plan describing projects. Funding is provided as a supplement to the operating budget for maintenance, following an assumption that approximately 50 percent of the formula amount is already contained in the operating budget.

Depreciation Accounting   Changes to depreciation accounting guidelines for higher education offer a potential solution to provide a constant funding source for capital renewal. Although not sufficient to fund substantial backlogs of deferred maintenance, maintenance depreciation entries in the balance sheet can provide a substantial source for renewal funding. The challenge is to create depreciation reserves from current revenues equal to the declining value of capital assets. Some institutions that were able to use "off–balance sheet" funding for capital renewal are currently creating depreciation maintenance reserve funds from revenues and including them in operating budgets.

The model created by Boston College in 1976 combined the annual operating budget and a separate capital budget for renewal funding. Boston College was able to rely on unexpended depreciation reserves for capital budgeting. Depreciation accounting and funding the depreciation charge through the operating budget was an innovative technique. Based on the concept that current users should pay for renewal and replacement, an equitable charge was included in the annual operating budget to develop a consistent source for funding facilities renewal. The retirement of long-term debt and a reduction in acquiring debt for new projects will improve the allocation of available funds for future renewal.

Quasi-Endowment Funds Conversion   Institutions with quasi-endowment funds or "funds functioning as endowment" have sacrificed interest earnings by designating the funds for use in capital renewal. This is a controversial action requiring approval of a governing board, but is a valid stopgap when current revenues are unavailable and the institution wishes to avoid incurring additional debt. Rensselaer Polytechnic Institute plans to partially fund $142 million in deferred maintenance by conversion of unrestricted quasi-endowment funds.

Plant Fund Reserves   Building up plant fund reserves by transferring income surplus offers a source for capital renewal funding. This decision is made in assigning priorities in the institutional budget-making process. Although not a guaranteed stream of funding, prudent financial management can create reserves allocated to fund deferred maintenance projects. Reserves can be drawn from a pool of funds as projects are defined and expenses incurred. Syracuse University used plant fund reserves to supplement maintenance operating budgets for funding more than $150 million in capital renewal and replacement over a 15-year period, beginning in 1973.

Fund Raising   Obtaining gifts for capital renewal represents a greater challenge than funding new construction. New or expanding programs and replacement of existing facilities have a greater appeal to donors than requests to fund deferred maintenance. However, as decaying campus facilities have become a high priority, attention has shifted from new projects to renovation of existing facilities as a target for designated gifts.

Strategies have varied from individual campaigns for specific facilities to an overall fund raising effort with unrestricted gifts channeled to capital renewal. These approaches afford alternatives for development programs and donor choices. Some campuses have prepared lists of capital renewal projects to be included in major fund raising programs. Changes in policies of foundations have seen the new priorities as valid reasons to award grants in support of renovation projects. By pooling challenge grants with gifts and other institutional resources, campuses can achieve a goal for a designated project not easily reached with a single gift.

Energy Conservation   Deferred maintenance projects for mechanical and electrical systems, utilities infrastructure, or central energy plants can be treated as unique capital renewal projects for energy conservation. Facilities audits have shown that 40 to 50 percent of deferred maintenance exists in these categories. The rationale that energy conservation will result from these projects is based on cost-benefit analyses identifying payback periods. Thus an investment in energy conservation can be considered self-financing.

Vanderbilt University finances energy conservation by a utility depreciation reserve created by a 14 percent "tax" added to the university's electric bills. The reserve has been supplemented by energy conservation grants. Syracuse University has obtained more than $6 million in energy conservation grants, some at 100 percent of project costs and others as matching grants. Sources include federal and state programs and programs offered by the local public utility to stimulate demand-side energy reductions. Incentives offered by private companies to participate in energy savings are an alternative method of funding a component of deferred maintenance.

Managing Capital Renewal and Deferred Maintenance Reduction Programs

Managing the long-range integrity of facilities in support of an institution's mission is a broader challenge than routinely responding to repair emergencies or requests for modifying offices or laboratories. Management of a capital renewal and deferred maintenance reduction program starts with clear understanding of the following:

- *View a facility as a collection of components and systems.* The deterioration of a component can cause breakdowns in other parts of a system. Evaluation of a repeated maintenance problem should consider the system nature of facilities. Facilities deterioration can be offset by maintenance management staff pooling knowledge of unsatisfactory conditions that are developing into major problems.
- *Keep track of facilities conditions.* An annual audit of physical conditions to note current problems and priorities should be a basic practice of facilities management. Familiarity with conditions enables the facilities manager to become aware of the most pressing needs. Lack of knowledge of conditions prevents anticipating major problems and avoiding budget surprises for overall campus fiscal management.
- *Maintain a five-year major maintenance and capital renewal program.* A five-year capital budget plan provides a level of confidence for senior administrators in the facilities management staff by regularly reviewing overall campus capital requirements. A level of capital requirements is established in long-range budget base planning, offering flexibility for emergencies or special situations that cannot be anticipated. Finally, the facilities manager has an operational

framework for maintenance management to direct staff, materials, and contractors to appropriate priorities.

- *Know the differences between maintenance, repairs, and major maintenance.* The categories may sound so routine that the important differences are not distinguished. A simple check on the practices of work management and control is the proportion of staff devoted to routine building and preventive maintenance and the proportion devoted to major maintenance. Although many facilities managers take justifiable pride in their staff's construction accomplishments, they are mistaken if they look at their labor pool primarily as a construction team. Unless the facility is located in an area where competition from contractors is unavailable, the wrong emphasis is being placed on work management if the majority of staff labor hours are allocated to major maintenance.

- *Manage maintenance as opposed to maintaining management.* Any size institution needs a work control center to assign tasks, control material purchases, and be a responsive service organization. Similarly, annual capital budget planning is part of a five-year capital budget plan that integrates all funding streams with work priorities. Failure to bring these basic management programs into routine operations is a sign of complacent management and of an organization prepared to complain about inadequate funds and a lack of appreciation for staff that is dedicated, loyal, and hard working.

- *View facilities management as a support service.* A central purpose of an academic enterprise is to maintain the quality of its academic programs. Maintaining the quality of campus life for all members of the campus community is a close second. In terms of allocation of resources, facilities come in at least third. Something else, hopefully, may qualify as last. Disappointment in the occasional short shrift in funding can be overcome by accomplishment in providing efficient service. Prompt responses to requests, explanations for delays and postponements, and attention to meeting the service needs of the specific requirements of a college or university are traits of the service organization.

Understanding these principles can create a fresh approach to practices and procedures. The facilities manager should take good stock of the condition of plant, including buildings, grounds, utilities, and equipment. Walk through the buildings, keeping in mind the operations and maintenance budget. Be candid in the self-evaluation of the maintenance management's effectiveness. Also, have a feel for the previous annual funding for major maintenance and the tempo of plant additions. (A quiet period or increase in activity? A shift from new construction to renovations?) Set aside frustrations from "inadequate" budgets and be self-critical of work control, staff performance, and the presentations of requests for increased base budgets and special capital appropriations.

As the facilities administrator walks the grounds and through buildings from basement to roof, note should be taken of deferred maintenance, especially for life safety problems and a building's exterior envelope—roofing, flashings, mortar, or other sealants—and places where deterioration permits moisture penetration. Check the operating records for failures of mechanical systems and complaints about heating and cooling. Finally, observe environmental safety conditions such as exits, toxic waste storage, sprinkler systems, and smoke detectors. Observe for any obstacles to disabled individuals.

This is an informal checklist to set the facilities manager onto the task of preparing a strategy for a major maintenance and capital renewal program. Formal aspects of the program begin with senior plant administrators evaluating the overall facilities management program for the institution. Later, tasks are defined for the administrator of the small campus to personally take on with available assistants, or, at a larger campus, for the delegation of staff.

A Plan of Action

To summarize, an action plan for comprehensive management of capital renewal and deferred maintenance reduction programs includes the following:

- Build a constituency of campus support.
- Develop a work plan.
- Inventory conditions.
- Select priorities.
- Determine funding requirements.
- Seek funding sources.
- Create public awareness of facilities conditions and funding needs.

Continue the cycle until results are produced. The actions are necessary to prevent further plant deterioration and protect capital assets.

The Management Renewal Challenge

The facilities officer should not be discouraged at the initial response to the magnitude of costs reported in a comprehensive audit for capital renewal/deferred maintenance reduction and the gap between current and required funding to maintain renewed facilities. Resource reallocation and supplementary funding will probably be required for renewal and replacement of facilities with a high proportion of deficiencies. Capital renewal is a long-term process, and programs should be designed with this in mind. The audit process is a key component of a capital renewal program that should be updated annually, reporting progress towards goals, identifying new priorities, and adjusting to programmatic changes affecting renewal and replacement.

A useful approach to consider is the revitalization of facilities staff to inspire confidence for funding deferred maintenance. Feelings of pessimism, frustration, and cynicism among the facilities staff at Santa Clara University stimulated a program of facilities management renewal to cope with lack of support for deferred maintenance funding. The concept centered on renewal of the management team, changing attitudes to gain credibility for funding deferred maintenance. Santa Clara's facilities management adopted a vision for their capital renewal program that included the following:

- Communicating, with credibility, the scope of the renewal and deferred maintenance needs and costs
- Proposing a strategy for achieving facilities equilibrium in a reasonable time frame
- Engaging in the budget decision process to ensure understanding of and advocacy for renewal projects
- Achieving measurable results, small and large, short-term and long-term

Facilities management team renewal can strengthen support for funding facilities renewal by increasing management's credibility through improved attitudes, actions, and accomplishments. The attitudes, visions, and strategies of Santa Clara University's management team are applicable for facilities managers throughout higher education.

NOTES

1. *The Decaying American Campus: A Ticking Time Bomb.* A joint report by APPA and National Association of College and University Business Officers (NACUBO) in cooperation with Coopers and Lybrand. Alexandria, Virginia: APPA, 1989.
2. Kaiser, Harvey H., and Jerry S. Davis. *A Foundation to Uphold.* Alexandria, Virginia: APPA: The Association of Higher Education Facilities Officers, 1996.
3. Boyer, Ernest L. *College: The Undergraduate Experience in America.* New York: Harper & Row, 1987.
4. National Research Council. *Committing to the Cost of Ownership: Maintenance and Repair of Public Buildings.* Washington, D.C.: National Academy Press, 1990.
5. See extensive discussion of backlog and funding projection models in NACUBO. *Managing the Facilities Portfolio.* Washington, D.C.: NACUBO, 1991, pp. 63–76.

CHAPTER 20

Facilities Condition Assessments

Harvey H. Kaiser
Harvey H. Kaiser & Associates

20.1 INTRODUCTION

Facilities condition assessment is a process of 1) developing a database of existing conditions of buildings and infrastructure by conducting a facilities audit, 2) assessing plant conditions by analyzing the results of the audit, and 3) reporting and presenting findings. This chapter provides the tools for conducting a facilities audit and assessing conditions and deficiencies identified by the audit. Guidelines are also included for reporting findings of the audit and assessment and making presentations for a persuasive case for the need to fund capital renewal and deferred maintenance (see Chapter 19).

20.2 THE FACILITIES AUDIT

Purpose of a Facilities Audit

The facilities audit systematically and routinely identifies building and infrastructure deficiencies and functional performance of campus facilities by an inspection program and reporting of observations. The audit process assists maintenance management and the institution's decision makers by recommending actions for major maintenance and capital renewal.

The audit is designed for use by facilities managers responsible for maintenance, capital renewal, and capital budgeting. Circumstances may differ between institutions that undertake a comprehensive survey of all facilities for the first time and those that have a specific set of goals for determining existing conditions. The basic principles presented here can be used for all levels of institutions, from a single structure to a facility consisting of multiple building complexes in dispersed locations. A continuous process of facilities audits, beyond a one-time program, provides up-to-date major maintenance priorities and can generate a significant portion of routine maintenance workloads. An effective audit program will extend the useful life of facilities, reduce disruptions in use of space or equipment downtime, and improve facilities management relations with facilities users.

Audit Structure

The audit is a process of collecting information on current conditions of a facility. The goals and objectives of a well-designed audit are as follows:

- Provide for a routine inspection of all facilities identifying deficiencies.
- Define regular maintenance requirements.
- Define capital renewal and replacement projects in order to reduce deferred maintenance backlogs.
- Develop cost estimates to correct deficiencies.
- Restore functionally obsolete facilities to a usable condition.
- Eliminate conditions that are either potentially damaging to property or present life safety hazards.
- Identify energy conservation measures.
- Inventory accessibility and disabled persons requirements.

The basic phases of the audit process are shown in Figure 20-1. The first three phases—designing the audit, collecting data, and summarizing the results—provide the database on maintenance deficiencies and functional performance. This is followed by presentation of findings. In this systematic approach the scope of the audit is first determined, the audit team is selected, and the inspection is planned. Next, data is collected through inspection of buildings and infrastructure and functional performance evaluations. Finally, the information from these inspections and evaluations is summarized, priorities are set, and results are presented.

Uses of the Audit

The audit is a method of collecting information on the current maintenance conditions and functional performance of a facility. It is designed to include the following:

- An inventory of facilities providing descriptions of characteristics
- Inspections of existing buildings and infrastructure conditions
- Evaluations of functional performance
- Recommendations for correcting observed deficiencies

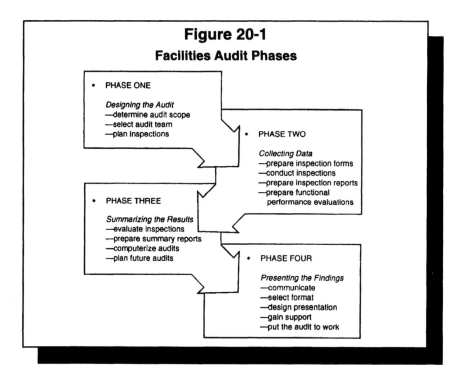

Figure 20-1

Facilities Audit Phases

- PHASE ONE

Designing the Audit
—determine audit scope
—select audit team
—plan inspections

- PHASE TWO

Collecting Data
—prepare inspection forms
—conduct inspections
—prepare inspection reports
—prepare functional
    performance evaluations

- PHASE THREE

Summarizing the Results
—evaluate inspections
—prepare summary reports
—computerize audits
—plan future audits

- PHASE FOUR

Presenting the Findings
—communicate
—select format
—design presentation
—gain support
—put the audit to work

The inspection process is accomplished by systematic inspections of buildings and infrastructure by components following the sequence of construction. Functional performance is evaluated for future planning purposes, capital renewal, and replacements. The methodology can be implemented directly or adapted to meet special conditions and needs of an institution without being rigidly followed in all aspects.

If an audit is designed in a comprehensive, easy-to-use format, it can be used 1) in the field without extensive training; 2) with or without consultant assistance; and 3) at any location, regardless of size.

Audit Information Users

The information gathered in the facilities audit has several intended users and serves many different purposes:

- *Senior Campus Administrators.* The facilities audit supports a consistent presentation of facilities deficiencies, leading to better priority setting when funding is limited. Results also provide documentation for capital renewal and replacement requests.
- *Governing Boards.* The facilities audit provides information to governing boards concerned with the condition of buildings and

infrastructure. Long-term capital budgeting and planning to eliminate deferred maintenance and maintain facilities in a functionally usable condition are also served by the audit.

- *Facilities Managers.* The facilities audit will provide information for routine maintenance, major maintenance, and capital renewal. The inspection procedures and forms can implement a routine process for observing deficiencies and incorporating them into maintenance planning. The facilities audit will also help the facilities management staff communicate with the administration; facilities problems that were avoided in the past can be objectively presented to senior administrators in the facilities audit.
- *Team Specialists.* The facilities audit will enable architects, engineers, and other specialists to gather data on building and infrastructure systems and components related to their disciplines. The information needs of the entire facilities team can be identified and used with more objectivity with the audit data; it should make it easier for the overall building and infrastructure needs of the institution to be studied.

Capital Asset Management

A facilities audit process using the methodology presented here is one element of a comprehensive approach to an institutional capital asset management program. This enables better planning of maintenance and capital expenditures and protects and extends the life of capital assets. Frequently, lack of planning and limited funds create maintenance backlogs and unattended facilities deficiencies. The results are facilities deterioration, deferred maintenance, and a significant financial burden for corrective measures. Therefore, it is necessary to have a process tool—the facilities audit—that clearly identifies and quantifies the condition and functional performance of the facilities organization and cost of various options for correcting deficiencies.

Elements of a comprehensive facilities management program are listed in Figure 20-2. The facilities audit is a component of the facilities management information database. Developed as a source for a strategic facilities development plan, an audit is essential in defining existing facilities conditions and is helpful in preparing a capital improvement plan.

Adapting the Audit

Although some information may be collected and analyzed by maintenance staff to ensure the performance of building and infrastructure components, planning data needed to assess long-term needs are not typically included. The audit goes beyond maintenance planning. By using the audit procedures, a database of information on facilities is produced that can be used as a baseline for future condition inspections.

Figure 20-2

Comprehensive Capital Asset Management Program

A. Strategic Facilities Development Plan
1. Physical development policy
2. Facilities management information database
3. Facilities improvement plan

B. Capital Budget Plan
1. Project schedule
2. Funding source
3. Impact of proposed space changes
4. Project priority selection guidelines
5. Capital project programming and budgeting

C. Facilities Management Plan
1. Operations and maintenance plan
2. Organization plan: facilities planning and operations
3. Space allocation procedures

The audit process is a thorough inspection of buildings and infrastructure, an evaluation of functional performance, and a reporting of observations. The process can be used as a guide for a self-evaluation program using in-house staff or as a guide for consultants retained for condition inspections. Facilities managers should incorporate the special characteristics of an organization and its facilities into an individualized facilities audit. Forms used for recording inspections can be adapted to unique building and infrastructure systems and components not included in standard forms.

Careful design of a facilities audit enables the facilities manager to predetermine the level of information to be obtained and ensure that the information gathered is appropriate for the projected application of the findings. However, the approach selected must be driven by the nature of an organization's facilities, budgeting methods, and organizational structure. Formats for reporting the audit findings should be tailored to match the input requirements for the maintenance work order system and the capital budgeting and planning processes.

As the facilities manager and staff gain experience with the audit, they will recognize its potential for enabling them to become familiar with building and infrastructure condition, provide information for maintenance planning, and restore and maintain the physical and functional adequacy of facilities. The audit contributes to overall effectiveness of the facilities management organization by developing condition inspections as a routine part of operating activities.

20.3   PREPARING FOR A FACILITIES AUDIT

Facilities Audit Program

To be successful, a facilities audit program requires the support of senior financial and facilities management. Planning for an audit program should incorporate management review of the process and form of results to ensure that requirements are met for capital asset management planning and allocation of resources. Senior management involvement in facilities audit planning results in reliable sources of information on the physical condition of the campus and in determining funding needs.

Conducting a facilities audit requires a clear set of objectives before committing staff and financial resources. Whether a campus has previously conducted facilities audits or is beginning one for the first time, there should be thorough preparation to ensure understanding and support of all staff involved in the process. Management and inspectors must make a commitment to collect accurate data, identifying deficiencies as objectively as possible.

An audit can be comprehensive, collecting information on building and infrastructure components and functional performance for all facilities, or it can be limited to a condition inspection program, eliminating the functional performance evaluation. It can also be selective for specific components of buildings, such as roofs, or a unique collection of information for safety or new regulatory requirements.

Audit results are used to plan routine and major maintenance, urgent and long-term measures to correct facilities conditions through a deferred maintenance reduction program, and capital renewal and replacement budgeting and planning. For example, audit forms can be used to provide a description of conditions for an auxiliary enterprise, such as residence halls or athletic facilities. The audit methodology can also be applied for a survey of one or more components for which information is needed because of legal compliance requirements to meet new regulations or codes.

Condition Inspections

Condition inspections are designed to provide a record of deficiencies and estimated costs to correct them. The deficiency–cost methodology requires training inspectors for a "self-audit" (or giving clear instructions to consultants) to produce an objective and consistent database for future reference. Incorporated into the training is the development of a process of continually observing and reporting deficiencies, flowing results into maintenance work and capital budgeting and planning. A successful audit program will introduce a culture of observing and reporting conditions, not on a one-time basis, but as a regular part of supervisory and tradespeople work performance.

The condition inspection part of a facilities audit is a visual inspection and recording of observations of buildings and infrastructure systems and components. Design of the inspection forms and methodology is based on how a building or infrastructure is constructed and how inspectors would logically proceed to make observations and collect data for deficiencies and costs of corrective measures. Building inspections begin with how a structure is placed in the ground; then travel upward to structural framing, exterior wall enclosures, and roof; and then move to the interiors. Each service system—heating, ventilation, and air conditioning (HVAC); plumbing; and electrical—is inspected separately. A comprehensive audit provides an inspection of architectural, civil/structural, mechanical, electrical, and safety components of each facility. Infrastructure inspections are conducted in a similar methodical manner.

Figure 20-3 outlines the systems and components for building and infrastructure inspections.[1] As shown, primary systems include the foundation, structural system, exterior wall system, and roofing system. Secondary systems include interior work that makes the facility usable: ceilings, floors, interior walls and partitions, and specialty work. Service systems include all operating systems, such as HVAC, plumbing, and electrical systems. Safety standards, including life safety and code compliance, are grouped together. Decisions on where to record data for a unique system or alternatives to the component definitions should be flexible and should be made during the initial organization of the audit and instructions to inspectors.

Major infrastructure components are listed in groups that can be inspected by appropriate specialized staff or consultants. These lists should also be reviewed, with flexibility retained until a final list is adopted.

Functional Performance Evaluations

Completing the database with an analysis of the facilities' functional characteristics provides a useful tool for renovating or planning alternative uses of facilities. Prior to summarizing the results of the condition inspection, the functional performance of a facility—suitability, adaptability, and use considerations—should be considered.

For example, in the condition inspections, a building may by found to have significant deficiencies, with costs of corrective measures exceeding replacement value. However, for historic, aesthetic, or other reasons, the building may be retained for remodeling and extended use. Major renovations resulting from the functional performance evaluation may include all identified priorities in a single improvement program. In contrast, a facility may be considered for remodeling, but the institution may want to recommend demolition because of conflicts with plans for future land use or sale as a source of revenues.

Figure 20-3
Building Component Descriptions

Primary Systems

1. Foundation and Substructure
   Footings
   Grade beams
   Foundation walls
   Waterproofing and underdrain
   Insulation
   Slab on grade
2. Structural System
   Floor system
   Roof system
   Structural framing system
   Pre-engineered buildings
   Platforms and walkways
   Stairs
3. Exterior Wall Systems
   Exterior walls
   Exterior windows
   Exterior doors and frames
   Entrances
   Chimneys and exhaust stacks
4. Roof System
   Roofing
   Insulation
   Flashings, expansion joints, roof
      hatches, smoke hatches, and
      skylights
   Gravel stops
   Gutters and downspouts

Secondary Systems

5. Ceiling System
   Exposed structural systems
   Directly applied systems
   Suspended ceilings
6. Floor Covering System
   Floor finishes
7. Interior Wall and Partition Systems
   Interior walls
   Interior windows
   Interior doors and frames
   Hardware
   Toilet partitions
   Special openings: access panels,
      shutters, etc.
8. Specialties (Examples)
   Bathroom accessories
   Kitchen equipment
   Laboratory equipment
   Projection screens
   Signage
   Telephone enclosures
   Waste handling
   Window coverings

Service Systems

9. Heating, Ventilating, and Cooling
   Boilers
   Radiation
   Solar heating
   Ductwork and piping
   Fans
   Fume hoods
   Heat pump
   Fan coil units
   Air handling units
   Packaged rooftop A/C units
   Packaged water chillers
   Cooling tower
   Computer room cooling
10. Plumbing System
   Backflow prevention devices
   Piping, valves, and traps
   Controls
   Pumps
   Water storage
   Water treatment
   Plumbing fixtures
   Drinking fountains
   Sprinkler systems
11. Electrical Service
   Underground and overhead service

Duct bank
Conduits
Cable trays
Underfloor raceways
Cables and bus ducts
Switchgear
Switchboard
Substations
Panelboards
Transformers
Variable frequency devices
Wiring
12. Electrical Devices
Lighting fixtures
Motor controls
Motors
Safety switches
Emergency/standby power
Baseboard electric heat
Lightening protection
13. Conveying Systems
Dumbwaiters
Elevators
Escalators
Material handling systems
Moving stairs and walks
Pneumatic tube systems
Vertical conveyors
14. Other Systems
Energy control systems
Clock systems
Public address systems
Sound systems
TV systems
Satellite systems
Security systems
Communications networks
15. Safety Standards
Asbestos
Code compliance
Egress: travel distance, exits, etc.
Fire ratings

Extinguishing and suppression
Detection and alarm systems
Disabled accessibility
Emergency lighting
Hazardous / toxic material
  storage

**INFRASTRUCTURE COMPONENT
DESCRIPTIONS**
1. Site Work
Roads
Walks
Parking Lots
Curbing
Fencing
Water retention
2. Site Improvement
Landscaping
Lighting
Furniture: benches, bike racks,
  waste receptacles, kiosks,
  signage
3. Structures
Bridges
Culverts
Retaining walls
Tunnels
4. Utilities
Central energy plants
Chilled water distribution
Compressed air
Distilled water
Domestic water
Electrical distribution
Energy monitoring and control
Fire protection
Irrigation
Steam distribution
Storm drainage
Sanitary sewage
Wastewater treatment and
  collection
Water treatment and distribution

20.4   DESIGNING THE AUDIT

The first phase, designing the audit, involves 1) determining the audit scope, 2) selecting the audit team, and 3) planning the inspection. The items that make up each of these steps are listed in Figure 20-4.

Building and infrastructure inspections can represent a significant commitment of resources. The audit design should be prepared to adjust staff schedules for in-house inspectors and allow adequate time for managing consultants when they are used. Direct costs must be budgeted for reproducing plans, preparing building histories, performing laboratory testing, and producing other information. If consultants are used, their costs must also be budgeted. Thorough preparation, including staff training, will ensure the usefulness of facilities audit results.

Determining the Scope of the Audit

An audit's scope is determined by 1) the goals and objectives for conducting condition inspections, 2) the methodology for inspections and assessment, 3) functional performance evaluations, and 4) the intended use of audit results. The choice of specific facilities to be inspected depends on

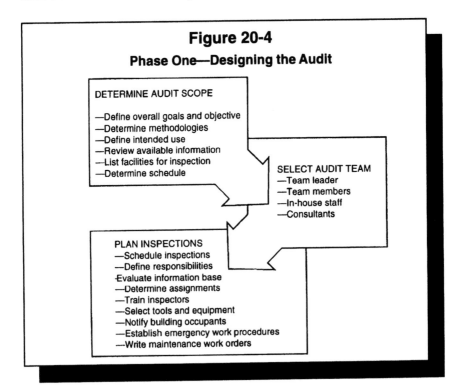

Figure 20-4

Phase One—Designing the Audit

DETERMINE AUDIT SCOPE

—Define overall goals and objective
—Determine methodologies
—Define intended use
—Review available information
—List facilities for inspection
—Determine schedule

SELECT AUDIT TEAM
—Team leader
—Team members
—In-house staff
—Consultants

PLAN INSPECTIONS
—Schedule inspections
—Define responsibilities
-Evaluate information base
—Determine assignments
—Train inspectors
—Select tools and equipment
—Notify building occupants
—Establish emergency work procedures
—Write maintenance work orders

the audit's goals and is often influenced by deadlines, available resources (staff and budgets), building access, and seasonal conditions. In determining the scope of the audit, keep in mind that inspection methodology produces quantitative information on building and infrastructure deficiencies and costs of correcting them. Information obtained from inspections should be tailored to staff ability and time to analyze results and implement corrections for observed deficiencies.

The following checklist should be reviewed when determining an audit's scope:

- Define the goals and objectives of the audit.
- Determine the methodology for inspections and assessments.
- Define the intended use of results, and determine report formats.
- Review available information on facilities to be inspected.
- Prepare a preliminary outline of the facilities and components to be inspected.
- Establish deadlines, availability of staff, and access to facilities.
- Coordinate the audit with building occupants.

The goals and objectives for conducting building and infrastructure inspections will determine the audit's scope. Objectives can include determining priorities for major and minor maintenance, evaluating deferred maintenance reduction programs, and providing cost estimates for capital renewal and replacement.

Initiating a capital renewal planning program on a campus requires establishing a database of conditions for all buildings and equipment. A specific request for information on a single building or building type (residences, laboratories, athletic facilities, etc.) will limit the scope of the audit. The need for information on one building or one infrastructure component (e.g., roofing, paving, or accessibility) will limit the scope still further.

Methodology for inspection and assessment includes choosing the forms to be used and deciding whether to use staff, consultants, or both. Defining the use of information and planning how to computerize inspection results at an early stage of the process can avoid the need to redo work. The intended results should be defined and final report formats designed to ensure that adequate inspection data are collected and analyses can be prepared.

A building inventory (Figure 20-5) lists buildings to be included in the audit and guides planning for staffing resources and overall scheduling of inspections. A decision is necessary to include all buildings in use (owned and leased) or only owned buildings (that is, only educational and general buildings). Forms should be prepared that include a master listing of all buildings and infrastructure that will eventually be inspected or omitted from the audit. The inspection requires basic information on all buildings and infrastructure for use in the inspection process and later in developing

Figure 20-5
Building Inventory List

Building No.	Building Name	Building Use	Location/ Address	Ownership O/L/O-L

summaries of information. Another element of the inventory is a building file, containing useful data on each inspected facility (Figure 20-6).

Next, the outline of building and infrastructure systems and components (Figure 20-3) should be reviewed to determine the scope of the audit. Items may be added or deleted from the elements included in each component. They may also be shifted from one component to another to fit the qualifications of inspection team members.

Finally, an overall completion date and/or any deadlines for completing inspections and assessing results should be factored into the scope to ensure timely completion of accurate reports. Thorough planning of inspections enables observed deficiencies and costs for corrective measures to flow directly into work orders and capital budgets. Functional evaluations of facilities should also be included in the audit; this assists in future planning and evaluation of improvements for program activities.

Selecting the Audit Team

Once the audit scope has been determined, the audit team leader and team members are selected. The audit team must be supervised to guide the process from start to finish. An audit manager should be identified to organize

---

Figure 20-6
Building File

1. Building # _____   2. Building Name _____
3. Address _____
4. Grid Location _____
5. Use _____
6. # Floors _____   7. Gross Area (sq. ft.) _____
8. Net Assignable Area (sq. ft.) _____
9. Ownership ____(O)   ____(L)   ____(O/L)
10. Book Value $ _____   11. Repl. Value $ _____
12. Age (Original construction, additions) _____

Land Data
1. Location _____
2. Ownership ____(O)   ____(L)   ____(O/L)
3. Book Value $ _____   4. Current Market Value $ _____
5. Year(s) Acquired _____
6. Area (acreage, square feet) _____
Notes: _____
_____
_____
_____
_____

---

the process, supervise preparation of a database of pertinent information, select audit team members, and schedule the audit. The manager should be able to prepare written reports, skilled in inspection techniques, capable of estimating and planning, and knowledgeable of maintenance practices and standards and should have good oral communication skills.

The number of members and qualifications of an audit team vary with amount of space and complexity to be inspected. Teams typically include architects and structural, mechanical, and electrical engineers. Complex inspections can require landscape architects and specialists in code requirements and conveying equipment. Familiarity with facilities operations and maintenance or facilities management is essential to ensure that inspections and assessments encompass the practical aspects necessary for application to work orders and capital projects. Knowledge or skills in installation and maintenance of a component are important considerations in selecting an inspector and evaluating the qualifications of a consulting team.

There are advantages and disadvantages to the use of in-house staff or consultants. In-house personnel are most familiar with the facilities to be inspected. Maintenance supervisors or facilities planning staff who work with facilities on a routine basis have access to operations and maintenance information and can readily use existing knowledge of deficiencies. However, the conflict of performing current assignments often places in-house staff in the role of part-time inspectors who must defer inspection assignments or interrupt scheduled tasks. Training can overcome lack of inspection experience and provide uniformity and consistency of inspections. Thorough training of in-house staff contributes to the success of an audit and should be done formally, covering the inspector's role and importance of the inspection, use of forms and results, inspection techniques, and completion of inspection forms.

Consultants are recommended when in-house staff with adequate inspection skills are not available. Consultants can complete an audit on a timely basis and produce objective inspections and reports while leaving in-house staff free to perform their regular assignments. Consultants can offer specialized technical knowledge for diagnosing problems, recommending corrective measures, and estimating maintenance and capital projects. However, the audit scope must be carefully defined to manage costs and ensure that results meet goals and objectives. A staff audit manager should be appointed to coordinate consultants' inspection techniques, access to facilities, reports, and findings. Consultants' work should be organized to ensure that future audits can follow a similar format, providing information usable as a database for comparison in future inspections.

Planning the Inspection

Thorough planning of the building and infrastructure component inspections is essential to produce accurate, timely, and useful results. There are several critical factors to be considered in planning inspections. These include the scope of the audit, scheduling inspections, responsibility for the audit, information requirements for inspections, inspection assignments, training, tools and equipment, notification to building occupants, and emergency work.

Scope of Audit   Specific buildings, infrastructure, systems, and components should be defined to ensure that inspections satisfy the audit's goals and objectives.

Scheduling Inspections   An overall timetable for inspecting facilities translates goals and objectives into a framework for individual buildings and infrastructure. A specific budget cycle may be the driving force; strategic planning with a need for financial and physical resource components can also be the impetus for a facilities audit. The level of staff effort or the need to retain consultants is determined by the following:

- The overall inspection timetable
- The number of facilities, their size and age, and the type of facilities to be inspected

- Staff availability
- Facilities access influenced by usage and seasonal weather
- Information available on the facilities
- The nature and extent of involvement with key building personnel, especially research or medical staff, if appropriate

An ideal audit program should include inspection of all buildings and infrastructure components on an annual basis to enable budget planning and to keep surveys of deferred maintenance current. However, resource limitations may make this goal unattainable. An initial audit should cover all facilities, regardless of resource limitations, and should be maintained on a cycle of at least two to five years. The initial audit can be extended over several years by inspecting some facilities each year; *inspections are a continuous process, not a one-time activity.*

An audit program performed on a continuing basis should become a routine activity of plant operations and maintenance that generates a large share of annual maintenance workload and provides an accurate listing of capital renewal projects. Management of the flow of audit information into maintenance work orders and capital renewal/deferred maintenance planning should also be considered for its ability to process and implement inspection data.

The final step in scheduling inspections is estimating required time for each facility. Each discipline involved should be considered, as should preparation of drawings, facility access, and pertinent operating and maintenance information. Enough time should be included for preliminary discussions with building occupants or a survey soliciting information on conditions. The primary use of a facility will determine the time required for an inspection. Guidelines for an inspection timetable are described in NACUBO's *Managing the Facilities Portfolio.*[2] Approximately equal time will be required for conducting inspections and preparing inspection reports.

Responsibility An audit manager must take overall responsibility for the inspection process, whether conducted by in-house staff, consultants, or a combination of both. The importance of this task is emphasized by allocating the majority of the manager's time to the assignment. The manager maintains control of the audit process by establishing him- or herself as the primary contact in facilities management for the inspection teams. The manager should inform building managers and the business office of the inspection schedule and the guidelines for inspectors. The manager also develops and supervises training for the inspectors, prepares the inspection schedule, assigns audit team members, monitors progress, and reports audit findings and recommendations.

Information Requirements for Inspections Planning the inspections includes developing an information base for each facility to be inspected. The audit manager is responsible for assembling information and supervising a central location that serves as audit headquarters, containing files, equipment, and access to copying machines and computers. The following is a suggested listing for an inspection information base:

- Facilities inventory
- Building file
- Maps with grids
- Building plans
- Infrastructure plans
- Project list and status of renovations, additions, and capital renewal
- Maintenance cycles (e.g., preventive maintenance, painting, equipment replacement)
- Inspection report information (e.g., priority criteria, craft codes, labor rates, and estimation sources)
- Codes and other regulatory requirements
- Building contacts (coordinators, managers, etc.)
- Access (keys, restricted areas, and safety precautions for hazardous conditions)
- Testing services (i.e., procedures for any specialized testing or laboratory services)

Inspection Assignments In-house inspectors are assigned building or infrastructure systems and components by the audit team manager. Disciplines should be matched with the inspection assignment. For example, primary systems inspectors may be a team with architectural, structural, mechanical, and electrical engineering skills. Two or more staff members should inspect a facility together for safety reasons and to improve communications between inspectors (i.e., identifying and diagnosing problems, selecting priority classifications, and estimating costs).

Training Training both in-house inspectors and consultants as to the inspection's purpose and schedule and proper use of the forms is essential to ensure uniform and accurate results. Knowing why, when, and how the audit is to be conducted contributes to the quality of inspection results and develops interest and enthusiasm for incorporating inspections and reporting of maintenance deficiencies into regular plant activities. The facilities manager should recognize that staff may consider the audit an additional burden to their normal assignments. The leadership of the facilities management department and the audit manager should impress on the inspectors the importance of their tasks and their contribution to the overall organization's operation and mission.

During training, objectivity in conducting inspections should be emphasized as critically important. Conscientious staff, naturally protective of their areas of responsibility, should be encouraged to reduce subjectivity in conducting inspections. When consultants are scheduled to conduct inspections, they should be sensitive to in-house staff during the process of identifying conditions.

Well-prepared training sessions in the use of the inspection forms assists in completing a successful audit program, guiding the thoroughness of an inspection, and reporting results. Team members should be encouraged to share inspection results with other audit team members and question maintenance personnel for any known building or infrastructure deficiencies. They should also be regularly informed of any management actions resulting from the audits, annual reports, or other summaries. A training program should include the following:

- Purpose of the audit
- Availability of information (drawings, maintenance histories, etc.)
- Inspector's responsibilities
- Maintenance standards
- Health, safety, and building codes
- Field notes and camera and video recorder use
- Use of forms
- Describing deficiencies
- Estimating maintenance and capital renewal project costs
- Use of results

Tools and Equipment An efficiently performed inspection requires inspectors to be prepared for a day in the field without returning to the base of operations unnecessarily. The following list of tools and equipment will ensure that the inspector is well prepared for assignments:

- Floor plans
- Report forms on clipboard
- Tape measure
- Screwdriver
- Pliers
- Adjustable wrench
- Flashlight
- Knife
- Portable tape recorder (optional)
- Camera with flash (optional)

Notification to Building Occupants Building occupants or users of infrastructure must be notified of a scheduled inspection. A preliminary discussion with users or a survey soliciting information can identify many problems prior to the inspection. Access to the building or any planned service disruptions for testing should be coordinated to avoid interrupting normal operations. Tours conducted while a building is occupied are useful uncovering deficiencies and considering corrective measures. Care should be taken to avoid inferring that inspections and identification of deficiencies will result in a specific program of remedial actions. Inspection results and any management actions should be shared with building managers.

Emergency Work Procedures should be in place for promptly remedying any emergency conditions observed during the inspection. Audit team members should be instructed in these procedures during training.

20.5 COLLECTING THE DATA

The second phase of the audit is collecting data by building and infrastructure components and preparing the inspection reports, as shown in Figure 20-7. This process is the detailed inspection, collection, and correlation of the information needed for the reports. Further detailed information on this process can be found in *The Facilities Audit: A Self-Evaluation Process for Higher Education*, published by APPA: The Association of Higher Education Facilities Officers.

20.6 FACILITIES CONDITION ASSESSMENT

The third phase of the audit assesses results of the condition inspections. The items that make up this phase are shown in Figure 20-8. The audit manager analyzes the inspection program and prepares summary reports. The assessment is structured for different purposes and levels of users within an institution, including the governing board, executive officers,

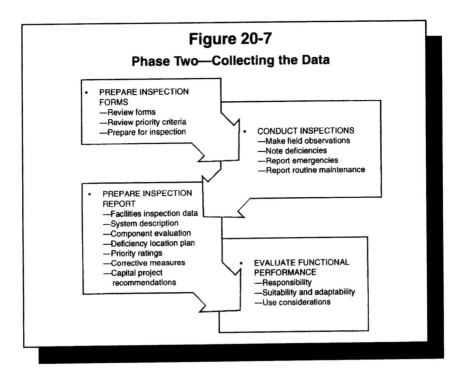

Figure 20-7

Phase Two—Collecting the Data

- PREPARE INSPECTION FORMS
  - —Review forms
  - —Review priority criteria
  - —Prepare for inspection

- CONDUCT INSPECTIONS
  - —Make field observations
  - —Note deficiencies
  - —Report emergencies
  - —Report routine maintenance

- PREPARE INSPECTION REPORT
  - —Facilities inspection data
  - —System description
  - —Component evaluation
  - —Deficiency location plan
  - —Priority ratings
  - —Corrective measures
  - —Capital project recommendations

- EVALUATE FUNCTIONAL PERFORMANCE
  - —Responsibility
  - —Suitability and adaptability
  - —Use considerations

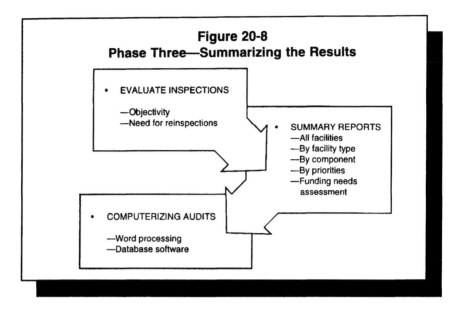

Figure 20-8
Phase Three—Summarizing the Results

- EVALUATE INSPECTIONS

  —Objectivity
  —Need for reinspections

- SUMMARY REPORTS
  —All facilities
  —By facility type
  —By component
  —By priorities
  —Funding needs
    assessment

- COMPUTERIZING AUDITS

  —Word processing
  —Database software

financial managers, the institution planning staff, and facilities management personnel.

Evaluating the Inspection Program

Thoroughness and consistency in the inspections are important considerations for the audit manager in reviewing inspection results. A random selection of inspections and visits by the manager can confirm results or suggest reinspections. The audit manager can judge the accuracy of cost estimates to a degree by independently performing sample estimates and comparing results and the methodology with those developed by the inspector. Subjectivity of the inspectors is a factor that should be considered in evaluating recommended corrective measures. The manager should suggest any needed improvements in inspectors' training programs.

Summarizing Inspection Results

The information base developed by the inspection reports can be the source for a wide variety of summary reports. Using building file data and inspection results, reports may offer information on the overall condition of facilities or facilities types as well as on individual facilities, systems, or components. Further sorting of data can identify projects by priority, cost ranges, and use of in-house personnel versus contractors. Reports can be designed for presentations for different purposes and are limited only by the needs determined by senior facilities and financial administrators.

Summary reports should be more than facts and figures. An executive summary should provide an overview and highlights of findings. A narrative should describe the audit process and objectives and priority selection criteria. Conclusions should include a summary of overall facility conditions, a needs assessment for capital renewal/deferred maintenance projects, and an assessment of the adequacy of operating budgets for maintenance. Audit summaries can be organized in several ways: by all facilities audited or by individual facilities; by all building systems or by individual building systems; and by all building components or by individual building components. A suggested list of reports is shown in Figure 20-9.

Deferred maintenance problems can be the impetus for starting the facilities audit process. A thoroughly prepared and summarized audit will provide the basis for evaluating deferred maintenance and costs for remedying deteriorated conditions. Results provide the basis for a deferred maintenance reduction program.

Figure 20-9
Facilities Audit Summary Reports

A.   All Facilities or Infrastructure
Report #1 - Summary of all facilities or infrastructure deficiencies
Report #2 - Summaries of all deficiencies by building or infrastructure type
Report #3 - Summaries of all deficiencies by building or infrastructure age
Report #4 - Summaries of all facilities or infrastructure deficiencies by systems
Report #5 - Summaries of all facilities or infrastructure deficiencies by components
Report #6 - Summaries of all facilities or infrastructure deficiencies by cost ranges
Report #7 - Summaries of all facilities or infrastructure deficiencies by priorities
Report #8 - Summaries of all facilities or infrastructure deficiencies by craft

B.   Individual Facilities or Infrastructure
Report #9 - Facility or infrastructure deficiencies
Report #10 - Deficiencies by building or infrastructure type
Report #11 - Deficiencies by building or infrastructure age
Report #12 - Deficiencies by systems
Report #13 - Deficiencies by components
Report #14 - Deficiencies by cost ranges
Report #15 - Deficiencies by priorities
Report #16 - Deficiencies by craft

C.   Expenditure Plan
Report #17 - Capital renewal/deferred maintenance priorities
Report #18 - Maintenance operating budget requirements

Comments from the audit are also helpful in producing feasibility studies of changes in building use, resulting in renovations or replacements. The inspections, reports of building and infrastructure conditions, and functional performance evaluation allow two critical questions to be addressed: 1) Is the facility suitable for its current use, or will it require remodeling? and 2) What is the actual cost of remodeling compared with a new building, and is relocation of programs to another building feasible and desirable?

Computer Applications

An overall goal of facilities management is an information system that integrates space management with maintenance work planning, major maintenance, and capital renewal projects. An interrelational database built on space characteristics has broad applications for facilities and financial management and other applications within an institution. Personal computers and commercially available services and products can be used to integrate a space database with the facilities audit process of inspecting and reporting building and infrastructure conditions.

The facilities audit process and methodology are designed for manual use but are readily adaptable to computer applications. A starting point for basic computerizing of audits is word processing and spreadsheet software for generating forms for inspector's use. Careful analysis of the information desired from the audit is necessary before designing individual screens for each form and report. More sophisticated programs allow faster preparation of summaries where computing capability is available to handle the data from a facilities audit. Finally, use of interrelational database programs allows integration of the audit into a facilities management information system.

Computerized applications of the facilities audit are a powerful tool for the facilities manager. Important benefits include readily retrievable information and capability for updating. Computerizing makes it easier to change the perspective of facilities management from collecting data for a one-time or occasional evaluation of conditions to updating facilities conditions on a continuing basis. Thus the audit can become an operational tool, useful for predicting needs, and a valuable component in decision making for managing maintenance and prioritizing improvement projects.

Future Audits

Although some information may be collected and analyzed by maintenance staff to ensure the working order of a facility, planning data needed to assess long-term needs are not typically included. The facilities audit process goes beyond maintenance planning. Following the audit procedures and using the suggested forms creates an information database that serves as a baseline for future condition surveys.

A database makes it easier to conduct special audits from the database. For example, surveys of hazardous materials, of conditions of components (e.g., roofs, roads), or of building or space types (e.g., housing, classrooms) take less start-up effort and will contribute to expanding a database. The audit process provides a powerful tool for budget planning and allocation of resources and offers a method for regularly incorporating information obtained with a high level of confidence into facilities management decision making.

20.7   REPORTING AND PRESENTING FINDINGS

The final phase in the facilities audit process is presenting findings and putting the audit to work. The steps that make up this phase are listed in Figure 20-10.

Communicating With the Audience

The facilities audit is one of the most valuable tools available to help facilities management perform its responsibilities, if the audit is developed and presented well. However, even a flawlessly performed audit is useless unless the information can be communicated to audiences in a readily understandable format. Careful consideration should be given to the audience, its interests, its knowledge of the subject, and the issues it faces as a decision-making group. Conclusions and recommendations should be able to stand on their own merits.

Presentation Format

The documentation provided by a facilities audit will be of more than passing interest because of its effect on campus financial conditions. Thorough preparation is necessary in the design of materials and presentations to establish and maintain the credibility of facilities management and the reliability of its information.

Before beginning the audit process, consider the presentation format. The wide array of summaries available from the inspections should focus on priorities and costs, with supporting material keyed to condensed presentations. If the report of facilities conditions is to be submitted in print

Figure 20-10
Phase Four—Presenting Findings

- Selective and designing communications
- Presentation format
- Gaining support

form only, without oral presentation, consider what graphic material would be helpful. The report itself may be presented as a brief statement of facts with graphics or as an extensive narrative that includes background, description of methodology, findings, and conclusions.

In all cases, the facilities officer should provide material that is concise, easily understandable, and attractively presented. It should be free of jargon, confusing terms, or acronyms that are not self-explanatory. Do not oversimplify for readability, but do design the documents for ease of cross-referencing.

Material should be developed in anticipation of the sharpest minds in the institution receiving the information. The documentation must be meticulous in detail and accuracy. Simple arithmetic errors and broad generalizations should be avoided through thorough checking of financial data, priority selections, and cost–benefit analyses. Expect the unexpected. Be prepared to answer the question, "What will happen if we postpone or don't do the work at all?" Organize the presentation so that the train of thought can be followed. Above all, keep the presentation simple and to the point.

Sample Presentations

Supporting printed material for a budget review session should be submitted in advance to all participants. The following format for budget review sessions, an example of which is shown in Figure 20-11, should be used to ensure the clarity and conciseness of the presentation:

1. Title sheet
2. Executive summary of the major conclusions and recommendations
3. A map of facilities with building names
4. Facilities age in periods of 5 and 10 years
5. Condition summary of priorities from the facilities audit building component form
6. Maintenance deficiencies summarized by categories, using cost estimates from facilities audit inspection forms
7. Maintenance deficiencies summarized by selected components from the facilities audit inspection forms
8. Project summaries identifying the building or facility and containing a short descriptive title and project budget; illustration of the funding source, separating operating from capital budgets, can be helpful
9. Detailed project descriptions, presented on individual sheets for each project

Gaining Support for the Facilities Audit

Once the facilities audit is complete, how do you gain support for a program to correct deficiencies? Essentially, this is done by developing an effective presentation, one that can sell the conclusions and recommendations. Consider the following items when presenting budget requests:

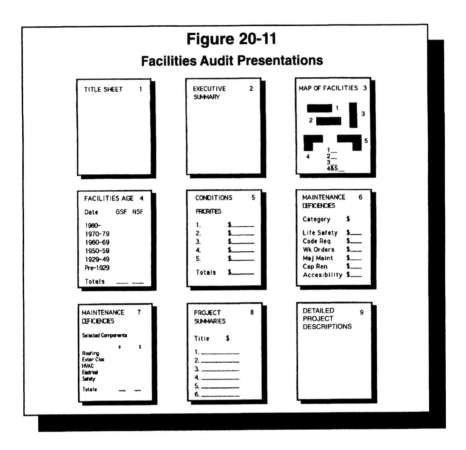

Figure 20-11

Facilities Audit Presentations

- *Overview.* Does the audit show a broad understanding of the budget mechanism and present a responsible fiscal position? Do the conclusions and recommendations fit into long-term policies and overall goals?
- *Credibility.* Does the audit show that previously allocated funds were well used? Does it take the initiative for the best use of new resources for new programs?
- *Competency.* Did the audit team and the implementing staff appear professional and competent during the audit process and in follow-up activities?
- *Thoroughness of Preparation.* Was the audit thoroughly researched and analyzed and professionally presented? Was there evidence of collaboration with budget managers and facility users in the preparation of the audit conclusions and recommendations? The form and content of the presentation must be an accurate presentation of the facts—objective, consistent, and capable of withstanding thorough scrutiny.
- *Supportive Senior Administrator.* The budgetary allocation process represents competition for limited financial resources. Without a strong

advocate, the facilities audit may be shelved. A senior administrator who understands the audit, its conclusions, and its recommendations is an invaluable ally in the funding and implementation process.

- *Preparation for Implementation.* The audit conclusions and recommendations must be in an immediately usable format for project planning and implementation. Administrators who will be involved in the implementation process should be included in the formulating of conclusions. Operational staff should also be involved when possible. These contributors improve the end result and ensure that the purpose of the audit is clear to all parties.

Putting the Audit to Work

Putting the audit to work means developing an ethic among maintenance staff to continually inspect, observe and report deficiencies, and maintain a timely and accurate record of facilities conditions. It also means developing funding alternatives, procedures for managing funding, and assuring that current use of resources is effective and efficient.

The facilities audit process provides the basis for determining capital needs to avoid future facilities deterioration. This enables development of short-term and long-term needs for dealing with the problem. The audit process also supports assessment of comparative facilities conditions and development of priorities. Too often, at this point, the process ends in frustrations with the governing board's lack of response to funding requests.

Many systems and institutions lack three important elements of a capital planning and budgeting process: 1) a project prioritization process, 2) a resource allocation model to formulate funding planning, and 3) a will to change the campus culture in favor of capital asset renewal. Although the number of successful examples is increasing, the evidence from national surveys shows that the rate of facilities deterioration also continues to increase. Despite declining public support and financial distress in higher education, the problem must be faced sooner rather than later.

NOTES

1. References used for developing this outline are the *Means Graphic Construction Standards* and *Means Cost Data,* annual editions, published by the R.S. Means Company, Inc. The reader is referred to these publications for an understanding of alternative component definitions and cost data for estimating costs of correcting deficiencies.
2. National Association of College and University Business Officers. *Managing the Facilities Portfolio.* Washington, D.C.: NACUBO, 1991. These guidelines are derived from U.S. Department of Navy standards (*NAVFAC Manual MO-322, vol. 2, Inspection for Maintenance of Public Works and Public Utilities*).

CHAPTER 21

Work Management Systems

Jay W. Klingel
University of Virginia

21.1 INTRODUCTION

Facilities organizations in higher education typically provide a wide range of services. Such services may include planning, design, construction, renovation, and maintenance of the institution's buildings, grounds, and infrastructure. Effective facilities organizations depend on an organizational component responsible for cognizance and oversight of the established systems for managing and controlling the flow of activities. This component in most facilities management organizations is defined as *work management*. This chapter will address the functional role of work management in a facilities organization and specifically relate the development and operation of the management systems normally assigned to the work management component.

As the business of higher education has changed in response to a more competitive environment, facilities support organizations have become a vital part of shaping this change. The requirement for adaptability and professionalism has never been greater in facilities management. Keeping pace with emerging information and communication technology, successfully marketing facilities services, and demonstrating creative and sound fiscal practices are but a few of the demands that require today's facilities managers to be leaders in the business of higher education. The management systems established to coordinate facilities services in support of the institution's mission are critical to the success of the facilities organization and to the institution itself.

The traditional activities of work management in facilities organizations—work reception, planning, estimating, and scheduling—continue to be important. However, in response to change, the approach to these traditional activities may be changing, and work management systems are emphasizing new activities. Customer relations, although always an important work management function, is becoming even more significant. Understanding the tools and capabilities of today's information technology—and predicting tomorrow's—has a tremendous impact on the effectiveness of facilities support, and as business operations in higher education move to responsibility center financing and a more entrepreneurial posture, management systems in facilities support must be prepared to accommodate these shifts.

This chapter will examine the fundamental components for an efficient work management program including work control, work order systems, financial systems, customer relations, computerization, preventive maintenance, service contracting, and a variety of maintenance management programs. The objectives and benefits of effective maintenance management programs will be described within the framework of policies and procedures designed to support the mission of the institution.

21.2　GENERAL CONSIDERATIONS

In exploring the essential elements necessary for an efficient maintenance management program, it is important to define the objectives and benefits of such a program and to consider the management policies and principles necessary to support the program. What is expected from a sound maintenance management program? How does it fit into the overall facilities organization, or better still, how does it fit into the institution itself? To determine these objectives, it is necessary to have a clear understanding of the mission of the institution and how the facilities organization supports that mission.

Many factors influence the institution's mission. Is it a public or private institution? Is the emphasis on two-year, four-year, or postgraduate programs? Is the institution experiencing rapid growth or downsizing to meet economic realities? The business of higher education is changing, responding to a dynamic customer base, and competing more than ever to attract students and staff for financial growth. The obvious mission of most institutions of higher education is undergraduate and graduate education—the business of teaching. However, there are other institutional missions. Scientific and medical research is being performed on college campuses as never before. Many universities have teaching hospitals, so health care programs have become a major part of the institutional mission. Community service programs, national collegiate athletic programs, and fine arts programs are significant parts of the higher education system across the United States.

The goal of the facilities management organization must fall within the overall framework of the institution's mission. An effective maintenance management program will provide leadership policies and decision making to promote effective use of the institution's resources. These resources include available budgets, personnel, materials, and equipment and the facilities organization itself. It is important to develop policies and procedures to ensure that all available resources are coordinated in supporting the institution's mission.

An effective maintenance management program will manage these resources to provide the optimal physical environment in support of the institution's mission. This physical environment includes the institution's grounds, classrooms, research facilities, residences, administrative offices, utility systems, and transportation systems—the real property of the institution. There is no better way for a facilities organization to support an institution's mission than by creating and maintaining a functional and desirable environment for teaching and learning.

Grounds maintenance is becoming an increasingly important part of an overall maintenance management program. It is often said that a prospective student's first impression of an institution is the condition and appeal of the grounds and landscape. A building maintenance program that provides comfortable, functional, and aesthetically pleasing classroom and residential space is a key factor in attracting future students. A well-maintained, modern infrastructure system that provides current technology and reliable utilities is also important. Finally, an ongoing effective custodial services program contributes significantly to the character of an institution. All of these programs must work in concert to deliver an environment conducive to the institution's mission—that is, higher education.

In realizing the goals of a facilities management organization within the framework of an institutional mission, it is important to explore the objectives of a sound maintenance management program. One could define maintenance management as a systematic approach to the maintenance and upkeep of facilities, grounds and infrastructure, in support of the institutional mission, applying such management principles as organization, planning, and control. There has been an increased emphasis in the past two or three decades in management approaches to maintenance in higher education. It is generally recognized that the post–World War II expansion in higher education institutions and facilities throughout this country has led to an increased awareness of the need for sound management principles in maintenance of facilities. These facilities built in the 1940s and 1950s have, in many cases, reached the end of their productive life, and systems are beginning to fail at a wholesale rate. Second, the escalating cost of energy consumption since the 1970s has led to a growing concern regarding expense of plant operations. As financial resources for plant operations and maintenance has proportionally decreased in recent years, an awareness of the high cost of deferred maintenance has led to an increased emphasis on professionalism in facilities management organizations. Finally, increased competition in higher education for tuition dollars has cast the facilities manager in a role of marketing of the institution's services, requiring today's manager to clearly understand the objectives of maintenance management programs.

21.3 OBJECTIVES OF MAINTENANCE MANAGEMENT PROGRAMS

One of the primary objectives of a sound maintenance management program should be the increase of plant utilization through a reduction in the rate of failure of equipment and systems. It is generally recognized that a well-planned preventive maintenance program can accommodate this objective. Facilities support of academic programs and research depends on providing both a comfortable and a reliable environment. Increased emphasis on classroom utilization standards continues to strain overused facilities and systems. Institutions cannot afford the lost use of facilities space that is due to equipment or system failure.

Another obvious objective is the reduction of maintenance costs—or stated otherwise, an increase in cost-effectiveness of the maintenance program. Available maintenance dollars have not kept pace with escalating requirements. Deferred maintenance in higher education facilities throughout the country continues to grow at an alarming rate. Facilities managers are constantly faced with the challenge to do more with less. A management program designed to increase the cost-effectiveness of maintenance activities is a requirement in the current financial climate. It is important to develop an effective, efficient, and reliable organization. Current managers must be cognizant of training opportunities, both technical and administrative, to enhance organizational efficiency. The facilities organization must be perceived as competitive, responsive, and reliable.

Another important objective of a maintenance management program should be to emphasize customer service-based management principles. The faculty, staff, students, and other stakeholders are customers of the facilities organization. Service must be provided with those customers in mind. The faculty and students are engaged in the overall mission of the institution and to best support that mission, the facilities organization must be constantly aware of customer relations.

Another objective is to develop a continual quality control program to evaluate maintenance services. Through management control, performance tracking, customer feedback, self-audits, or other methods, the policies and procedures of maintenance management programs should be constantly and continually evaluated.

The benefits of a positive maintenance management program should lead to an improved level of maintenance and reliability, improved productivity, and protection of the institution's capital investments. The value of the facilities organization of an institution is often its most notable endowment. This facilities endowment requires continuous reinvestment, which will lead to a comfortable and convenient educational facility, aesthetically pleasing surroundings, and a safe and secure environment for the customers of the institution—that is, the students, faculty, staff, visitors, parents, and general community.

21.4   WORK CONTROL SYSTEMS

The first basic element in a sound maintenance management program is the work control function. Work control essentially comprises three processes:

1. The work order system, which identifies and categorizes work
2. Work authorization, which cites availability of resources
3. Work planning, scheduling, and reporting

Institutions approach the work control function differently according to the size and business philosophy of the institution, but whether the institution is a small college or a large university, a sound maintenance management approach depends upon a recognized process of work control.

Work Order System

In most college and university facilities organizations, the work order system is the basis for identification of requests for services, or work. When an activity is required of an organization—whether a facilities management organization or almost any business—the work order identifies that a service is required or a sales transaction is to take place. For any business, an order—or in our case, a work order—does basic things for the organization. First, it alerts the responsible unit of a requirement for services. An effective work order system should be designed such that the work order itself clearly indicates which part of the organization will perform the requirement. Second, the work order describes the work or services to be performed. Whether the requirement is a preventive maintenance task, installing a lock, replacing a roof, or constructing a facility, the work order is the primary identification and building block for the accomplishment of the service. Third, in the facilities management area, the work order authorizes fund expenditures for the described work. An approved work order that has assigned responsibility and describes the work will also inherently authorize that funds are available for such work. Finally, an effective work order system will allow the organization to track performance in the accomplishment of such work. The organization is able to know when the work was required, when it was completed, who performed the work, and the cost of performing that work. Thus the work order system is the backbone of a work management program. The ability to plan for effective resource utilization depends on a system of identifiable and quantifiable units of work.

The development and management of a work order system in a college or university setting must begin with work requests. For the facilities organization to be responsive to the institution's community, there must be a recognized way for work to be identified and requested. Where do these work requests come from? Who wants the work to be done? Who are the customers of the facilities organization? How does the facilities organization receive these requests?

Work required of the facilities organization can originate from virtually every corner of the institution. The obvious sources of most requests are students, faculty, and staff. Most of the requests from students will involve requirements in the residence halls, and the largest number of requests come from faculty and staff for services in classroom, administrative, and research facilities.

Many institutions have a network of facilities coordinators throughout their campuses. In this system an individual representing a department or several departments in a facility is assigned to be the liaison with the facilities organization in most matters regarding facilities services. Building occupants let the facilities coordinator know of required services, and the coordinator, in turn, identifies the requirements to the facilities organization. Likewise, when the facilities organization needs to communicate facilities information to the users of a facility, the facilities coordinator is in a position to communicate to the building constituents. Depending on the size and complexity of a facility, the role of a facilities coordinator can range from a full-time position to an administrative assistant with several other duties. It is important to note that frequently the facilities coordinator is an employee of the user department instead of the facilities organization. The facilities coordinator can be an excellent source of clear, consistent work requests, avoiding confusion and redundancy for both the user department and the facilities organization.

The facilities organization and its employees are another significant group of people identifying work requests. On most campuses the housekeeping crews and grounds maintenance crews are invaluable sources for identifying maintenance requirements. These groups are in the buildings and on the grounds on a daily basis and should be encouraged to identify this work. Facilities inspection teams are to identify maintenance requirements and, in fact, should identify major maintenance and deferred maintenance requirements as part of their primary duties. Certainly tradespeople and other maintenance crews are a significant source of work identification. In the normal course of preventive maintenance, tradespeople will identify corrective maintenance for equipment they are inspecting, and of course, management by walking around (MBWA) nearly always results in a list of items to be accomplished.

Other sources of work identification include the auxiliary businesses on a college campus. The food services unit, the bookstore, and the athletic department have significant requirements of the facilities organization on most campuses. The security or campus police force, the campus events scheduler, and even the general public are included in the many sources who identify and require services from the facilities organization.

Because of this diverse group of people requiring services, an effective work order system must accommodate the communication of these services. How are work requirements communicated?

- *Telephone calls.* This is by far the most common method of identifying work. Why? Because it is the easiest method of reporting work needs. Most customers of the facilities organization would rather pick up the telephone, dial a familiar number, describe their needs, and rely on the facilities organization to satisfy their requirements. Telephone communication works quite well for most routine maintenance, service calls, and other types of minor work. It is quick and easy, and there is two-way oral communication.
- *Preprinted forms.* Most facilities organizations have developed work request forms. These forms are effective in that they generally provide all the information one needs to proceed with the work. In addition, the preprinted form provides signed authorization of the work. For larger work orders that may result in significant expenditure of funds, the consistency of information and the signed authorization are important parts of the documentation trail.
- *Oral requests.* These include walk-in customers to the service desk from the institution's community, as well as requests from the facilities staff.
- *Inspection results.* The two types of formal inspections prevalent in most facilities organizations are the preventive maintenance program and the facilities assessment program. Both of these programs are normally designed to include a feedback loop that identifies maintenance requirements.
- *Electronic requests.* In recent years, electronic mail has become an increasingly popular method of communicating work requests to the facilities management organization. Many college campuses have developed the ability for customers to enter on-line work requests in the facilities organization's automated work order system.

This discussion of who the customers are and how they communicate work requirements to the facilities organization is an important part of the application. Regardless of the size of an institution, the facilities manager must develop and maintain a system through which work identification is communicated between the institutional community and the facilities organization.

The *work order* contains certain pieces of information that are necessary to describe the required work. A simple but basic work order might identify only a few pieces of information, as follows:

- Date
- Requestor's name
- Building name
- Location of work
- Description of work

An example of a simple work order format is shown in Figure 21-1.

Depending on the amount and the type of information to be managed, a work order could become more complex. In today's world of computerized work order systems, many work management systems maintain a database with a significantly expanded work order format. Information required on a more complex work order could include:

- Date and time
- Requestor name, phone number, fax number
- Building name and number
- Exact location of work required
- Nature of work or problem to correct
- Type of work (category)
- Urgency of work date or time to be completed
- Equipment numbers identifying individual pieces of equipment
- Source of funds

Figure 21-1
Short Work Order Format

Work Order 666666   Requestor KLINGEL   Phone 555-5555   Date 1996-05-3
(Press Ctrl-Entr or Double Click on Description Field to Enter Additional Data)
Request Date
Description BROKEN GLASS IN WINDOW/ROOM 101/TALBOT HALL
Equipment BU0210   Talbot Hall

Details

Work Type 1R
Area U   Supervisor ROGERS   Status WAPPR   Date 1996-05-17-16.19.00.
Location 0210   Lead Craft   Old Group/Zone   Priority 0
Cost Center   Crew   New Group/zone   Entered By E0008

Scheduling information			Master Work Order	
	Start	Completion	Master WO	
Target			#Sub WOs	0
Scheduled			Modified	
Actual			By E0008 Date 1996-05-17-16.19.0	

- Priority
- Responsible person
- Date complete

Figure 21-2 is an example of an expanded work order format.

Another important element of the work order system is the ability and flexibility to *categorize* work. Different categories of work will be accomplished in different ways by an organization and by different components of the organization. Because an organization responds differently to various types of work, it is easier to identify and track progress through a simple means of categorization. Most facilities organizations categorize their workload into five distinctive types.

1. Emergency work
2. Preventive maintenance
3. Routine maintenance service
4. Minor work
5. Major projects

Emergency work is unplanned work that requires an immediate response. Normally life or facilities are at risk, and all necessary resources are brought to bear on the situation. The necessary response may include suspension of certain rules and procedures and the coordination of many departments or agencies. This type of work is often tracked separately to be able to identify costs and to provide information for future management decisions.

Preventive maintenance work orders are issued as part of a planned and controlled program of continuous inspections and corrective actions on identified building systems and equipment. The principal idea of preventive maintenance is that through a program of continuous and regular servicing, the service life of equipment can be extended, and equipment failure can be prevented. A detailed description of preventive maintenance systems follows later in this chapter. Most facilities managers believe that, with the exception of emergencies, preventive maintenance work orders carry the highest priority of the categories of work.

Routine maintenance service normally comprises day-to-day minor maintenance activities. The most common types of routine service work include rooms that are too hot or too cold, leaking pipes, malfunctions in door hardware, and lights that are burned out. Most organizations respond to these types of services on a first in/first out basis. Zone maintenance concepts and batching work orders by geographic areas are other approaches to accomplish routine service work.

Service work necessary to eliminate a hazard threatening life or property, to prevent significant disruption to scheduled activities, or to avoid significant financial loss is performed as emergency service work and normally carries the highest priority in any organization. The use of two-way

Figure 21-2

Expanded Work Order Format

Work Order 454545 | PAINT EXTERIOR/ROTUNDA |
Equipment BU0001 | Rotunda |
Requestor | KLINGEL | Phone | 555-5555 | Date | 1996-05-17-16.30 | Priority | 0 |
Location | 0001 | Status | WAPPR | Date | 1996-05-17.16.30.00 | Work Type | 3R | Area | U |

Responsibility	Job Details
Cost Center [39] Lead Craft [ ] Supervisor [ ]	Job Plan [ ]
Project [ ]	PM Master [ ]
Manager/Phone [WEEKS 555-6666]	Service Contract [ ]
Construction [ ]	PM Trade [ ]
Supervisor/Phone [ ]	Old Group/Zone [ ]
Construction [ ]	New Group/Zone [ ]
Inspector/Phone [ ]	Meter ID [ ]

Project Information	
Project # [ ]	Master Work Order
Contractor/Phone [XYZ PAINTERS 555-7777]	Master WO [ ]
Consultant/Phone [ ]	#Sub WOs [ ]
Problem Failure Class [ ] Problem Code [ ]	Page Down for More Details

Totals			Scheduling Information		
	Estimated	Actual		Start	Completion
Labor Hours	[ ]	[ ]	Target	[ ]	[ ]
Labor Cost	[ ]	[ ]	Scheduled	[ ]	[ ]
Material Cost	[ ]	[ ]	Actual	[ ]	[ ]
Equip. Cost	[ ]	[ ]	Estimated Duration [ ]	Remaining Duration [ ]	
Total Cost	[ ]	[ ]	Downtime Required? [ ]	Interruptable? [ ] Crew [ ]	

Accounting Information		Special Flags	
Cost Contract [ ]	Fiscal Customer Code [ ]	History? [ ]	
Cost Allocation [ ]	FM GL Code [ ]	VP/Unit [ ]	
Surcharge % [ ]	W.O. Budget [ ]	Utility Dist. Code [ ]	
Fund Code [ ]	Contract Amount [ ]	Department Code [ ]	
Object Code [ ]	Final Bill Flag [ ]	Calendar [ ]	
		Entered By [ ]	

Modified   By | E0008 | Date | 1996-05-17-16.30.00. | End of Screen 2

radios, cellular telephones, or pagers is essential in dispatching appropriate personnel to respond immediately to emergency situations. A well-trained and knowledgeable work reception staff can recognize an emergency situation and quickly notify the appropriate personnel to respond.

Minor work is normally described as discretionary nonmaintenance work, generally of nonurgent scope. This category of work might include moving furniture, making signs, or adding an electrical outlet. The minor work may have a minimum dollar limit or hourly limit, is normally not estimated, and is usually accomplished on a first in/first out basis.

Major work includes renovation projects, maintenance repairs of significant scope, new construction, or other projects exceeding an established minimum dollar limit. Major work is normally prioritized by urgency of need rather than by the order in which it was requested. Most facilities organizations plan and estimate major work and in general place a higher emphasis on management control of that work.

These particular categories of work that have been described may not fit every facilities organization, but the notion of categorization of work to provide a more effective means of managing the flow of the different types of work should fit within the framework of any work management system. It is important to communicate to the facilities organization as well as the institutional community about the different categories of work and the different methods in which the facilities organization responds to each category.

Work Authorization

The next step in the work control process is work authorization. Once the services of a facilities organization have been requested and then identified by a work order, a decision must be made as to whether to accomplish the work. This decision is essentially *work authorization*. Authorization should be based on several considerations:

- Scope of work
- Estimated cost
- Availability of funds
- Availability of workforce
- Impact if not done

After determining these factors, work may be authorized and assigned to the appropriate facilities management unit for performance. However, the decision may be made not to authorize the work. In fact, when the magnitude of the current deferred maintenance on college campuses across the nation is considered, there is evidence that a significant amount of work is unauthorized. Lack of funding or available workforce has resulted in a growing trend of identification of maintenance requirements without authorization to perform.

The responsibility for authorization of work should be clearly established and communicated within the facilities organization. The importance of this step necessitates the appropriate level of review and approval. Although work authorization has traditionally been a management function, there are many who argue that trends in empowering the workforce are shifting the decision on work authorization to field personnel. Again, the size of the institution and the facilities organization will determine the most appropriate level of work authorization within the organizational structure. Because the final decision on authorization ultimately becomes a funding issue, those responsible for budget management are normally responsible for work authorization.

Once funding is identified, the work order is assigned to the organizational component responsible for completion of the work. Choices may include a work center or shop within the facilities organization, a service contract administrator, or the procurement officer responsible for competitively bidding work. In most large facilities organizations, the responsibility for assigning work falls in the work management area. Depending on the size and organizational structure, assigning work might be the responsibility of the work control unit, the scheduling office, or, in the small organization, the facilities director. Regardless of how a particular institution decides how to assign work, the responsibilities and procedures for this work management function should be clearly established and communicated.

Planning

Planning simply is decision making—deciding on the best strategies for the effective use of available resources to meet goals of the organization. Planning precedes action and must consider the goals and objectives established by the facilities organization in order to decide on the best alternatives to meet those goals. Planning can include such activities as establishing goals and objectives, developing policies and procedures, setting priorities, estimating, and scheduling. Most facilities organizations think in terms of long-range planning programs, short-range planning, and individual job planning.

Long-Range Planning   Considering the goals and objectives of the facilities organization within the mission of the institution, managers must think in terms of where their programs should be in future years. Long-range planning in a maintenance management program considers the multiple-year, cyclic maintenance activities such as:

- *Paint program.* How often should one plan to paint the exterior of facilities? Interior spaces? Is there an established frequency, and if so, what is the resource requirements in future years?
- *Roof program.* Roof replacements, based on the life of the roofing system, are fairly predictable. Long-range planning permits the facilities manager to program funds and resources in future years.

- *Mechanical systems audits.* Through periodic audits, mechanical systems such as elevators or major heating, ventilation, and air conditioning (HVAC) equipment can be programmed in future years for replacement or major upgrades.
- *Capital renewal.* Through long-range planning of capital renewal programs, the facilities manager can make better short-range maintenance decisions.
- *Building systems replacements.* Interior building systems such as finishes, lighting retrofits, or fixture replacements can be programmed on a future multiyear cycle.
- *Road and walk program.* How often will roads and walks be resurfaced or replaced? How much is budgeted each year for road and walk maintenance?
- *Space administration.* What is the institution's goal regarding growth? Are there standards regarding classroom utilization?
- *Staffing/training.* How will the facilities organization change in the next 10 to 20 years? What is the anticipated attrition? What types of training are required to prepare the staff for future technology?
- *Shop equipment/vehicles.* What type of replacement program is in place for vehicles and other equipment? How will these replacements be funded?

Long-range planning encompasses those predictable recurring activities that allow the facilities manager to make better decisions in programming resources. Ideally, long-range planning sets the stage for more effective short-range planning.

Short-Range Planning Although long-range planning is based on organizational direction over several years, short-range planning considers a horizon of three to six months. It is the process of balancing workload—both current and anticipated—and resources. To determine an effective use of resources, one must realize the organization's workload in quantifiable terms. Typically, organizations measure workload in either number of hours required to accomplish work or number of dollars required to accomplish work. Through measurement of the current workload and anticipating future workloads, managers are able to make informed decisions on resource levels required to accomplish that work. Figure 21-3 as an example of a monthly workload based on predicted work hours.

The facilities manager has a variety of resources available to accomplish work. These resources, however, must be balanced with the workload so that there are enough people and materials to accomplish the work, but not more people and materials than needed. Most facilities organizations combine productive hours spent by their facilities organization staff with

Figure 21-3
Sample Monthly Workload

REPETITIVE/PREDICTABLE	HOURS/MO	NONPREDICTABLE	HOURS/MO
Preventive maintenance	2400	Major work	3000
Routine service	2000	Emergencies	250
Minor work	1800		
Total	**6200**		**3250**

hours spent by contracted resources to be consistent with workload. This combination allows an organization to be more flexible in adjusting to peaks and valleys of work.

Individual Job Planning   The third level of planning is that of individual job or project planning. Job planning usually includes such items as scoping, estimating, scheduling, and procurement—the activities that precede execution of the project. Which jobs require this planning effort? The majority of activities performed by a facilities organization include preventive maintenance and service work requiring small amounts of labor and material. Job planning is not usually desirable on those types of work orders. Individual job planning should be approached on a needs assessment basis. Managers must weigh the cost versus the benefit of tight planning and control on individual jobs. Many organizations establish an hourly cutoff point or a minimum dollar volume as the decision-making factor for determining whether to plan a job. An institution must decide for itself what criteria it will use for that distinction. It is becoming apparent in today's world of providing excellent customer service that it is important to invest some planning effort into *all* jobs being requested and funded by customers of the facilities organizations.

Some of the activities included in individual job planning are as follows:

- Determine the customer's expectations. What does the customer think the cost should be? When does the customer need the service?
- Determine necessary materials, equipment, and time frame. What materials will be needed and how soon? What equipment will be used, when does it need to be available, and for how long will it be used?
- Determine necessary labor and time frame. Which crafts will be required to accomplish this work, and how long will they be occupied?
- Determine necessary contractual support.
- Estimate cost.
- Determine current backlog of work. How does this particular job fit into the current backlog? What is its priority?

- Determine method of accomplishment. Should the entire job be contracted?
- Determine proposed schedule. What is the target start and the target completion of work?

The elements of job planning lead to a discussion of two of the more important elements in an effective work management program: estimating and scheduling.

Estimating

Estimating is a fundamental tool of a work management system. Whether estimating the cost of maintenance work to be performed within the facilities organization budget or the cost of projects to be funded by customers of the facilities organization, the success and reputation of the organization largely depends on the ability to accurately assess the cost of work. Large facilities organizations often have a component of the organization staffed by people whose primary duties are planning and estimating work. Other organizations may rely on shop supervisors to estimate work as necessary. In smaller organizations, perhaps the director or senior manager provides most of the estimating services.

There are many methods used for providing estimates for facilities related services:

1. *Predetermined time standards.* One of the most useful and accurate methods of estimating is application of predetermined time standards. The Navy's Engineered Performance Standards (EPS) has been a popular source of maintenance task estimating in facilities organizations for many years. There are other organizations producing labor hour standards for maintenance and construction tasks that have been used effectively. Most of these companies offer training programs on the use of their standards, and the standards books are updated on a periodic basis to reflect economic times or changes as a result of work sampling studies. Most of the companies offering estimating standards packages have developed computer software applications, making the estimating task much faster and easier.

2. *Historical cost of similar work.* Many facilities organizations have developed histories of cost per square foot for various types of project work. These historical records can be an effective tool in determining the anticipated cost of similar work. In order for historical cost to an effective source of accurate estimating, the sample of projects must be fairly large.

3. *Experience.* An estimator's experience can be an accurate estimating method for many types of projects. However, most estimators have extensive experience in only a small number of trades.

4. *Professional estimators.* Many facilities organizations have contracts with professional estimators for larger projects. One advantage of a consulting estimator is objectivity.
5. *Contract data.* Extrapolation of recent contracted services can be an effective estimating tool. Unit rate contracts and delivery order contracting can also be used effectively to estimate the potential cost of a project.

Most facilities organizations use a combination of all of these sources of estimating tools. No single estimating method is the right one for every project. The size of a project, its complexity, its priority, the skills of the assigned estimator, and other factors may influence the determination of which estimating methods would best serve the purpose.

The typical large project estimate and job plan will include the following:

1. Detailed list of materials required, prices, delivery time, and potential vendor
2. Estimated labor hours by craft
3. Necessary equipment
4. Subcontract costs
5. Basic sequence of work
6. Estimated duration/project schedule. The duration should include the procurement stage, construction stage, and occupancy stage
7. Project responsibility assignment

The estimates are used for many purposes, some requiring more accuracy than others. The importance of estimating can be explained by considering how estimates are used:

- To help determine the size of the workload
- For scheduling purposes
- To measure performance on an individual project or an individual shop
- To compare in-house costs with contract costs
- To plan future staff levels
- For maintenance budget forecasting
- As a means to quote costs to customers requesting facilities services

Offering to perform work funded by customers through a chargeback system has become more prevalent on college campuses since the 1980s. An emphasis on responsibility center accounting and charging requirements in facilities as a result of communication and research technology have contributed to the frequency in which facilities organizations are performing services to customers on a contractual basis. Thus, the importance of accurate estimating has significantly increased. It is one thing for an organization to estimate its own work and find out later that the work cost more than had been planned—that is often disappointing. It is quite another matter to estimate costs for a customer who is providing funds and then realize the work cost significantly more—this can be disastrous. For years facilities organizations have prepared estimates for customers funding work. Those estimates were exactly that: an estimate of what the work would cost. There were no guarantees. The fund provider would raise funds based on the estimate and hope the job came in under or on budget.

In more recent years, however, facilities organizations have become more businesslike and more entrepreneurial and have begun to offer work to customers under a variety of contract options:

- *Time and materials.* This is the oldest and most frequent type of contract between facilities organizations and institutional customers. It costs the facilities organization what it costs the customer. Estimates are prepared and offered, but again, there are no guarantees. The fund provider assumes all the risk. If the cost of a project comes under the estimate, everyone is happy—the estimator, the work crew, and the customer. However, if the cost of the work exceeds the estimate, no one is happy. The customer is outraged, and the estimator and work supervisor are pointing fingers at each other. For many projects, however (e.g., small projects and projects of indefinite scope), this contract method is often the preferred option.
- *Fixed price.* For a project with a definite scope, detailed estimates can be prepared and offered to fund providers on a fixed price basis. Regardless of what the facilities organization spends in performing the work, the cost to the customer is the agreed-upon fixed price. In this option, the facilities organization assumes the risk. The customer is generally happy because the cost of the project is known before the project is begun. There may be a tendency, however, for estimators to "pad the estimate" to avoid an unpleasant review if the facilities organization realizes a significant loss. Customers may become aware of this tendency in a noncompetitive arena. An explanation of the estimate and a negotiation of a final price agreement usually results in a fair price and a satisfied customer.
- *Guaranteed maximum price.* The guaranteed maximum price format offers the advantages of time and materials for a customer with the security of a maximum price. Essentially, a guaranteed maximum price proposal will offer a customer billing for time and materials up to, but not exceeding, the agreed-upon guaranteed maximum price. This type of contract works well on projects of indefinite scope or fast-track projects. Often a general scope and price can be agreed upon and procurement and construction can start while the design is being finalized. As an incentive for both the facilities organization and the funding customer to complete the work under budget, a formula of shared saving can be derived: if the completed project costs are under the guaranteed maximum price, the savings will be shared based on a negotiated percentage.

- *Unit rate contracting.* Another type of contracting used by facilities organizations is unit rate contracting. For well-defined or repetitive type services, the facilities organization may estimate and propose a unit rate for those services. That rate is then used as the basis for final costs depending on the number of units required.

Scheduling

Another important aspect of the planning component of the work management system is scheduling. Scheduling is the assignment of labor resources over a period of time through a specific project or work process. A formalized scheduling system in a facilities organization can provide the following benefits:

1. It provides a systematic, coordinated approach to matching resources with requirements. If a facilities organization has quantified its workload and labor resources, a scheduling system will help coordinate efforts.
2. It levels the workload. The facilities organization's workload can be a series of peaks and valleys. One month it seems there are more projects than one can ever complete; other months some trades are looking for work to keeping them productively employed. A scheduling system can help anticipate these peaks and valleys and assign resources to avoid overloaded and nonproductive times or to allow planning for any outsourcing needs.
3. It provides a basis for evaluating actual work versus planned work. A scheduling system can help managers and the workforce determine how well work is being executed in comparison with work plans.
4. It satisfies organizational priorities. A formalized scheduling system can ensure that the staff is performing work in the order that management intended. The tendency in a decentralized unscheduled organization is to work on projects that are enjoyable or that are for favored customers. Scheduling systems will ensure that work is performed in accordance with the organization's procedures and priorities.
5. It is an effective communication tool. An effective scheduling system requires frequent communication between management and staff and among the separate trades and work centers.
6. It establishes commitments/customer service. A scheduling system can be an effective tool for customer communication. A schedule will establish start dates and completion dates for customer-required work. The risk in establishing these commitments is outweighed by the value in customer service.

The size and complexity of an organization should determine how formal and complex the scheduling system should be. A scheduling system can be a sophisticated computerized system, capable of a detailed plan of assigning individual labor, materials, and equipment. In some settings a manual schedule assigning work crews to generalized types of work may be the most effective. We will explore in this section the basic foundations of a scheduling system and its application to long-range, weekly, and daily scheduling.

The first step in establishing an effective scheduling program is planning—that is, determining resource availability. This is accomplished by measuring the number of productive hours during a given time period for different types of work. To measure available productive hours, one must determine the percentage of nonproductive time anticipated by employees. This nonproductive time could include supervision, leave, training, administrative activities, and other miscellaneous activities requiring employees to be away from the work site. Productive work usually ranges between 1600 and 1800 hours per person per year.

The productive hours will be applied toward the accomplishment of the different types of work categories established by the organization. A decision can be made on how to allocate work hours to service work, preventive maintenance, or major work orders. A determination of resource availability can influence decisions on method of accomplishment of work. Zone maintenance concepts, batch maintenance, and on-demand contracts are examples of methods of accomplishing work that will impact resource availability.

The second step in an effective scheduling program is the planning process of determining the organization's workload. There are essentially two separate classifications of work in most facilities organizations. First, there is the repetitive, predictable, somewhat constant level of workload made up of preventive maintenance, service work, and minor work. Although this type of work is generally not estimated, this level of workload can be fairly predictable through trends of previous months and years. It is relatively simple to establish the anticipated hours per month the organization will spend in this class of work (see Figure 21-3). The second major classification is that of major work. This is the less predictable, sometimes seasonal work requirement that can lead to the peaks and valleys of workload previously described. By effectively planning and estimating major work, however, this workload can be translated into a predictable, orderly component of the organization's overall workload.

Most work management programs approach scheduling on a long-range, weekly, and daily basis. There is a different emphasis and a different set of responsible players for each of these levels.

Long-Range Scheduling   Most long-range scheduling programs, sometimes known as master scheduling, have a time horizon of three to six months and concentrate on major work. Long-range scheduling is an attempt to break down estimated projects by number of hours per craft per month and distribute that commitment over monthly increments. The scheduler makes decisions based on overall work hour availability, craft availability, priority of projects, material delivery schedules, and customer and seasonal needs. Figure 21-4 is an example of a typical master schedule.

In most organizations the scheduler, with input from both management and shop supervisors, develops and maintains the master schedule. The master schedule becomes an effective communication tool for the organization describing the plan for the workforce in the coming months. It is also an important customer relations tool, allowing the organization to be more effective in making commitments to customers for future projects. Of course, a master schedule is never cast in stone. Priorities change, emergencies arise,

Figure 21-4
Typical Master Schedule

LONG RANGE SCHEDULE

SHOP:
Carpentry
WORK HOURS

JOBS or PROJECTS	Aug	Sep	Oct	Nov	Dec	Jan
6387 ABC Hall	190					
8522 XYZ Dorm	150					
3710 Bldg 8						
2232 Bldg 10-Room 201	40	160	160			
6757 Bldg 11-Room 10	40					
8700 DEF Hall						
New Addition-XYZ Hall		100	50	50	50	
Dean Dokes Office			80			
Alumni Bldg Roof		180			120	
Admin Bldg Renovation						
Stairway West Parking		30	160	160	180	180
Bldg 8-Rooms 101, 102, 103				160		40
Total load (workhours):	**420**	**470**	**450**	**370**	**350**	**220**
WORKHOURS CALCULATION						
Capacity (workhours/workday)	32	32	32	40	40	40
Number of workdays in period	23	20	22	21	16	20
Planned workhours available	736	640	704	840	640	800
Total load (workhours)	420	470	450	370	350	220
Net workhours available:	**316**	**170**	**254**	**470**	**290**	**580**

and actual labor differs from estimates. The master schedule must be a dynamic document, constantly adjusting to the changes in the organization's plans. It also serves as the basic work sheet for an organized approach to weekly scheduling.

Weekly Scheduling The weekly schedule is the plan for assigning labor resources to specific work categories or jobs in the coming week. The weekly schedule is normally developed on a Wednesday or Thursday of the preceding week in a negotiation process by the scheduler and shop supervisors. The assignment of labor resources is based on several factors:

1. Available work hours. This can be affected by planned leave, holidays, attrition, and other factors.
2. Available materials and equipment. To accurately schedule, materials planners must communicate realistic delivery dates for necessary materials.
3. Rate of success in the current week's schedule.
4. Priorities. The overall plan of the master schedule becomes a guide in developing priorities for the weekly schedule.

A weekly schedule does not necessarily define individual mechanics' daily work, but rather the number of hours by a particular shop or work center to be spent each day on specific work orders. In the case of preventive maintenance and service work, because each work order may be too small to effectively schedule, most weekly schedules assign a number of hours per day by each shop for that type of work category. Figure 21-5 is an example of a typical weekly schedule.

As mentioned, the weekly schedule is developed through a negotiation process. Work supervisors discuss progress by their particular trade on multicraft projects, adjustments are made for changes in the current week, and the scheduler represents management in prioritizing efforts. Through this weekly interaction, the scheduler is able to assemble plans from each work center and produce a plan for the next week's efforts that will be available for all in the organization to work toward. Through these weekly negotiations, the scheduler is also able to update the master schedule as necessary. Significant changes in the master or weekly schedule enable the scheduler to notify customers of changes in commitments and management in changes in priorities.

Daily Scheduling The daily schedule is actually a by-product of the weekly scheduling process. The foreman, or shop supervisor, based on the weekly schedule, makes daily decisions regarding the assignment of labor resources. Using the weekly schedule as a guide, the foreman adjusts to absences, emergencies, changes in scope, materials delays, or other factors that change the weekly plan. Supervisors are able to use the master

Figure 21-5
Typical Weekly Schedule

WEEKLY WORK CENTER PROJECT LOG

Shop: *Carpentry*
Week ending: *8/1/96*
Number of personnel: *4*

		DAILY ACTIVITY					
	MO	TU	WE	TH	FR	SS	TOTAL
Total hours	32	32	32	32	32	32	192
Overhead: Leave	16	0	0	0	0	0	16
Other	0	8	0	0	8	0	61
Fixed assignments	0	0	8	8	8	0	24
Available for scheduling	16	24	24	24	16	32	136

Description	Work Order #	Hours Scheduled						Weekly Total
		MO	TU	WE	TH	FR	SS	
Bryant Hall, doors	9282	16	10	0	0	0	0	26
Ruffner, install casework	9566	0	10	10	0	0	0	20
Police station, mailbox	9578	0	0	0	6	0	0	6
New Cabell, floor tile	9666	0	0	0	0	8	0	8
Slaughter Rec, lights	9872	0	0	8	0	0	0	8
Mem Gym, painting	9884	0	0	0	12	6	0	18
Olsson, roof and gutters	9955	0	0	0	0	0	24	24
Jag School, renovation	9987	0	0	4	4	0	0	8

and weekly schedules to maintain priorities and understand how changes in their shop's schedule may affect others, resulting in enhanced communication.

An effective scheduling program requires a management commitment. A proactive approach to planning and scheduling will increase the effectiveness and organization of work by the entire work unit. Management guides the organization through staffing levels and priority settings. Shop supervisors are able to carry out the policies and procedures of the organization through an organized, coordinated fashion. Customers are more informed on realistic commitments, and expectations become more consistent with the ability to perform. The number of reactive crisis projects, forgotten work orders, and miscommunication between shops are greatly reduced through an effective scheduling system. Obviously, the larger and more diverse a facilities operation is, the more important it is to have a modern scheduling operation. The use of computers and scheduling software has made the task much simpler and easier to manage. In past manual systems, a change by one trade could ripple through the entire schedule requiring countless manual changes. In current automated scheduling systems, one keystroke can accommodate that same change in only seconds.

Tracking and Reporting

Tracking work progress and reporting on work progress is another important part of the work control process. In tracking the progress of work an organization is assessing major milestones of a job: when did or when will the job start, and when was or when will the job be completed. Reporting on work is the process of communicating with management and customers the current progress and the current plans for a project.

Tracking work comes in several forms. Weekly schedule compliance is an effective method of tracking progress. For instance, how close was the actual weekly execution of the work in relationship to the plan developed in the weekly schedule? Another method of work tracking, exception reporting based on performance standards, is particularly effective for preventive maintenance and service work. An organization may set performance standards for the number of days for completion of minor work or a maximum allowable backlog of work. Through exception reporting on incomplete work orders, management is able to track performance versus plans. For major projects it is beneficial to track progress on an individual project basis. During the job planning process, a series of milestones are normally established. As the project is accomplished, variances in scheduling and tracking should be reviewed.

Reporting to customers on major projects accomplished is an important communication tool for facilities organizations. The work management system should develop methods for up-to-date reporting on the progress of major projects to customers on a routine basis. In large institutions, it is not uncommon to have literally hundreds of major projects under way at any given point in time. Accurate information on project status is a fundamental requirement for healthy customer relations. Even when there is bad news, customers will appreciate that more than no communication and an eventual unpleasant surprise. Figure 21-6 is an example of a typical progress report sheet for a major project.

Customer service and customer relations are playing an ever-increasing role in the facilities business in higher education. Issues such as privatization, budget distribution, and evolving technology make it increasingly important that the facilities organization not only performs, but is perceived as an able performer by the institutional community. Therefore, facilities organizations across the country are investing more time and management energy in customer relations and customer communications. Reporting on the status of customers' work is one of the

Figure 21-6
Typical Project Information Report

Run Date: 06/13/96   INTERNAL PROJECT MANAGEMENT REPORT   Page: 2

Project Number: 045056
Project Desc.:   PROVIDE ACCESSIBILITY FOR ROUSS HALL
Requestor: BILL BOHN   Building: ROUSS HALL HANDICAPPED
Request Date: 08/12/95   Consultant:
Project Manager   BILL BOHN   Construction Supt.: Bob Thompson

Standard Task	Planned	Current	Actual	% Complete
PIR Requested				
PIR Completed	05/10/96	05/13/96	05/20/96	10%
PIR Approved	05/10/96	05/25/96	05/30/96	
Contract Awarded	06/01/96			33%
Preliminary Drawing Design Approved				60%
Working Drawing Design Approved				
Construction Start				
Construction Completed				
Closeout				

Status Comment: Project on hold, awaiting approval of capital funds.

most important ways to communicate and will be expanded on in the next section.

We have examined the work control process and discussed its major elements: work identification and categorization, work authorization, planning and scheduling of work, and tracking and reporting on work. In most organizations the majority of people are in the business of performing work. The work management system allows managers to manage what, when, how, how much, and how well the organization performs this work. The work management system can be complex and computerized, with full scheduling and tracking controls, or more informal, with a minimum of control. An organization must find the right balance of control to enable it to meet its goals and objectives in supporting the institution's mission. A work management system should work for the customers of the organization and for the staff performing the work of the organization. It should be a support system designed to make the organization's production more organized, more customer responsive, and more cost-effective.

Customer Relations

The emphasis on customer service in the facilities management arena has never been stronger. The growing competition for students, quality faculty members, and financial support for educational programs has required institutions to become effective in marketing their educational services. This also applies to facilities organizations. Facilities managers must develop a keen sense for who their customers are and how to best meet their expectations. It is no longer adequate to be technically competent and cost efficient. To be successful, a facilities organization must be perceived by its customers as an organization dedicated to providing services with the customer in mind. Chapter 5 covers this topic in detail. This chapter will explore the functional role of work management systems in supporting and enhancing customer relations.

Ask members of an institution's faculty and staff to name the three most important ways their facilities organization could improve, and the most likely response will be, "improve communications, improve communications, and improve communications." Establishing and building a timely and meaningful dialogue between the facilities organization and its customers is the first step in effective customer relations. The work management component of an organization is in an excellent position to develop systems for customer communications.

Facilities Representatives As mentioned earlier, establishing an effective network of facilities representatives, or building managers, is an excellent means of two-way communication between a facilities organization and building occupants. Regardless of the services being provided (e.g., preventive maintenance, service work, major projects), people working, teaching, and engaging in research want to know as much as possible about the facility. If they have asked for a service to be provided, they want to know when it will be provided, who will provide it, how it will be done, and how long it will take. The facilities representative concept is an effective means of communicating this information to a representative of the building occupants. Just as the work management component should be the primary source of contact for the facilities organization, the facilities representative should be the primary source of contact for a department or a group of departments within a building. Although it is important for all parts of a facilities organization to support the facilities representative concept, it should be a priority for the work management component to encourage the community to endorse a network of facilities representatives.

Customer Survey Another effective means of customer communication is through a customer survey program. Such a program, designed to assess levels of customer satisfaction on facilities services, has several benefits. Foremost, it provides an avenue for customers to tell an organization which parts of the service they are happy with and which parts of the service need improvement. By assessing customer satisfaction levels on items such as quality of work, responsiveness, communication, and cost, an organization can learn more about the expectations of its customers and how to better meet those expectations. Second, a survey program allows a facilities organization to influence the expectations of its customers by describing the organization's abilities or performance standards as part of the survey instrument. The facilities organization is able to both educate the customer and seek customer input. Third, a survey program is normally seen in the eyes of the customers as a genuine effort on the part of an organization to increase communication and feedback. Customers appreciate the opportunity to be heard, and assuming an organization reacts positively to constructive criticism, customers become part of the improvement process of the facilities organization.

Customer Service Representatives Another example of increasing communications is the assignment of customer service representatives to groups of customers. The customer service representative, although an employee of the facilities organization, acts as a true customer advocate. Different departments and programs throughout higher education have different sets of needs. Needs that are important to the chemistry department and its research programs may be completely different from those that are important to the intramural recreation department. An interrupted utility system, altered cleaning schedules, or interrupted access to classrooms can have different effects on different educational programs. The customer service representative can learn the needs of different customer groups and educate the facilities organization on ways to best respond to the various needs of these different groups.

Through an effective work control process, the work management component of an organization is in a position to provide information to customers on the status of work in their facilities. Work reception, scheduling, and reporting enable the work management component to establish a regular pattern of communication regarding the status of services. Different customers may require different levels of communication. Some customers may not be interested in routine day-to-day work such as preventive maintenance, but will be interested in major construction or repair work. Other departments may want to know every time a representative of the facilities organization enters the building. It is the responsibility of the work management component to learn its customers' expectations, establish an effective network of communications, and respond accordingly to the various requirements throughout the community.

21.5 PREVENTIVE MAINTENANCE

For a management program to be most efficient and meet the stated goals and objectives of the maintenance management effort, a significant emphasis must be placed on preventive maintenance efforts. Preventive maintenance can be defined as a planned and controlled program of continuous inspections and corrective actions taken to ensure peak efficiency and minimize deterioration. Merely reacting to equipment breakdowns is not an acceptable method of managing the maintenance effort. A proactive approach to systems and equipment maintenance in which certain precautionary maintenance activities are performed will enhance the life and dependability of such equipment and systems. Regardless of the size and resource base of an institution, the preventive maintenance program is a key to efficient maintenance management.

The analogy of preventive maintenance on an automobile can be helpful in applying preventive maintenance concepts to an overall maintenance management program at an institution of higher education. Oil changes, tune-ups, and state safety inspections are examples of preventive maintenance that add to the life and efficiency of an automobile. These same principles apply to equipment and systems in facilities. Typical preventive maintenance work includes inspection, cleaning, adjustment, and lubrication of equipment; minor parts replacement; analysis and testing; and minor repairs. In addition, reporting on the results of the preventive maintenance inspection or consideration of further maintenance activities is an instrumental part of a preventive maintenance program.

Preventive Maintenance Objectives

A well-planned preventive maintenance program is instrumental for the facilities organization to meet the objectives of its maintenance management effort. A primary objective of a preventive maintenance program is to significantly reduce the frequency of unscheduled breakdowns and downtime of critical equipment and systems. Emergency generators, elevators, and HVAC equipment are critical to the operation of a facility and its activities. For educational and other programs to be successful, facility support of critical equipment and systems must be reliable. A preventive maintenance program in which these critical pieces of equipment are tested and inspected on a routine, scheduled basis greatly enhances performance reliability. It is up to the institution and the maintenance manager to determine which equipment is most critical and therefore requires the most preventive maintenance attention.

Second, a preventive maintenance program greatly enhances maintenance planning efforts. Through estimating and scheduling planned preventive maintenance activities, the maintenance workshop is engaged in more predictable, planned activities. The very notion of a preventive maintenance program encourages the development of accurate and comprehensive equipment inventories. A well-planned and effective preventive maintenance program significantly reduces emergencies, thereby avoiding the need to divert workforce to unplanned activities. Finally, through the reporting process inherent in preventive maintenance, future maintenance activities can be planned on a systematic basis.

Another objective of a preventive maintenance program is the reduction of overall maintenance costs. By reducing emergencies and unscheduled breakdowns, less overtime is required. Although the cost of a preventive maintenance program is considerable, it is generally recognized that these preventive measures result in a lower overall cost.

With these objectives in mind, the benefits of a planned preventive maintenance program are too great to ignore. These benefits include the following:

- Reduced equipment downtime
- Extended service life for equipment, systems, and the facility
- Improved overall appearance of the facility
- Reduced energy consumption
- Improved safety
- Improved reliability and thereby increased customer confidence
- Reduced cost

Establishing a Preventive Maintenance Program

All preventive maintenance programs contain the same basic elements and are developed in three basic phases: planning, implementation, and continuous review. The *planning phase* is by far the most time consuming. It requires a considerable investment of labor and financial resources but is essential to an effective program. The planning phase includes the following tasks:

- *Identifying the equipment inventory.* What pieces of equipment and building systems require maintenance on a routine basis?
- *Determining tasks.* What are the preventive maintenance activities for each piece of equipment and building system identified?
- *Determining frequencies.* How often will the work task be repeated?
- *Setting schedules and estimates.* When will the work be accomplished, by how many people, and at what cost to the institution?

Once the planning phase has been completed, the shorter *implementation phase* is carried out. The implementation phase consists of distributing the work orders, establishing shop scheduling controls, and starting the process of performing assigned preventive maintenance activities.

Finally, the *continuous review phase* is the process of performing, evaluating, and modifying the actual preventive maintenance work. As preventive maintenance activities are performed, management monitors both labor productivity and equipment productivity. Continuous feedback from shop personnel performing preventive maintenance activities provides management with valuable data for planning future maintenance activities. This continuous evaluation of both process and results allows for continuous improvement to the preventive maintenance program.

Equipment Inventory

The most important element in planning a preventive maintenance program is developing and maintaining an accurate comprehensive equipment inventory. The most accurate method of obtaining a complete and up-to-date list of equipment is through a physical inspection of all facilities. These inspections can be performed by tradespeople or planners from the facilities organization staff or by engaging a consultant specializing in development of preventive maintenance inventories. Often equipment inventories can be developed from as-built construction documents, but such inventories will normally be accurate only on relatively new facilities. It is important to prioritize the type of equipment to be included in inventories and to develop procedures for updating those inventories. The policies and procedures must be in place to recognize changes in inventory as a result of alterations, new construction, and deletion of equipment.

Once the equipment inventory is complete, an equipment record is developed for each individual piece of equipment or building system. The equipment record, at a minimum, should include the following data:

- *Equipment number.* For computerized tracking, an equipment number should be established for each piece of equipment requiring preventive maintenance activity. Depending on the type of computerized system used, the number can be "smart," signifying characteristics of the equipment, or merely sequential or random. If sequential or random numbers are used, other fields in the equipment record will provide necessary identification.
- *Description.* Each equipment record has a description signifying the type of equipment or building system.
- *Location.* Each equipment record indicates the facility in which the equipment is located and the location within the facility.
- *Nameplate data.* The nameplate data further describe the equipment by listing the manufacturer, model, size, voltage, and other distinguishing characteristics.
- *Parts list.* Equipment records should include major components and parts along with corresponding suppliers.
- *Manufacturers' operation and maintenance data.* The equipment record should include pertinent information on the manufacturers operation and maintenance manual for a quick reference. Figure 21-7 shows an example of a typical equipment record.

Preventive Maintenance Work Orders

The next element in the planning phase of a preventive maintenance program is establishing the preventive maintenance work orders. The work order in this case provides the actual written instructions for the preventive maintenance activities to be performed on a piece of equipment or a building system (Figure 21-8). This can be one of the most difficult and time-consuming tasks in developing the program. One needs to determine which preventive maintenance tasks the technician should actually perform and with what frequency. On the one hand, too many

Figure 21-7

Typical Preventive Maintenance Equipment Record

```
06/12/96              UVa Facilities Management              PAGE: 1
16:04:59              4P  Preventive Maintenance    WEEK    WOPRINT
====================      Maximo W/O:169112      ====================
PM Trade: 16   Date Scheduled: 11/07/95
Department: ZM Day Frequency: 365
FM Cost Center: 84     Next Scheduled: 11/07/96
Location - Bldg: 0256 Group/Zone:
Job Plan(s): U8630
Equip/Machine: EU030328                                            |
Equipment: FAN COIL UNIT (256, FLR 1, W. ALCOVE, NO 02)
Equipment Long Description (Nameplate Data):

SIZE                  (U) SR1203S
INVENTORY NO.         (U) 1362.00
QUANTITY              (FILTERS)) 2
MODEL                 (M) CR6307A
MOTOR                 (M) 1/6 HP, 115/60/1, 1050 RPM, 2.4 AMPS
CAPACITY              (U) 5.0 GPM, 1010 CFM
W.O.                  270256
UNIT NO.              02
MANUFACTURER          (U) MARLO COIL
LOCATION              256-CHEMISTRY BUILDING, WEST ALCOVE, FLOOR 1
REMARK                -FILTER 1'' X 9'' X 37 5/8''
                      -HOT & CHILLED WATER SYSTEM
LUBRICATION           (F, FM) SEAL
DRIVE                 DIRECT
STARTER               MANUAL
             ***  End of Equipment Long Description  ***
```

tasks or too high a frequency will overload a preventive maintenance program and make its implementation impossible. On the other hand, eliminating or overlooking critical tasks could lead to unnecessary maintenance failures.

In writing preventive maintenance instructions, also called a *job plan,* it is helpful to consider the following:

- *Standardize for each class or type of equipment.* Preventive maintenance instructions should be standard to common components when possible and should be developed according to like items or groups of items. Once a set of preventive maintenance instructions has been developed for a built-up roof on one facility, for example, that same set of instructions should apply to other built-up roofs throughout the institution. In the same vein, air handling units, regardless of manufacturer, will likely contain the same basic set of instructions. Standardizing instructions for common types of equipment is an

Figure 21-8
Typical Preventive Maintenance Work Order

```
06/12/96           UVa Facilities Management              PAGE: 1
16:04:59           4P Preventive Maintenance    WEEK      WOPRINT
===================   Maximo W/O:169112    ===================
Job Plan Number: U8630
Job Description: FAN COIL UNIT.
Job Crafts:
Job Materials:
Job Tools (Equipment):
Job Operations:
          1- CHECK MOTOR.
             1-CHECK CONNECTIONS AT TERMINALS.
             2-CHECK INSULATION RESISTANCE . . . . MEGOHM(S).
             3-CHECK TEMPERATURE.
             4-CHECK BEARINGS WITH STETHOSCOPE.
             5-CHECK MOUNTING SOLIDITY.
             6-LUBRICATE (IF APPLICABLE).
             7-CLEAN.
          2- CHECK MANUAL STARTER.
             1-CHECK CONNECTIONS AT TERMINALS.
             2-CHECK THERMAL RELAY (IF APPLICABLE).
             3-CHECK VOLTAGE . . . . VOLTS.
             4-CHECK AMPERAGE . . . . AMP(S).
             5-CHECK GROUND.
             6-CLEAN MOUNTING.
          3- CHECK (IF APPLICABLE).
             1-PILOT-LAMPS.
             2-PUSH-BUTTON(S) MECHANISM.
             3-CONTROL(S).
             4-THERMOSTAT.
          4- CHECK FAN.
          5- CHECK PLUMBING COMPONENTS.
             1-CLEAN COIL (USE CARE TO PREVENT FLATTING OF FINS).
             2-CHECK COIL CONDITION.
             3-CLEAN STRAINER.
             4-CHECK STOP-VALVES.
          6- CHECK FILTER(S).
             1-CLEAN OR REPLACE IF NECESSARY.
          7- CHECK GENERAL FUNCTIONING.
          8- CLEAN UNIT.
          9- CHECK AND CLEAN CONDENSATE PAN.
             1-REMOVE OBSTRUCTIONS FROM AROUND DRAIN.
             2-CHECK CONDENSATE DRAIN LINE.

           ***  End of Equipment Job Operations  ***

******************  End of PM W/O :  169112  ******************
```

effective means of establishing the developmental stage of preventive maintenance work orders as long as feedback systems are established to allow the modifications that apply to unique pieces of equipment.

- *Use manufacturers' operation and maintenance manuals.* Most operating manuals or manufacturers' brochures will recommend some type of preventive maintenance schedule. If manuals are not on hand, they can be obtained from the manufacturer or other institutions that have already implemented a preventive maintenance program and have the same type of equipment. It is important to know that in some cases an institution's risk management program may require that manufacturers' operation and maintenance manuals be followed in order to qualify for insurance reimbursements.
- *Consult with tradespeople and supervisors.* One of the best sources of information is located within the facilities management organization. Supervisors and technicians who have been maintaining equipment over a long period know the tasks they perform when the equipment fails and usually have a good idea of what causes failure or, better yet, what might prevent it.
- *Emphasize preventive activities.* Preventive maintenance instructions should emphasize activities such as cleaning, adjustment, lubrication, alignment, and investigative actions such as checking for heat, vibration, noise, and overall operation.
- *Separate job plans.* Equipment often requires different levels of preventive maintenance activities throughout the maintenance cycle. For example, an emergency generator may require a complete overhaul annually, a less intensive check quarterly, and merely a test start-up monthly.

Determining Frequencies

Once a preventive maintenance job plan is developed, the frequency in which this work is accomplished must be considered. In determining frequencies, the following factors should be considered:

- *Impact of downtime.* Some equipment breakdowns are merely a nuisance and can be repaired with little or no impact on the operation of a facility and its programs (e.g., clocks, bathroom exhaust fans, locks, etc.). Certain other equipment breakdowns, however, can result in serious consequences. Elevators, emergency generators, HVAC systems, and fire and life safety systems are examples of equipment in systems that carry considerable consequences as a result of unscheduled breakdowns.

- *Equipment type.* The type of equipment has a bearing on the required frequency of preventive maintenance inspections. Dynamic equipment (e.g., pumps, air handling units, or other equipment with moving parts) requires more frequent inspections than static equipment.
- *Operating hours.* Certain pieces of equipment are more suited to frequencies based on operating hours as opposed to frequencies based on calendar days.
- *Environmental factors.* Equipment exposed to extreme heat, dust, moisture, or other environmental factors may require more frequent preventive maintenance inspections than normally expected.
- *Age.* Frequencies may vary according to the age of equipment.
- *Safety and code requirements.* Local code requirements or established safety rules may dictate the frequency of preventive maintenance inspections on some equipment. Elevators, sprinkler systems, and smoke detection systems are required by code to be inspected at prescribed intervals.
- *Operation and maintenance manual.* Most major equipment and building systems carry manufacturers' operation and maintenance manuals recommending the frequency of maintenance activity.
- *Cost factor.* The cost benefit of preventive maintenance must be considered when determining preventive maintenance frequencies. The cost of replacing a piece of equipment should be compared with the cost of preventive maintenance activities to extend its life.

Estimating and Staffing for Preventive Maintenance

Another task in the development of a preventive maintenance program is the establishment of the total workload required for the program. Each work order, or job plan, is estimated based on industry standards, predetermined time standards, the mechanics input, or other methods. By estimating all preventive maintenance work orders and factoring for frequencies, the annual workload for a preventive maintenance program is determined.

Staffing for the preventive maintenance effort varies from institution to institution. Some institutions have a dedicated preventive maintenance work center that works solely on preventive maintenance, whereas other institutions successfully mix preventive maintenance and other facilities services throughout their workforce. Often, designated preventive maintenance crews are scheduled to perform preventive maintenance during evening or night shifts to avoid disruption to normal institutional activities. Preventive maintenance is not normally customer interactive, so this type of scheduling has merit. When deciding to assign a designated preventive maintenance crew, one must consider that the service crews and preventive maintenance crews are not familiar with each others' activities. Finally, because a preventive maintenance program, by its very nature, is easily defined and predictable, outsourcing for preventive maintenance services is also a viable alternative.

The final step in the implementation of a preventive maintenance program is scheduling the work orders. The key to a good preventive maintenance schedule is balance. Preventive maintenance tasks should not be grouped during any particular period but should result in an even workflow over a given period of time. There are some preventive maintenance work orders that are seasonally dependent. Some equipment can be taken out of service only when it is not needed, so major overhauls are normally scheduled during Christmas or spring break. Heating equipment should receive major service in the summer, and cooling equipment should receive major service in the winter. Most computerized preventive maintenance systems are designed to aid the preventive maintenance manager in developing a complete and balanced schedule of preventive maintenance work orders.

To ensure that the required tasks are performed, there must be a strong commitment by management to the preventive maintenance program. Individuals assigned to preventive maintenance work must not be pulled from their schedule except in emergencies. There will always be some preventive maintenance tasks each month that are not completed for one reason or another. The system developed must monitor these incomplete tasks and reschedule or cancel them. If similar problems continue month after month, management must decide to either increase the staff assigned to preventive maintenance or compromise the scope of the preventive maintenance program.

Management Controls

A successful preventive maintenance program requires constant performance review by management and supervisors. Many organizations review performance reports that indicate the percentage of preventive maintenance work orders completed in the scheduled time frame. Another common performance indicator measures actual work hours versus estimated work hours for preventive maintenance activities. These performance reports can be helpful not only in determining productivity, but also in providing valuable information leading to adjustments in the preventive maintenance program and annual workload.

Another indicator of the effectiveness of a preventive maintenance program is measuring building system reliability. By developing trend data on major system breakdowns, an institution is able to analyze the effectiveness of the preventive maintenance program. A similar trend analysis would involve contrasting the percentage of maintenance funds budgeted for preventive maintenance with that budgeted for major maintenance and repair. Ideally, an efficient maintenance program would result in an increasing percentage budgeted for preventive maintenance as requirements for major maintenance and repair decreased.

A preventive maintenance program does not establish itself or become more effective without support from all levels of the organization. There must be a strong and continued management initiative for a preventive maintenance program. Management must impress on the facilities organization, and throughout the campus, the high priority of the program. There is always a temptation to neglect preventive maintenance work. It is not glamorous. Rarely does a customer, or for that matter a supervisor, praise or thank an employee for carrying out a preventive maintenance work order. Therefore, management must establish and maintain the high priority of preventive maintenance work and, through performance indicators, communicate appreciation for the effectiveness of the program to the people performing the work.

For a preventive maintenance program to be successful, it also must be accepted by the workforce as a beneficial program. By including the workforce in the development of the program and responding to their feedback regarding the quality of the program, the workforce is able to develop a sense of ownership and acceptance for the program. An aggressive training program for the mechanics and tradespeople involved in preventive maintenance is another method of maintaining acceptance of a preventive maintenance program at the working level. Such training can include technical methodology, predictive maintenance techniques, computerized applications such as bar coding, and management aspects of successful preventive maintenance programs.

21.6 FINANCIAL SYSTEMS

For a work management system to provide a comprehensive program of maintenance management, it must be integrated with a policy of fiscal control. Budgeting for maintenance activities and an accounting of expenditures incurred while performing maintenance activities are keys to financial systems to support the work management process. Budgeting for the different maintenance categories that correspond to categories developed in the work order system is an effective method of breaking down the maintenance budget. Forecasting expenditure plans for the fiscal period provides a means for assessing financial performance throughout the fiscal period. A cost accounting system provides financial information for effective decision making. What is work costing? Are available budgets being used most efficiently? What are the life cycle costs of different pieces of equipment, systems, or programs?

Budgeting

The three most common types of budgeting processes are incremental budgeting, formula-based budgeting, and zero-based budgeting.

1. *Incremental budgeting.* One of the most frequently used concepts for budgeting in higher educational facilities organizations is incremental budgeting. This concept is based on using the previous year's budget as a target. The upcoming annual budget is modified by increasing or decreasing the target by an incremental percentage. The target or base budget may be amended for new facilities, new programs, regulatory compliance, or other justified additions or deletions to the budget.

2. *Formula-based budgeting.* Some institutions use a formula-based budgeting process in which maintenance budgets are established depending on asset units. The most common form of formula-based budgeting is based on a maintenance cost per square foot of facility space. This type of budgeting becomes more accurate if the age and use classification of facilities are factored into the formula. Another form of formula budgeting is based on the ratio of maintenance dollars in relationship to facility replacement value. This concept of a reinvestment rate on facilities assets is an effective business-based approach to maintaining the value of capital investments.

3. *Zero-based budgeting.* Perhaps the most accurate, but also the most difficult, method of maintenance budgeting is zero-based budgeting. This concept requires a detailed inventory, justification, and prioritization of budgetary requirements for all maintenance programs. By starting with zero and building the entire budget through needs justification, the final budget can be negotiated with an understanding by all parties on which programs will be funded and which requirements will be deferred.

To adequately manage maintenance budgets, it is helpful to break the total budget into categories or subaccounts. The previous section on work control discussed several categories of maintenance work orders. By dividing the maintenance budget in accordance with these maintenance work categories, managers are better able to plan and account for the financial approach to maintenance management. A common method of this budget differentiation is through establishment of separate fund accounts. By budgeting for each major category of maintenance effort, the work control system, including work identification and authorization, is integrated into a financial plan.

A typical maintenance budget could be subdivided into the following categories:

1. *Preventive maintenance.* The preventive maintenance budget should be a predictable budget. It normally is labor intensive, with a well-defined inventory of equipment, job plans, and frequencies.

2. *Service work.* Somewhat less predictable than preventive maintenance, the service work account funds routine minor maintenance services.

3. *Painting services.* Another predictable maintenance budget is that of painting services. Through an effective work control program of long-range

and short-range planning, annual painting budgets, and expenditures can be accurately predicted.

4. *Major repairs.* It is important to reserve a certain amount of the operating budget to fund major maintenance and repairs. These projects are identified through a combination of facilities assessment programs, corrective measures recommended through a preventive maintenance inspection, and unscheduled breakdowns.

5. *Grounds maintenance.* A considerable component of the operating budget should be dedicated to an adequate grounds maintenance program. This category can be difficult to plan because of the unpredictability of weather and its influence on grounds maintenance requirements. Drought conditions necessitating irrigation, storm cleanup, snow and ice control, and other weather-related extremes can all lead to higher than expected expenditures in grounds maintenance activities.

6. *Custodial services.* The budget for custodial services should be relatively predictable. Costs per square foot and historical budget performances can aid in preparing accurate budget requirements for custodial services.

A majority of the maintenance operating budget will normally be consumed by the routine maintenance categories of service work and preventive maintenance. It stands to reason that the more effective a preventive maintenance program, the fewer funds that will be required for contingency repairs in a major repair budget category. A typical maintenance budget at a large university is shown in Figure 21-9.

Frequently more detailed information is also useful. Many times the tasks within the broad categories of service work, grounds maintenance, or custodial work are reflected in the work order accounting system. This information can provide detail for future management decisions on services provided and their costs. For example, the work order system may allow differentiation among the ground maintenance tasks of flower bed maintenance, mowing, and tree pruning.

Most institutions differentiate in the operating budget between true maintenance expenses and those facilities services requirements that would be construed as improvements. These improvements could be safety improvements, landscape projects, fire and life safety improvements, or asbestos or other regulatory compliance issues. Again, effective long-range planning can prepare an institution to budget adequately for those improvements that would meet the institution's mission. It is desirable to separate and establish these improvements accounts. In many organizations there is considerable pressure to compromise preventive and routine maintenance services to respond to improvement projects. Although the importance of improvements at an institution should not be minimized, such improvements should not occur at the expense of maintenance requirements. By distinctly budgeting for improvements, managers are better able to protect and fully use maintenance budgets for their intended purpose.

Figure 21-9

Maintenance Budget

ANNUAL MAINTENANCE BUDGET

Account Code	Category	Amount
12556	Preventive maintenance	850,000
12362	Corrective maintenance	1,200,000
11454	Grounds maintenance	980,000
13745	Maintenance painting	400,000
14452	Major repairs	800,000
15642	Custodial services	2,200,000

Expenditure Plans

Just as it is important to separate the maintenance budget into manageable maintenance subaccounts, it is also important to plan for expenditure rates throughout the fiscal period. Most expenditure plans are devised by breaking the annual budget into monthly spending allotments. Managers should avoid the tendency to simply divide the annual budget into twelve equal monthly allotments and instead should take into account the seasonal requirements of different maintenance categories—as well as holiday periods, number of work days per month, and other factors—in developing a realistic expenditure plan. Through use of such a planning process, managers are able to analyze current budget situations throughout the year. A typical spending plan is shown in Figure 21-10.

Effective budgeting is essential to sound maintenance management programs. The comprehensive approach to maintenance budgeting is a fundamental step in influencing the amount of funds dedicated to the maintenance program. A budget planning process that coordinates the institution's budget office, the facilities management staff, and the facilities organization's customers leads to improved communications through a unified approach to the overall institutional budgeting process. Budgeting for maintenance services is a competitive process. Many of the competing programs include faculty salaries, capital improvements, academic programs, new equipment, administrative initiatives, and other programs, all at the institutional level. In public higher education institutions, there is also competition at the state level—road systems, prison systems, primary and secondary education, and a variety of other essential programs. Through open communication and sound management techniques,

Figure 21-10
Spending Plan

1995-96 Spending Plan

BUDGET CATEGORY	JUL	AUG	SEP	OCT	NOV	DEC
PREVENTIVE MAINT	$65,399	$76,045	$47,909	$56,274	$71,483	$47,148
CORRECTIVE MAINT	$91,040	$106,005	$103,511	$112,241	$125,960	$86,052
GROUNDS MAINT	$62,204	$74,025	$62,684	$69,583	$68,389	$59,244
PAINT	$18,989	$26,767	$28,891	$28,761	$47,892	$17,036
MAJOR MAINT	$52,820	$64,001	$112,580	$124,018	$96,387	$83,535
TOTAL	$290,452	$346,843	$355,575	$390,877	$410,111	$293,015

BUDGET CATEGORY	JAN	FEB	MAR	APR	MAY	JUN	TOTAL
PREVENTIVE MAINT	$49,429	$69,201	$58,555	$73,764	$70,723	$74,524	$760,454
CORRECTIVE MAINT	$77,322	$92,287	$88,546	$110,994	$124,713	$128,456	$1,247,127
GROUNDS MAINT	$36,024	$60,300	$70,493	$76,985	$101,128	$125,333	$866,392
PAINT	$19,987	$18,801	$23,443	$37,366	$43,268	$36,913	$348,114
MAJOR MAINT	$83,535	$67,599	$121,062	$162,187	$158,845	$158,587	$1,285,156
TOTAL	$266,297	$308,188	$362,099	$461,296	$498,677	$523,813	$4,507,243

the facilities manager is able to influence the amount of funds available for facilities services in support of the institutional mission.

Cost Accounting

To help budget more effectively, managers need to know where and how funds are spent. A cost accounting system in support of the work management policies and procedures can assist a manager in the analysis of fiscal activity. There are three primary components of a cost accounting system in a facilities management system: the billing system, work order accounting, and budget accounting.

Billing System   The *billing system* is simply the method of accounting for expenditures. Every resource has an associated expense. Labor expenditures are measured by hourly rates, materials and contracted services have associated costs, and certain indirect administrative costs are part of the overall cost of services. A billing system is the method of charging or accounting for those expenses.

Work Order Accounting   Work order accounting is the method of determining the actual charges on a specific work order. To determine the cost-effectiveness of accomplishing certain tasks, the cost accounting system must capture and report the actual cost. The level of detail in the cost accounting system is consistent with the level of detail established in the work control system. In other words, if the decision is made to issue a specific work order for a minor maintenance task such as adjusting the thermostat or changing some light bulbs, the cost accounting system should be devised to provide financial details at that same level.

Budget Accounting   Budget accounting is the financial report of cumulative, actual expenditures and encumbrances against overall annual budgets. This component of cost accounting reports on how much has been spent, spending in relationship to spending plans, and how much remains in each of the maintenance fund categories. Figure 21-11 gives an example of a monthly budget report.

The budgeting and accounting process is quite different for public and private institutions. Public institutions are often unable to carry over balances at the end of the fiscal period. In addition, the approval process for public institutions is much more stringent. Often public institutions are required to work within state accounting procedures. Overall, public institutions have much less flexibility than private institutions in budgeting and accounting areas.

Most facilities organizations have developed the ability to bill for services. The first step in establishing this ability is the development of labor billing rates. The distributive labor rate is the combination of direct costs (e.g., salary and benefits) and indirect costs (e.g., training and administrative support). The labor rate is computed by adding the total direct costs and indirect costs in a years time and dividing by the projected number of hours per year. An example of computation of labor rates is shown in Figure 21-12.

The development of distributive labor rates is a principle of effort-based accounting that permits a reliable system of charge-backs. This process allows the facilities organization to bill maintenance budgets, bill university customers for nonmaintenance activities, bill auxiliaries for services provided, and in some cases, to bill noninstitutional customers for services provided. As responsibility center accounting and privatization issues become more prevalent in higher education, facilities organizations must develop the means to accurately account and report on the total cost of providing services.

Responsibility Center Accounting

Most facilities organizations carry on their business much like an auxiliary. An auxiliary bills out its services for a fee, is fully recoverable, earns interest

Figure 21-11
Monthly Budget Report

ACADEMIC DIVISION BUDGET PERFORMANCE REPORTS

For the 10-month period ending April 15, 1996

EXPENDITURE CATEGORY	APRIL BUDGET	ACTUAL	VARIANCE	YTD BUDGET	ACTUAL	VARIANCE	%	ANNUAL BUDGET	BALANCE
CORRECTIVE MAINT	$116,967	$70,213	$46,754	$1,137,852	$1,255,548	($117,696)	-10%	$1,406,153	$150,605
GROUNDS MAINT	$116,267	$52,534	$63,733	$781,240	$755,474	$25,766	3%	$1,089,053	$333,579
PREVENTIVE MAINT	$78,284	$85,084	($6,800)	$699,167	$797,188	($98,021)	-14%	$884,704	$87,516
GENERAL PAINTING	$54,208	$60,927	($6,719)	$310,377	$328,669	($18,292)	-6%	$402,581	$73,912
MT. LAKE FACILITY	$4,825	$7,306	($2,481)	$46,115	$41,935	$4,180	9%	$57,511	$15,576
MAJOR MAINT	$91,870	($18,582)	$110,452	$657,076	$591,728	$65,348	10%	$862,674	$270,946
TOTAL ACADEMIC	$462,421	$257,482	$204,939	$3,631,827	$3,770,542	($138,715)	-4%	$4,702,676	$932,134

Figure 21-12
Labor Rate

LABOR RATE DEVELOPMENT
Based on a sample four-person shop

Direct costs (salary, benefits)	$120,000
Indirect costs (training, tools, supervision, uniforms, etc.)	$35,000
Total cost center costs	$155,000
Revenue year per person	
Total hours paid	2,080
Unproductive time (holidays, leave)	−330
Productive hours	1,750
Four people	x4
	7,000
Cost center expenses	$155,000
Projected productive hours	7,000
= Billable rate of $22 per hour	

on its carry-forward balances, and is able to establish reserves and, in some cases, issue bonds. However, it is unusual for a facilities organization to be designated as a bona fide auxiliary enterprise. Responsibility center accounting attempts to structure the financial activities of an organization so that each subunit is responsible for recovering its expenses, both direct and indirect, through billing for their services. The objective is to assign responsibility for controlling the cost and managing the bottom line to the supervisor at the point where costs originate. Decision making is decentralized at the responsibility center level.

Financial Performance Reporting

Financial performance in the provision of facilities services is an indicator of the effectiveness of a maintenance management program. One method of determining financial performance is comparative analysis. There are several sources of published performance indicators, such as *Comparative Costs and Staffing Report,* published by APPA: The Association of Higher Education Facilities Officers; and an annual benchmarking report, published electronically by the National Association of College and University Business Officers (NACUBO). Through such comparisons, organizations are able to compare their performance against other institutions or against recognized industry standards. Comparing financial performance with internal standards can also be an effective form of analysis. Establishing performance standards as part of the work control process can provide a basis by which to measure financial performance. A variance analysis process, which measures the planned versus actual results on cost estimates and schedules, can help an organization become more competitive and cost-effective. Labor productivity reports that measure backlog of work, overtime levels, absenteeism, and facilities utilization can also be a useful method of financial performance reporting.

Financial Management Summary

An effective maintenance management program combines an organized approach to work control with a sound policy of financial budgeting and accounting. The trends in today's climate of a businesslike, entrepreneurial approach to facilities management requires the successful facilities manager to be a strong financial manager. Through accurate forecasting and budgeting and meaningful reporting, managers are in a better position to determine the cost-effectiveness of services provided. It is no longer adequate to be a strong technical manager and leader of facilities services. One also has to develop the ability to be a financial manager as part of the total package.

21.7 COMPUTERIZATION IN WORK MANAGEMENT

Perhaps the most dramatic change in the facilities management business in recent years has been the emergence of computers and automated data technology. About 1980, computerization first found its way into facilities management organizations. Although most organizations at that time depended on mainframe computer systems to manage financial, human resource, and student records, such central systems were not normally operated with the facilities organization in mind.

Currently, however, it is unusual to find a facilities organization in higher education that is not using computerization to enhance management of facilities services. The affordability of hardware and software, the growing popularity and flexibility of personal computers, and the computer literacy of young people entering the job market have all contributed to a rise in popularity of computerization in facilities management organizations. It stands to reason that because institutions of higher education are in the business of preparing students for careers in a technocratic society, administrative and support functions would be in a position to benefit from modern computerized systems.

Many facilities organizations are moving toward a concept of integrated maintenance management systems in which many of the related computerized applications are integrated into a single system. There are a variety of applications in databases typical to most facilities organizations. These may include the following:

- Work order system
- Cost accounting
- Personnel
- Inventories
- Purchasing
- Utilities
- Computer-aided design (CAD) drawings
- Building and infrastructure data
- Word processing
- Scheduling
- Key control
- Estimating
- Fleet management
- Space and real estate management
- Project management

With current software and hardware technology, the integration of information from separate databases is an opportunity not to be missed. Through the use of local area networks (LAN), an integrated maintenance management system can provide more information to more people, faster than ever.

The most obvious use of computers in maintenance management systems in higher education facilities organizations is the work order system. The maintenance manager with current information technology is able to store and retrieve data, at a detailed level, for all work performed. Estimating and scheduling have become much less labor intensive as software continues to be developed to support these processes. There are several commercial software programs available for estimating and scheduling.

Communications through electronic mail (e-mail) has also been a valuable technological advance for facilities organizations. Most institutions of higher education have a campus network linking academic, research, student, and administrative units. Through the campus LAN, facilities organizations are able to communicate easily with a larger group of customers on a more frequent basis. Likewise, customers are able to communicate with the facilities organizations through electronic media. Whether through e-mail or the ability to submit on-line work requests, the computer opens another avenue of communication between the facilities organization and its customer base.

Investments

An organization must consider the investment in committing to an integrated maintenance management system. The cost of hardware and software can be considerable depending on how many users will be on the system. There are a variety of software packages available designed to support an integrated maintenance management system. An institution considering how to proceed on a purchase of maintenance software should benefit from the experience of institutions that have used the software. There is also expense involved in purchasing computer hardware—processor, monitor, keyboard. Technology is changing rapidly, and one must be sure to purchase hardware that is able to support current powerful software packages. Most personal computers range in cost from $1,500 to $3,000, with comparable laptop systems ranging from $4,000 to $5,000.

Although an experienced, reliable computer support staff can be worth their weight in gold, they represent another considerable investment in a computerized management system. The savings realized in avoiding a large staff for performing manual duties far outweighs the expense of hardware, software, and support staff costs, but the investment in dollars and time when first establishing a computerized system can be taxing on an organization.

The modern facilities organization cannot afford to ignore the value and importance of information technology, especially in the business of higher education. Higher education has been in the forefront of developing and using the Internet, particularly the incredibly large collection of information within the Internet by means of the World Wide Web. Through electronic mailing groups, facilities managers across the country are able to communicate, describe problems, share solutions, and generally exchange information on-line, within minutes. The computer can be the most valuable tool in the entire maintenance organization. Chapter 12 contains a more thorough discussion of information technology in the facilities organization.

21.8 MATERIAL MANAGEMENT AND PROCUREMENT

Another important element of administrative support to a maintenance program is materials management. To effectively plan and schedule labor resources, materials and equipment support must be planned and scheduled with equal attention. As with most management systems, the difference between public and private institutions with regard to their approach to materials management is considerable. The lack of flexibility in most public systems adds to the challenge in a materials management system, but with proper planning, even the most stringent procurement systems can be managed to provide the necessary support to the facilities maintenance function.

The procurement function in most institutions is handled by a central procurement office. This is especially true in public institutions. The reason for a centralized function is to manage a system that is both legal and responsive. With a centralized procurement function, the institution can be assured that procurement laws and regulations are followed. In addition, a centralized procurement office is able to coordinate the purchasing requests of a variety of departments and is in the position, therefore, to save the institution considerable funds through bulk purchasing. The disadvantage of a central procurement function is the compromise of flexibility by the end user. Action on material requests deemed important by the facilities manager can be low on the priority list of a procurement officer with a backlog of material requests. A decentralized procurement system in a large institution can obviously be more responsive to the end user but in some cases can compromise legality and coordination. Progressive institutions are seeking the appropriate balance between centralized and decentralized procurement functions. Through open lines of communication, proper procurement planning, and training in procurement policies and procedures, institutions are restructuring to meet the demands of their consumers while serving the institutional mission.

Regardless of how a facilities organization procures its goods and services, the managers of the work control system should be linked directly to the procurement office. The work control activities of planning and scheduling depend on a responsive and reliable procurement function. Together, decisions must be made regarding inventory, prioritization, methods of contracting services, and developing adequate plans of communication.

Establishing and maintaining an adequate inventory of commonly used materials and equipment is a major component of a responsive procurement system. Everyday items such as light bulbs, fasteners, pipe and fittings, cleaning supplies, and drywall are inventoried and on hand at most large facilities organizations. Although warehouse or general stores space can be expensive to operate, the advantages of on-hand materials are too important to overlook. However, the tendency by many is to inventory far more than necessary. Most procurement operations have determined the cost-effective stock usage levels and have combined on-site warehousing and inventories with "just-in-time" delivery contracts. Procurement officers have found that it is normally less expensive to have vendors maintain inventories of frequently used materials for as-needed or same-day deliveries.

Many large facilities organizations are engaged in major projects—alterations, small construction, major maintenance and repair—requiring purchase and delivery of large pieces of equipment. Adequate warehouse space is useful for staging materials and equipment until they are needed for installation at the job site. Again, the cost of operating warehouse space can be considerable. Institutions must make decisions regarding the cost-benefit ratio in determining the appropriate use of warehouse space.

In addition to goods and equipment, service contracting is another procurement function providing a considerable level of support to the maintenance management function. Service contracts for many specialty areas such as pest control, paving, concrete, ceiling and floor finishes, and painting are the most prevalent forms of service contracts in facilities organizations. These contracts are most effective when procured on a long-term, open-end basis, priced according to unit rates. Many organizations complement their staff with hourly rate service contracts. Contracting for services can often be the most economic way of accomplishing certain categories of work. Effective work management systems should balance the capabilities of its staff with the use of service contracts.

Regardless of whether an institution is public or private and whether the procurement function is handled under central or decentralized basis, the maintenance manager must integrate the work control process with the procurement system. It has been said that materials are the lifeline of the productive workforce. The first thing a work sampling or productivity study specialist will scrutinize is the availability of materials and supplies to the productive workforce for the accomplishment of work. The staff that spends its time driving to and from supply houses all day is certainly not a productive staff. Too many times we have heard, "I could have completed the work, but I didn't have the right materials." More detailed discussion of procurement systems can be found in Chapters 16 and 17.

21.9 MAINTENANCE MANAGEMENT PROGRAMS

Work management systems normally comprise a variety of maintenance management programs. Although many work management programs are based on detailed procedures for managing work, other maintenance programs are on a macro level, designed to provide decision-making information on a system-wide basis.

Facilities Assessment Program

Periodic facilities assessments are necessary to determine the condition of each facility by identifying maintenance deficiencies and building systems renewal requirements. Identified major replacements, code compliance issues, and substandard conditions form the basis for developing programmed major maintenance and specific capital renewal projects. Through a comprehensive facilities assessment program, an organization is able to establish a baseline of facilities conditions through a detailed and structured inspection process. Short-range and long-range renewal needs can be maintained and reports can be produced to assist in space administration, financial management, and corrective actions for those responsible for maintaining facilities. A detailed discussion of facilities assessments can be found in Chapter 20.

The backbone of a facilities assessment program is a continuous and cyclic series of maintenance condition inspections. A team of trained inspectors, through visual inspections, identify and quantify maintenance deficiencies. A report of the deficiencies can be developed, sorted by building trade with an estimated value of repairs, and prioritized by need. Many institutions use their own staff of planner/estimators for inspections, whereas others engage consultants to perform inspections and provide reports.

Whereas maintenance condition inspections concentrate on maintenance and repair requirements, facilities audits are normally performed to identify nonmaintenance requirements, such as code compliance requirements, energy improvements, environmental and other regulatory requirements, and adaptations for changes in future use.

Through a comprehensive approach to facilities assessments, an organization is able to determine the dollar volume of unfunded maintenance for the entire facilities organization. Through an annual report on the overall conditions of facilities, a manager is able to develop past and future trends relating available maintenance funding to the volume of unfunded or deferred maintenance.

In addition to being a valuable management tool for maintenance programming, a facilities assessment program can be a useful tool of communication with occupants of facilities. Prior to performing inspections, the inspection teams should alert facilities coordinators of the schedule for inspections and the intent of the program. Often facilities coordinators and the various staff in the facility are anxious to participate in the process. Once the report is developed, a discussion with the facilities coordinators about their facility can be quite educational. Including customers in the decision-making process can pay valuable dividends in customer relations.

Quality Control Programs

Quality control is defined as a system for ensuring the maintenance of proper standards by periodic inspection of the product or service. A work management system inherently exercises some level of quality control. The most common form of quality control is through supervisory inspections. The importance of inspecting work, both during its accomplishment and once it is complete, cannot be overemphasized. This is the most common form of quality control in almost every industry.

Another form of quality control is through customer satisfaction surveys. As discussed earlier, the customer survey can be a valuable customer relation tool, but it can also be of value as a quality control instrument, providing management with data and trends leading to enhancement of policies and procedures for more efficient delivery of service.

Performance measurement is yet another common and effective method of quality control. Through the establishment of standards, and measuring performance in relation to the standards, an organization is able to determine how well it is attaining its goals. Common performance measures include the following:

- Shop schedule compliance
- Preventive maintenance tasks completed
- Service work complete within time frame
- Major work completed on time/within budget
- Building cost per square foot
- Overtime and absenteeism rates
- Productive hours
- Energy usage
- Safety/workers' compensation loss time
- Equipment failure
- Amount of rework

Benchmarking has recently become another useful form of evaluating the performance of an organization. Through the development of an institutional profile and comparing performance with similar institutions, management is in a position to better determine an organization's strengths and weaknesses.

Insurance Program Management

It is important to have a single coordinator of insurance-related activities in the facilities organization. Depending on the institution's insurance policy, many facilities-related expenses may be reimbursable. Costs related to flooding, vandalism, equipment breakdown, and third-party claims are all examples of expenses that could be reimbursed through the institution's insurance program. Early identification of a potential insurance reimbursement is the key to successful risk management coordination. The consistency of claims information and effective communication could lead to considerable financial return to the facilities organization.

CHAPTER 22

Emergency Management

Richard S. Fowler
University of Virginia

With additional material from Nadesan Permaul, Jill Finlayson, and Treacy Malloy, University of California, Berkeley, and from Robert Collins, Davidson College.

22.1 INTRODUCTION

The college or university facilities organization can expect to assume a leadership role in dealing with unforeseen facilities challenges affecting the campus. Colleges and universities are vulnerable to emergencies created by flash flooding, hurricanes, thunderstorms, earthquakes, hazardous material incidents, resource shortages, or other disasters. Although emergencies are usually unforeseen situations that cannot be handled routinely, planning for their possible occurrence can lessen their impact. Planning includes anticipating possible effects and prescribing actions to minimize the damage and to restore normal operations.

22.2 OBJECTIVES OF EMERGENCY PREPAREDNESS PLANNING

A facilities organization's primary mission in the event of an emergency should be to protect lives, minimize danger or loss to university or personal property, restore vital services as quickly as possible, and support the institution's emergency plan. A facilities emergency operations plan (EOP) will provide a sound basis for emergency preparedness. The plan should establish the organizational and operational concepts and procedures necessary to minimize loss of life and property and to expedite recovery from any disaster situation confronting the facilities organization and the campus.

What are the potential emergencies for a given facility? How do managers prepare for the unforeseen? Managers must adapt and exploit the strengths of the existing organization to cope with disasters. To gain an appreciation for the magnitude of a disaster, try visualizing the last time you personally had a really bad experience, and then multiply that occurrence a hundred- or thousand-fold. Think of parallel experiences, and then apply the things your organization does best in coping with these catastrophes.

The contemporary campus plan is an outline of the management responsibilities of the campus and the lines of authority. It should contain a description of how business as usual changes during an emergency, checklists with organizational charts, summaries of roles and responsibilities, lists of essential service providers, and a functional matrix. An outline for a typical emergency plan is given in Appendix 22-A. Some useful definitions are provided in Appendix 22-B. The EOP may also contain a series of annexes. Detailed plans for each essential service provider should be provided in these attachments. To achieve some key emergency functions, such as building assessment or mass care and shelter, several departments may have to coordinate the response activities outlined in their respective annexes and determine who will lead the effort. These annexes should be updated at least annually. This responsibility should be formally assigned and managers held accountable.

A campus-wide plan should coordinate the efforts of support activities under the institution's control. For a small college, this may be the only plan for the institution. For a larger university, the plan may be only a part of the entire university's plan. Resources outside the institution's control, such as city fire service or utility companies, may ask for or be asked for assistance when an emergency occurs.

22.3   EMERGENCY PLANNING

A simple six-step planning model follows:

1. Plan for the unexpected.
2. Communicate roles.
3. Foster cooperation with outside units.
4. Write the plan.
5. Practice the plan.
6. Revise the plan.

The purpose of a facilities EOP is to establish the organizational basis for operations within the facilities organization to respond to any type of disaster or large-scale emergency situation. It should assign broad responsibilities for emergency preparedness, mitigation, response, and recovery. These responsibilities are generally extensions of normal, day-to-day functions involving the same personnel and material resources. Experience has shown that it is best to keep the structure for emergency operations as close to normal day-to-day operations as possible, while emphasizing simplicity.

Supporting annexes identify the concepts and procedures by which the facilities organization can effectively apply available resources to minimize casualties and property damage and restore essential services following an emergency or disaster.

The chief facilities officer and staff must ensure that the policies and procedures established for emergency operations are consistent with the institution's EOP. The plan should support the college or university and should consider the external community. In large metropolitan areas, the plan may be an integral part of a much larger effort, including city, county, and state disaster management organizations. During emergencies, the chief facilities officer and staff control and direct the functions of the department in support of the institution. A typical responsibility matrix is provided in Appendix 22-C. Lines of communication must remain open, and emergency supplies and procurement agreements should be in place. A specific individual, usually the chief facilities officer, should function as the director of facilities emergency operations. The director of facilities emergency operations should activate an emergency operations center, conduct an analysis of the emergency on a continuing basis, and take the steps necessary to minimize the effects of the disaster.

Situation and Assumptions

Emergencies of many types, sizes, intensities, and duration may occur within the boundaries of the university, with or without warning. Such emergencies can affect the safety, health, and welfare of the university population and cause damage or destruction to property. Appendix 22-D shows typical stages of emergency warnings. An emergency may be localized, such as in a transportation incident, or it may cover a large geographic area, as in weather-related events.

During emergencies, a college or university may be required to be self-sufficient for a period of time. Outside services may be interrupted, so the resources of the institution must be organized to provide essential services. Routine operations should become secondary to emergency operations.

In the event of an emergency situation that exceeds the facilities organization's response capabilities, outside assistance may be sought through other university organizations or by mutual assistance agreements with outside organizations.

Organization

The facilities emergency operations organization should be structured around the existing facilities organization, functions, and order of succession. The more the EOP resembles day-to-day operations, the more likely it is to succeed. Figure 22-1 shows an example of a typical emergency operations organization for a large university; Figure 22-2 shows a typical small college organization.

22.4   FEDERAL EMERGENCY ASSISTANCE

The Federal Emergency Management Agency (FEMA) is dedicated to emergency management. FEMA has published the *Emergency Management Guide for Business and Industry: A Step-By-Step Approach to Emergency Planning, Response and Recovery.* For a copy of the guide, managers can call 1-800-480-2520 or visit the FEMA website on the Internet at http://www.fema.gov. FEMA is prepared to assist local organizations in their efforts to plan and train for emergencies. Consideration should be given to having one or more managers receive FEMA training in disaster plans and operations.

FEMA is requiring state agencies to use the Standardized Emergency Management System (SEMS) as a guide for developing their plans. All other agencies that would like to be reimbursed for disaster-related expenses should also be able to demonstrate that they utilize SEMS. Through this standardized approach, agencies will be able to work together more smoothly by using the same command structure and terms and by having clearly defined memoranda of understanding. The Incident Command System is an important part of SEMS. This system is used by fire agencies in many states and municipalities across the United States.

22.5   FACILITIES EMERGENCY MANAGEMENT TEAM

The key to maximizing immediate response and minimizing prolonged effects of an emergency situation resides in control and communications procedures. The establishment and training of a facilities emergency management team (FEMT), headed by the director of facilities emergency operations; an emergency operations center (EOC); and designated coordinators and team leaders are critical to achieving successful results in an emergency situation.

The FEMT assists the chief facilities officer on matters fundamental to the facilities organization's responsibilities. The FEMT is composed of staff charged with developing plans, techniques, policies, and controls for managing major disruptions to the university. For any emergency situation, the responsibilities of the FEMT should be as follows:

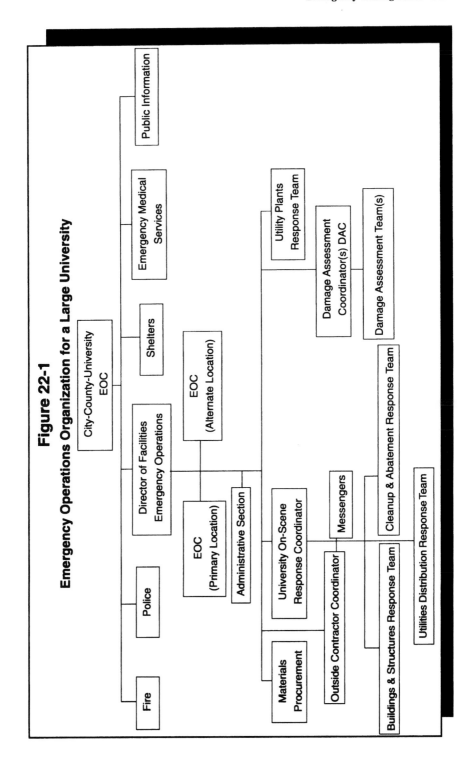

Figure 22-1

Emergency Operations Organization for a Large University

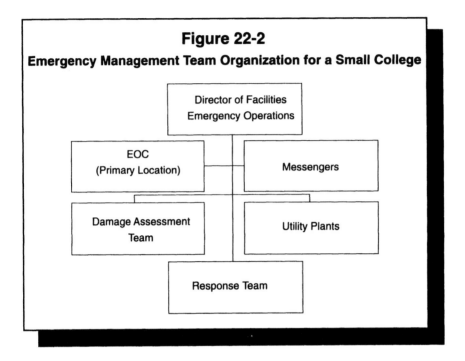

Figure 22-2

Emergency Management Team Organization for a Small College

1. To ensure that responsibilities are clearly delineated for emergency management, training, and reporting
2. To establish and test contingency plans for emergencies
3. To provide the initial response to an emergency situation
4. To minimize loss of life and property
5. To provide for continuing operations

Activation   The FEMT is activated, on order of the chief facilities officer, to enable personnel to train in and practice the plan, to facilitate planning, and in response to a significant natural or man-made disaster.

Staff Recall   A current recall roster should be provided to the EOC at least on a quarterly basis. A copy of all recall rosters should be provided to the director of facilities emergency operations. Recall rosters should include the following information for each employee:

- Name of employee and specialty
- Radio call number, beeper number, and telephone number
- Home address and telephone number

Remember that in the case of a major disaster, the staff may also be concerned with problems at home. Managers should plan accordingly.

Alert Drills Exercises should be conducted periodically to ensure that all personnel are familiar with and competent to execute their emergency assignments. Each unit head should prepare written instructions for use by personnel in emergencies.

Units involved in emergency preparedness operations should conduct joint mock drills. A campus-wide committee should establish criteria and act as the evaluator of performance in emergency drills. Timely recall is an essential prerequisite to an effective FEMT. Alert drills should be conducted periodically, perhaps on a semiannual basis.

Readiness Conditions All FEMT members and their alternates should be thoroughly familiar with weather condition warnings.

FEMT Checklists Each potential EOC participant should have a checklist of the key responsibilities of the position and the required initial actions on arrival in the EOC. Those identified to be in the EOC and their alternates should have the authority to initiate the actions set forth by the director of facilities emergency operations.

22.6 THE EOC: CALM AT THE EYE OF THE STORM

The EOC may be a command post for the entire college or for only the facilities organization's use within a larger integrated program. An effective EOC is lean, organized, and practiced. Depending on the character, scope, and magnitude of an emergency incident, a variety and number of EOC participants may be mobilized. However, staffing should be kept to the minimum number of participants necessary. The purpose of the EOC is fourfold:

1. To provide rapid and continuing coordination of all elements of a facilities management organization
2. To provide for the rapid collection and dissemination of new information and developments
3. To communicate with allied emergency operations
4. To communicate with the university administration and local community

Equipment and Setup in the EOC

The location of the institutional EOC can be a designated facility or a facility converted to an EOC when there is an emergency. A designated facility is preferable, because setup time is less. However, most campuses will have to identify a shared-use space. The space selected should be near the campus communications center, usually the fire or police department. This is where much of the information providing an initial assessment of the impact of the emergency will be received. The facility should be appropriately chosen based on security, seismic, and other safety considerations.

A typical listing of equipment for the EOC is described in Appendix 22-E. The EOC should have access to important materials, but it should not be cluttered with unnecessary tools. Some key considerations when equipping the EOC include an uninterruptible power supply and extra fuel for a backup generator, good ventilation, sufficient computer and communications hook-ups, telephones and telephone lines, televisions and radios, fax machines and copy machines, and dry erase boards and maps for tracking events and resources. Be prepared to set up and staff an emergency hot line telephone bank.

EOC Orientation and Training

EOC participants should receive an orientation to the facility. They need to know how to set up the facility and how their role fits into the facility's operations. Discuss how communication will travel and how decisions will be made. A tabletop exercise in the EOC will help to familiarize participants with the room and setup procedures ahead of time.

22.7   COMMUNICATIONS

Successful communications are vital to the containment and control of emergency situations. The communications plan should be tested periodically, normally in conjunction with EOP training or recalls. The concept of communications is designed to support appropriate response to emergency situations ranging from natural disasters to major accidents.

The telephone will normally be used as the primary means of communications during crisis situations. If time permits, portable cellular telephones should be obtained prior to the onset of an emergency, for flexibility. In storms or other emergency situations, telephone lines may be down, requiring backup communications. Radios may be used as the secondary means of communication. Citizens band radios may be used as a third means of communication should telephone service be down during emergency situations. Messengers may also be used.

22.8   SUCCESSION OF AUTHORITY

The succession of authority for the facilities organization should be clearly spelled out. During an emergency, it is anticipated that key personnel will sometimes be unavailable. Again, it is important that the existing organizational structure be reflected in the emergency management organization.

22.9 DECLARATION OF EMERGENCY

The administrative head (president or chancellor) should be responsible for the initial declaration of a university or college emergency, as well as for the declaration of the end of the emergency. After the emergency is declared, the director of facilities emergency operations is responsible for directing all utilities, facilities, and grounds emergency operations. The declaration of an emergency should be made first to the director of facilities emergency operations and then to the campus community and the general public.

Building occupants normally function through a chain of command, with a building supervisor and other individuals in authority. The building organization will develop procedures for identifying and reporting conditions requiring facility attention.

22.10 CONCEPT OF OPERATIONS

The EOP forms the basis for training, disaster preparedness, and emergency response to a natural or nonnatural emergency that threatens or occurs within or near a university. The director of facilities emergency operations has overall responsibility for maintaining and updating this plan. All facilities directors should be encouraged to recommend appropriate improvements and changes at any time to the director of facilities emergency operations.

The director of facilities emergency operations exercises direction and control through the FEMT and the EOC during disaster operations. The EOC will be prepared for operations and may be partially or fully staffed, depending on the type and size of the disaster.

Available warning time will be used to increase readiness measures to ensure maximum protection of people, property, and supplies from the effects of potential disasters. For example, with an impending hurricane, time and effort might be well spent in boarding up windows, securing loose construction materials, and filling sandbags. The effects of disasters can be substantially mitigated with advance planning and action.

All appropriate personnel and resources should be fully committed before assistance from outside organizations is requested. Requests for assistance should be made through the EOC.

Appropriate operations procedures should be carried out and actions taken to improve emergency operating capabilities. Training and test exercises should be conducted on a recurring basis. Training should consist of periodic exercises and drills to test and practice emergency plans and procedures. When disaster threatens, all personnel with emergency operations responsibilities should take those actions necessary to maintain control and conduct emergency operations.

Activation of the EOP

The director of facilities emergency operations will activate the EOP when a disaster, or threat of one, is of sufficient severity and magnitude to warrant coordinated action to prevent or alleviate damage, loss, hardship, or suffering. Requests for information by news media or others should be referred to a specific, qualified individual.

Emergency Management Assignments

A detailed listing of emergency task assignments should be prepared. Files of current plans and standard operating procedures should be available in the EOC to ensure the capability for prompt implementation. These plans and procedures should be reviewed and updated at least semiannually. One copy of all such plans should be filed with the EOC.

Direction and Control

A designated individual, usually the chief facilities officer, will function as the director of facilities emergency operations, whereas another may be responsible for the development of the EOP and continuity of operations to accomplish emergency responsibilities effectively. Others should be responsible for the development of emergency plans and standard operating procedures to accomplish their tasks effectively.

Financial Considerations

Although subordinate to emergency and recovery operations, procedures should be established to record expenditures in the execution of emergency operations. Implementation of the EOP may require unplanned expenses in both preparation and execution of emergency operations. These expenses should be identified separately from those required for normal operations for potential subsequent reimbursement from federal or state sources. A financial manager should have overall responsibility for fiscal planning, control, and operation under the requirements of the plan. All disaster-related expenditures should be documented to provide a basis for reimbursement.

Damage Assessment

During emergencies, there may be a need to assess damage to facilities and determine the corrective work required. Drawings of utility lines may be critical to emergency planning and operations. The facilities organization usually has skilled personnel capable of assessing damage to property. The type and degree of support needed from other organizations will be determined by the amount and types of property or facilities damaged. The task of a damage assessment unit is to expeditiously evaluate grounds, buildings, facilities, structures, and utilities within an affected area to determine whether operations can continue safely in specific damaged facilities. Specific responsibilities may be as follows:

- To support the EOP by providing timely assessment of any disaster-related damage incurred
- To control and coordinate all damage assessment activities
- To advise fire, police, construction, maintenance suppliers, and others regarding damage to physical facilities and to recommend corrective actions
- To assist police and fire departments in taking necessary actions to rescue or protect persons and property
- To survey campus buildings and infrastructure and report the findings to the EOC
- To make surveys to determine the status of buildings for the safety of all concerned. When action must be taken, this unit prepares the necessary documentation to conduct the appropriate function (repair or condemnation)
- To coordinate all utility use and interruption
- To remove utility service from buildings as necessary

Damage Assessment Organization and Functions Facilities resources should be employed as necessary during emergency operations to allow for the most efficient means of assessing damage. A qualified individual should be designated to conduct damage assessment operations. Appropriate personnel should be designated to serve as members of damage assessment teams.

Preemergency Operations The damage assessment coordinator (DAC) will alert, assemble, and instruct damage assessment team members regarding the extent and nature of operations to be performed and provide to team members required specialized or protective equipment.

Emergency Operations To ensure the safety of personnel and the protection of university property, the damage assessment coordinator should collect and record damage information as it is reported and formulate a prioritization schedule for assessment. Should a disaster occur without prior notice, all damage assessment team personnel should report to the damage assessment coordinator for instructions and specific tasks. The damage assessment coordinator should relay pertinent information to the EOC.

Postemergency Operations   Outdoor damage assessment activities should take place when conditions are safe. The DAC should compile all damage assessment results and forward them to the EOC.

Logistics Considerations

It is extremely important to identify available equipment and resources. Individual operating units should develop complete listings of available equipment and resources and their location. All costs incurred from purchasing supplies for an incident should be consolidated and reported to the financial manager for appropriate action.

A facilities organization will normally have standing contracts that can be used for emergency responses. These may include the following:

- Fuel
- Emergency tree services
- Roofing
- Paving
- Snow removal
- Refuse collection
- Trash disposal
- Equipment and operators

A list of vendors who will accept approved emergency purchase requests should be established. A listing of vehicles available to the facilities organization should be available to the EOC.

22.11   REPORTS

Brief situational reports should be made by the DAC to the EOC on arrival of teams at affected areas and thereafter as conditions require. A brief written report should be completed and submitted to the EOC as soon after attendance to emergency situations as conditions permit.

22.12   SOME LESSONS LEARNED

Major disasters experienced in recent years by member institutions of APPA: The Association of Higher Education Facilities Officers provide some important lessons learned.

Hurricane Hugo (1990): The Citadel

Colleges and universities have the resources of small cities and more. Leverage those resources, and you will recover more quickly. You have transportation and access to resources and credit while others do not. You can go get what you need or have it delivered.

With regard to FEMA, document all losses and provide an audit trail. Use reasonable unit costs and do not expect payment for what cannot be replaced (e.g., a 300-year-old oak tree will be replaced by a sapling). Because in-house labor costs are frequently challenged, consider outsourcing. Ask for reimbursement for productive efforts only; do not try to collect for time wasted or excess management. Partial losses and code upgrades in place of direct replacement require clarification and negotiation. Be prepared for an audit by FEMA or by the state. You have the right to appeal any denied claims. You may have an opportunity to submit information to mitigate denial of claims. Negotiation is also a FEMA option on disputed claims. Most FEMA field staff are temporary who are poorly trained and will not bend to your view. They can respond only to documentation and regulation.

Look after your employees. Employees needed in recovery operations will not be productive or will not report to work if their basic needs are not satisfied. Provide what they need and cannot get while they are working. Feed them and house them. They may be prepared to work around the clock, as they may have nowhere else to go. Accept help from volunteers only if you can support and supervise such help. Consider day support only.

Maintain the ability to prepare, serve, and store food. If you want employees, volunteers, and contractors to work for you, you may have to feed them. Maintain the ability to store water, and provide a dry, safe location for housing, with appropriate bedding. Provide suitable bathing facilities and provisions for waste disposal. People will need money after a disaster. Banks, automatic tellers, and auto deposit facilities may not be open. Have access to pay records. Computers may be unreliable or may not work at all.

Before and after the event, the first to react is the first to get the necessary materials and equipment and the right price!

Top off the fuel level of storage tanks to prevent floating during a flood. Fuel will be needed after the event. Maintain the ability to pump fuel with generators and operable pumps. Make sure vehicles are fueled and safe. Disperse and protect spare vehicles.

During Hurricane Hugo disaster recovery operations, the following actions were taken:

- Housed families and contractors
- Fed families and contractors
- Provided temporary child care services for workers
- Provided gasoline for private automobiles for employees to get home and back, because gas stations were closed
- Provided lumber and roofing, bedding, and food to workers for temporary repair and patching of their homes
- Paid cash for payroll because banks and instant tellers were closed for weeks
- Bartered services for dump trucks and generators

Loma Prieta and Northridge Earthquakes (1989 and 1994): University of California at Berkeley, Stanford University, and California State University at Northridge

Have a recovery plan. Disaster preparedness includes knowing how to get your multi-million dollar business up and running as soon as possible. Even though the Berkeley campus of the University of California did not suffer significant structural damage, the cost of repairs and the time to implement them suggested that in a significant disaster, planned recovery is essential. Knowing which buildings are key to the effective delivery of academic and student services, targeting those structures for reinforcement, and planning alternatives if a program of reinforcement will take years to accomplish are all part of the necessary planning.

Even though Stanford University has not accomplished all the repairs to its campus from damage caused by the Loma Prieta earthquake, they have accomplished the goal of keeping the institution viable. In the case of California State University at Northridge, significant damage to campus buildings remains, but the campus is operational. The use of tents and other alternative sites for services and instruction made it possible to resume operations in a timely manner. On the Northridge campus, there was land available to use for tents and the relocation of services. However, land-locked campuses such as University of California, Los Angeles, and the University of California, Berkeley, have to make other provisions for services if buildings cannot be made operational in a timely manner.

22.13   SUMMARY

Emergency preparedness planning must be unique to each college and university. Much useful experience and information are available from outside sources. Although this information is transferable to some extent, it must be utilized in light of the unique conditions of the individual institution. Through this analysis, an action plan for emergency operations can be developed that provides for the orderly handling of an emergency and efficient recovery to normal conditions, at minimal cost. The key to successful emergency operations is preparing the facilities organization in an orderly, logical way to support the unique college or university environment involved.

Appendix 22-A

Outline of a Typical Emergency Operations Plan

Table of Contents
Preface
Distribution List

Basic Plan

Purpose
Situation and assumptions
Organization and assignment of responsibilities
Concept of operations
Authorities and references
Definitions

Functional Annexes

Annex A: Direction and control

Appendix 1: Emergency services organization chart
Appendix 2: Matrix of responsibilities chart
Appendix 3: Emergency services organization telephone listing
Appendix 4: Succession of authority
Appendix 5: Local situation report
Appendix 6: EOC layout chart
Appendix 7: EOC procedures
Appendix 8: EOC staffing
Appendix 9: EOC message form
Appendix 10: EOC log form
Appendix 11: EOC message flow

Annex B: Communications

Annex C: Emergency public information

Annex D: Law enforcement

Annex E: Facilities management

Appendix 1: Facilities management responsibilities
Appendix 2: Facilities management organization
Appendix 3: Facilities management personnel resources
Appendix 4: Facilities management equipment resources
Appendix 5: Buildings with emergency power
Appendix 6: Facilities management vehicle resources
Appendix 7: Facilities management recall procedures
Appendix 8: Transportation resources

Annex F: Medical and health

(Appendix continues)

Annex G: Fire service

Annex H: Emergency medical services

Annex I: Evacuation

Appendix 1: Shelters
Appendix 2: Primary EOC and shelter location map
Appendix 3: Secondary EOC
Appendix 4: Facilities and transportation for persons with
            special needs

Annex J: Special facilities

Annex K: Auxiliary workforce

Appendix 1: Registration form for volunteers
Appendix 2: Resource distribution centers
Appendix 3: Donations management

Annex L: Damage assessment

Appendix 1: Damage assessment teams
Appendix 2: Initial damage assessment

Annex M: Federal disaster assistance

Appendix 1: Federal response plan emergency support
            functions
Appendix 2: Map of potential federal response field facilities
Appendix 3: Potential federal response disaster field facilities

Annex N: Hazard mitigation

Hazard-Specific Annexes

Annex O: Flooding

Annex P: Hazardous materials incidents

Annex Q: Dam safety

Annex R: Resource shortage

Annex S: Airport emergency

Annex T: Water contamination

Appendix 22-B

Definitions

Emergency: A sudden and unforeseen occurrence or condition, either as to its onset or as to its extent, of such disastrous severity or magnitude that facilities organization intervention beyond normal activities is required to coordinate the actions of employees and other organizations in accomplishing emergency operations.

Emergency Communications Center: The communications center that is staffed 24 hours a day to handle routine fire, rescue, and police emergencies in city, county, and university communities (usually 911).

Emergency Operations Center (EOC): Centrally located command center, equipped with communications, information, and emergency power for coordination of emergency response and emergency public information.

Emergency Operations Management: The planned response of the facilities organization to contain an emergency situation and to minimize the loss of life and property; to provide emergency recovery assistance and limit the residual effects of an emergency situation.

Emergency Services: The execution of functions, other than normal tasking, to prevent, minimize, and repair injury and damage resulting from natural or man-made disasters. These functions include warning; communications; evacuation; resource management; plant protection; restoration of utility services; and other functions related to preserving the health, safety, and welfare of the university population and property as well as transient visitors in the vicinity of the university during emergency conditions.

Federal Emergency Management Agency (FEMA): A federal agency whose mission is to provide leadership and support to reduce the loss of life and property and protect the nation's institutions from all types of hazards through a comprehensive, risk-based, all-hazards emergency management program of mitigation, preparedness, response, and recovery.

Hazardous Material: Substances and materials in quantities and forms that may pose an unreasonable risk to health and safety or

(Appendix continues)

to property when transported. Hazardous materials include explosives, radioactive materials, etiologic agents, flammable liquids or solids, combustible liquids or solids, poison or poisonous gases, oxidizing or corrosive materials, irritants, compressed gases, and hazardous waste.

Major Disaster: Any natural or man-made disaster in any part of the United States that, in the determination of the president of the United States, is or is thereafter determined to be of sufficient severity and magnitude to warrant disaster assistance above and beyond emergency services by the federal government to supplement the efforts and available resources of the several states, local governments, and relief organizations in alleviating the damage, loss, hardship, or suffering caused thereby and is so declared.

Natural Disaster: Any hurricane, tornado, storm, flood, high water, wind-driven water, earthquake, landslide, mud slide, snowstorm, drought, fire, or other natural catastrophe resulting in damage, hardship, suffering, or possible loss of life.

Nonnatural Disaster: Any industrial, nuclear, or transportation accident; terrorist action; explosion; conflagration; power failure; or resource shortage that threatens or causes damage to property, human suffering, hardship, or loss of life.

Resource Shortage: The absence, unavailability, or reduced supply of any raw or processed natural resource or any commodities, goods, or services of any kind that bear a substantial relationship to the health, safety, and welfare of the university population.

Severe Weather Warning: Severe weather conditions that could cause serious property damage or loss of life and have been actually observed or reported. For example, a flash flood warning means that heavy rains have occurred and low-lying areas are likely to be flooded.

Severe Weather Watch: Atmospheric conditions indicate that severe weather is possible but has not yet occurred.

State of Emergency: A condition declared, usually by a state governor, in which, in his or her judgment, a threatened or actual disaster in any part of the state is of sufficient severity and magnitude to warrant disaster assistance by the state to supplement local efforts to prevent or alleviate loss of life and property damage.

Emergency Management   593

Appendix 22-C

Typical Responsibility Matrix

	President/chancellor	Director, facilities emergency operations	Police department	Fire department	Auxiliary services	Medical department	Rescue squads	Public information office	Public transportation system	Emergency operations center	Facilities organization	Shelter/evacuation coordinator	Utilities manager
Direction and control	P	P		S									
Emergency public information	P	P						P					
Law enforcement			P										
Traffic control			P										
Communications		S								P			
Warning and alert	S	S								P			
Fire response				P									
Hazardous materials response		S	S	P								S	
Search and rescue		S	P	S									
Evacuation		S	P	S					S				
Reception and care		S				S						P	
Mass feeding					P							S	
Utilities services													P
Street maintenance											P		
Debris removal											P		
Damage assessment		S									P		
Resource and supply											P		
Medical services					S	P	S						

P = Primary responsibility

S = Secondary responsiblity

Appendix 22-D

Typical Stages of Emergency Warnings

First Stage: Warning Alert

1. This stage indicates the possibility of a toxic gas or other emission and will be signaled by a continuous high tone signal from one or more local sirens. In the unlikely event that there is a sudden unexpected release without warning and personnel detect or have unusual physical reactions to toxic materials, they should *immediately* proceed inside to the nearest assembly area and begin to execute the first stage of the warning alert.
2. People are to go inside, close windows and doors, and shut down air conditioners and other ventilation equipment. No smoking will be allowed inside buildings. Tune radios to local radio and television stations.
3. Students in classrooms or dormitories will move to assembly areas in each of the main campus buildings. Designated supervisors will issue further instructions.

Note: If the alert is sounded during sleeping hours, physical facilities and security personnel will awaken people in dormitory and housing areas.

Second Stage: Shelter in Place

1. In the event that evacuation is not necessary or feasible, personnel will be directed to remain indoors in the designated assembly areas until the hazardous condition has been eliminated.
2. Supervisory personnel at all levels will assist in informing and directing people to seek shelter inside and remain there.

Third Stage: Prepare to Evacuate

1. The third stage warning alert will be broadcast by radio and television and indicates that preparations for evacuation are to be made.
2. People must remain calm, gather any needed items such as medications and clothing, and await further instructions.

Fourth Stage: Evacuation

1. The evacuation order will be issued by radio and television. Supervisory personnel will provide further instructions.
2. People will be directed to proceed to automobiles in preparation for evacuation. Cars will be loaded to seating capacity.
3. People needing transportation should proceed to the bus stop near the student union, where vehicles will be loaded.
4. Emergency personnel, such as state police and sheriff's department personnel, will be at key points to direct traffic.
5. Traffic will be directed away from hazardous areas. Automobile radios should be tuned to the designated station for further information.

Appendix 22-E

Contents of a Typical Emergency Operations Center

Typical Emergency Equipment

I. Communications Equipment
   A. Pagers
   B. Radios
   C. Cell Phones
   D. Chargers
   E. Receivers (TV, Radio)

II. Electrical Equipment
   A. Emergency generators/fuel
   B. Portable heaters
   C. Lighting
      1. Flashlights
      2. Interior/exterior floodlights
   D. Video/still cameras
   E. Batteries
   F. Fans

III. Office equipment
   A. Calculators
   B. Computers/Printers
   C. Copy machine

IV. Personnel
   A. Trained assistance
   B. Crowd control
   C. Emergency medical

V. Protective Equipment
   A. Chemical
      1. Cartridge respirators
      2. Chemical protective equipment
      3. Chemical spill cart
      4. Containment systems
      5. Gas dispersal systems

  B. General
   1. Hard hats
   2. Masks
   3. Hearing protection
  C. Fire fighting
   1. Clothing
   2. Extinguishers

VI. Food/Housing
  A. Emergency feeding
  B. Bottled water
  C. Portable shelters/tents
  D. Portable toilets
  E. Water purifiers
  F. Air ventilators
  G. Housekeeping supplies
  H. Beds/sleeping bags/cots/blankets

VII. Clothing
  A. Rain gear
  B. Protective garments

VIII. Tools and Repair Equipment
  A. Chain saws
  B. Pole climbing equipment
  C. Lifting equipment (for bracing)
  D. Necessary tools
  E. Flood suppression
   1. Sandbags (empty)
   2. Water pumps

IX. Transportation Needs
  A. Appropriate vehicles
  B. Fuel, oil, fluids
  C. Parts supplies

X. Miscellaneous
  A. First aid/medical supplies
  B. Petty cash

SECTION II-B

BUILDING SYSTEMS

Editor:
John D. Houck, P.E.
Oklahoma State University

INTRODUCTION

Building Systems

The following entry was made on July 7, 1958 in Eric Hoffer's personal diary, which later became one of his published works, *Working and Thinking on the Waterfront: A Journal, June 1958–May 1959.* Hoffer was a longshoreman for more than 20 years in the San Francisco Bay area. From time to time I think of the following passage and reflect on its content.

> Same ship, same place. Six hours [worked]. In the morning I took the Key-System bus to Encinal. As I walked down the several blocks from the bus stop to the docks I was impressed by the gardens in front of the houses. The houses of average size, are fairly old, yet in excellent shape. The people living here are mostly workingmen. The sight of the gardens and houses turn my mind to the question of maintenance. It is the capacity for maintenance which is the best test for the vigor and stamina of a society. Any society can be galvanized for a while to build something, but the will and the skill to keep things in good repair day in, and day out are fairly rare. At present, neither the Communist countries nor in the newly created nations is there a profound capacity for maintenance. I wonder how true it is that after the Second World War the countries with the best maintenance were the first to recover. I am thinking of Holland, Belgium, and Western Germany. I don't know how it is in Japan. The Incas had an awareness of maintenance. They assigned whole villages and tribes to keep roads, bridges, and buildings in good repair. I read somewhere that in ancient Rome a man was disqualified as a candidate for office because his garden showed neglect.[1]

These are remarkable words. Although nearly 40 years have passed since his thoughts were committed to writing, much of the content of his message remains the same. The question of maintenance of building systems is covered in the following chapters, which explore such topics as general building systems, mechanical systems, electrical systems, and other major elements of a physical facility. The level to which fcailities managers aspire to keep their existing inventory of buildings in good repair is the "best test for the vigor and stamina" of their college or university. Facilities managers should read, study, and enjoy the following chapters to the benefit of their institutions.

—John D. Houck, P.E.

NOTE

1. *Working and Thinking on the Waterfront: A Journal, June 1958–May 1959.* New York, Harper & Row, 1969.

CHAPTER 23

Architectural and Structural Systems

Warren Corman
Kansas Board of Regents

23.1   INTRODUCTION

Thousands of buildings exist on college campuses and all appear to be different. Their outward appearance is limited only by the designer and the builder. Buildings also vary according to their primary structure. The following sections describe these basic types of structures.

23.2   BUILDING FRAMING TYPES

The four basic structure or framing types are 1) wall-bearing, 2) reinforced concrete, 3) structural steel, and 4) a combination of these types.

Wall-Bearing

Wall-bearing refers to a building type that contains thick exterior masonry walls that support the floor and roof joists. Such structures are usually only one or two stories. Before the widespread use of skeletal framing, it was not unusual for buildings to have many stories, but the walls at the first floor were necessarily thick (3 or 4 ft.) to withstand the unit pressures on the masonry. A typical one-story wall-bearing structure might consist of a 12-in. exterior wall made up of 4 in. of face brick and 8 in. of concrete block, with the roof framing constructed of steel open-web bar joists.

Wall-bearing construction is usually found in fairly simple structures in which no major modifications are anticipated. They are rather easy to construct but are not as flexible in floor plan as the other building types because of the heavy bearing walls. Because masonry and mortar cannot be laid in wet weather or freezing temperatures, cold or inclement weather can slow masonry construction, thus delaying building completion.

Reinforced Concrete

Reinforced concrete framing is different from wall-bearing construction in that it consists entirely of freestanding columns braced by horizontal beams at each floor level. Skeletal framing is necessary for high-rise construction, and the columns and beams can be designed to carry almost unlimited loadings from the dead load of the building weight and the live load of contents, occupants, and the wind. Skeletal framing provides wide flexibility for future change to the layout, as there are few, if any, solid masonry bearing walls to remove or relocate.

Reinforced concrete frames are usually formed on the job and poured floor by floor as the structure rises. If beams are a real problem, either functionally or aesthetically, a flat slab floor system can be designed that eliminates beams by thickening the floor construction.

Reinforced concrete is an ideal material for structures, because it is fireproof and does not have to be protected. This is especially important in high-rise structures, where fire safety and exit codes are more stringent.

Precast concrete framing is ideal for some situations. The beams, columns, and even units of the floor slab may be cast in a factory and delivered to the job already cured and ready for erection. Quite commonly, these units are prestressed with wires or cables under tension. The connecting joints are usually welded to steel plates precast into the concrete. One of the most common uses of precast concrete structures is in the construction of multistory parking garages.

Structural Steel

Steel-framed buildings have characteristics similar to concrete-framed buildings in that the structural frame is freestanding and constructed before the exterior walls are erected. Steel is erected more quickly and is easier to work with in winter weather than is concrete. Steel-framed buildings range from the relatively simple Butler and Armco type prefabricated metal buildings to the more complex and sophisticated "superdomes," the sports arenas and high-rise structures. Steel structures are shop fabricated and field erected using rivets, bolts, welds, or combinations thereof.

Steel framing must be fireproofed for certain types of occupancies and for multistory construction. If a steel structure is not properly insulated, it will suddenly collapse in a fire if the temperature of the steel rises well over 1,000° F. Steel construction is lighter in weight, more quickly erected, and usually costs less initially than concrete framing.

Combinations of Types

Some of the more simple buildings are combinations of framing types. It is not uncommon for one- or two-story structures to have exterior walls that consist of wall-bearing masonry, while the interior supports are all steel columns with steel beams or trusses.

Some areas of the country are highly suitable for wood frame construction and wood siding. This is particularly true in California, Oregon, and Washington, where the native woods are abundant, and the products of the giant redwoods, firs, and cedars weather well.

23.3 FOOTINGS AND FOUNDATIONS

Spread Footings

Spread footings are simple concrete footings bearing on the ground to support concrete foundation walls or grade beams above them. They are commonly used on simple and low-rise structures, but they are usually not adequate for tall buildings or for poor or unstable soil conditions. Tall buildings create large unit pressures that spread footings usually cannot accommodate. Poor soils cannot safely support spread footings without settlement and cracking.

Pilings

The most common and best substitute for spread footings is piling. Piling can be made of treated timber, steel, or concrete. Concrete piles may be precast or cast in place. Piles are also either friction or point bearing. Friction piles carry their loads from the friction generated along their surface between the pile and the soil. Point-bearing piles usually sit on a stiff stratum of shale, rock, gravel, or other bearing strata capable of carrying large unit pressures.

Piles are usually placed in groups and capped with a heavy concrete top. Some piles in each group are driven at a slight angle from the vertical to provide batter, thus creating more stability within each individual pile grouping. They are driven into the soil by a large free-falling weight or a double-acting hammer weight. The pile driver guides the pile as it is driven.

Poured-in-place piles are created by drilling large holes into the ground and filing them with concrete after the bottom of the hole has reached a suitable bearing stratum. Pilings vary in diameter from 12 in. to as much as 5 or 6 ft. and sometimes reach 60 ft. in depth. In shifty soils or where water is present, the pile drilling may have to be lined with a steel casing as it is being drilled, and the hole may have to be pumped full of concrete as the steel wall is withdrawn.

Pilings are more expensive than simple spread footings but are sometimes justified because the footing is a critical part of the building design and must last for the life of the building. It cannot be revised, replaced, or even maintained without a huge expenditure of funds; thus it is not an item to gamble with in the overall design of the building.

23.4 FLOOR SYSTEMS

Monolithic Concrete

Many concrete framed buildings have concrete floor systems that are poured in place as a part of the overall structural system. This is called a *monolithic* concrete system. Monolithic systems are usually solid and stable and tend to reduce sound and vibration transmission problems.

Precast Concrete

Many companies manufacture a precast floor unit that can be set in place by a crane and anchored to the supporting beams. Precast units have the advantage of quick erection in all types of weather and the same solid and noise-reducing characteristics as monolithic concrete. Some units are also prestressed and may have hollow cores to reduce weight. A typical unit might be 6 in. thick and 12 in. wide. Precast units do not have quite the same flexibility as poured-in-place concrete when designing special openings through the floor.

Steel Systems

Steel systems consist of lightweight steel joists or trusses with steel decking spot-welded to them. Sometimes a thin layer of concrete is poured on top of the decking to add strength and stability, reduce noise and vibration transmission, and improve fire resistance.

Steel units can be erected quickly and are usually more economical than concrete. Many steel deck units are designed with compartments or cells to carry wiring for power, computers, telephones, and other equipment.

Composite Systems

Many types of floor systems are on the market and in use. One such system is called a *composite* system because it uses steel beams for the main support and a poured concrete floor on top of the beams. The concrete and steel are designed to act together as a single unit to carry the loads by rigidly securing the top flange of each beam to the concrete slab. This anchoring is achieved by a series of steel studs welded to the top flange of the beam and encased into the concrete when the slab is poured. This type of system is lighter in weight yet provides all the advantages of a concrete floor.

In general, floor systems with mass reduce noise and vibration transmission. In some cases specific design needs may have to be met by the floor system to accommodate specialized research equipment. It may be necessary to locate some equipment on a slab-on-grade floor with an appropriately designed mounting pad or piling.

23.5 EXTERIOR WALL TYPES

Solid Masonry

Exterior walls composed of masonry units have been used for centuries. Still in common use are facings of brick and native stone with a backup material made of concrete block or lightweight cinder masonry units. Such walls are usually 12 to 16 in. thick and are relatively easy to maintain. They have poor insulation value and should be insulated in the core or on the interior face with high-quality, permanent insulating materials.

Masonry Veneers

One of the more common exterior wall systems consists of an exterior wythe (facing of one thickness) of brick or stone masonry anchored to wood or metal stud framing behind it. This system can be heavily insulated in the stud cavities, but extreme care must be taken to ensure that the exterior veneer is solidly and permanently anchored and that water entering the core of the wall is swiftly and surely routed back to the exterior.

Curtain Walls

The term *curtain wall* is used to describe any exterior wall suspended from floor to floor on the structural frame of the building. This is a direct opposite of the term *wall bearing*. One type of curtain wall, popular since the 1950s, is a system composed of metal (usually aluminum) extrusions anchored together to form an exterior grid of vertical and horizontal mullions. The spaces formed by these mullions are filled with windows and opaque insulated panels. The number of designs, shapes, colors, and materials are almost unlimited.

High-quality curtain walls are relatively easy to maintain and have performed well. The major concern involves keeping them well caulked to avoid leakage in heavy rains and strong winds.

Precast Units

Curtain walls faced with precast concrete units provide a good, permanent exterior building face. Concrete units can be formed into any shape and texture and are usually of a size that can be easily transported and erected. Steel anchors are embedded into the concrete units so that the units can be welded or bolted to the building structure. The precast units can be backed with steel studs and gypsum board for insulation. This assembly makes an excellent exterior wall and is extensively used for university building projects. The stone and brick industries have developed and promoted prefabricated panels of brick or stone that can be erected in large sections similar to precast concrete units.

Wood Facades

Wood facades are normally used on wood-framed structures. Redwood, cedar, and fir weather well in their natural state and can be utilized in many shapes (e.g., lap siding; shakes; board-and-batten, tongue-and-groove, and shiplap siding).

Miscellaneous Types

Tilt-up wall construction is sometimes used for warehouse and other simple utilitarian-type structures. Tilt-up wall panels are made of concrete and poured in a flat plane on the job site, preferably next to the wall location. After curing, each panel is tilted up into place and anchored to the building structure. This is an economical method of construction but is not used widely on college campuses.

In certain parts of the country attention must be given to the expansion/contraction characteristics of the materials that are chosen. Widely varying expansion coefficients among materials that are not properly accounted for in the design will cause serious structural problems later. Also, sealants are used on some wall construction types, and these must be carefully chosen and properly installed so that they accommodate the expansion characteristics of the material they are joining.

Seismic design is also a major consideration. Wall construction and its integration into the structure will need to meet appropriate seismic design standards. These design standards vary substantially by state and will not be covered here, but the facilities manager should ensure that all design professionals working for the university are familiar with the applicable seismic design standards.

23.6 ROOFING

Nothing in the building industry has gained more attention than roofing and the problem of leaking roofs. Roofs generally fall into two basic categories: flat (or nearly flat) or steep sloping. Older buildings tend to have steep roofs or roofs with an adequate slope. Newer structures, which generally accommodate larger floor plans, are often topped with dead-level (flat) roofs or roofs with little slope.

Steep roofs are usually made of shingles (slate, asphalt, wood, asbestos, or clay tile) or standing seam metal sheets. Dead-level roofs are built up with layers of felt and hot moppings of asphalt or coal tar pitch and then capped with a flood coat of the hot liquid material.

Built-up Roofs

A built-up roof is constructed by placing alternate layers of saturated felt paper and moppings of hot asphalt or bitumen; it is then flood-coated on the top surface of the last layer of felt with the hot liquid asphalt or bitumen. Sometimes rock or gravel aggregate is embedded onto the hot pour to form a protective surface. Usually four or five layers are used, and such a roof is expected to last at least 15 to 20 years.

A built-up roof should have at least a $1/2$-in.-per-foot slope to readily allow water drainage. One should expect to receive at least a 10-year guarantee on the roof against leakage. The guarantee should include the flashings and counterflashings in addition to the roof itself. Flashings and counterflashings are usually made from galvanized steel or copper.

On older roofs with a slope of less than $1/2$ in. per foot, it is necessary to use coal tar pitch instead of asphalt. Pitch flows at a lower temperature and tends to seal itself in warm weather. Thus, it will flow off a sloping roof in hot weather and clog the gutters and downspouts. Four grades of asphalt exist; their use depends on the roof slope, which can vary from almost flat to 6 in. per foot.

The installation of a built-up roof is extremely labor intensive in the field and requires strict control of workmanship and attention to the weather. Built-up roofs are flexible and can adapt to almost any roofing problem.

Single-Ply Roofs

The term *single-ply* describes a factory-made sheet system from a single material or a laminated material. The sheets are shipped in a large continuous roll and are cut to fit field conditions on the roof. The top surface of a single-ply sheet may be factory coated or field coated. Single-ply roofs can be grouped into many classifications according to type of installation, material type, chemical composition, or manufacturing process. The usual installation types are adhered or externally ballasted. Joints between sheets are sealed with contact cements or welded using solvents or heat.

Single-Ply Adhered A fully adhered system is usually attached to the top surface of the roof insulation by contact cements spread by hand or sprayed. The partially adhered system uses mechanically attached plates spaced over the roof surface or other types of individual mechanical fasteners. These mechanical fasteners can be a source of failure. Adhered sheets are fairly easy to maintain, as rips, tears, or holes are apparent and can be repaired.

Single-Ply Ballasted   A different installation method involves laying the single ply onto the roof deck without any direct adhesion except at roof edges or penetrations. The loosely laid membrane is then held down by rounded, smooth, clean rocks with a diameter of about 2 in. Because the ballast tends to cover any problem areas, such areas are less easily maintained.

Other Roofing Types

Steep roof slopes are readily compatible with various types of shingles or with standing seam metal sheets. Conventional shingles are composed of asphalt, clay tile, cement tile, wood, or slate. Standing seam roofs are made of long, narrow sheets of metal and are joined by a raised, interlocking watertight joint (the standing seam). Also on the market are foamed coatings that spray on and provide insulation and waterproofing. These are usually used for reroofing and not for new construction.

Roof Protection

It is important to protect roofs from damage from pedestrian traffic. If equipment on the roof requires regular maintenance, roof walkways or stepping stones must be installed for the maintenance people. Special units are made for this purpose. In any case, the number of penetrations should be kept to a minimum and may require close coordination with the mechanical system design.

23.7 WINDOWS

The basic material for window frames is primarily metal or wood, although new high-strength plastics are appearing on the market. The most popular windows for university buildings are of aluminum or nonferrous alloys. They are long lasting, do not rust or rot, can be extruded into intricate shapes to receive good weather stripping, and require little maintenance. Factory-applied permanent finishes are popular. The raw aluminum is coated with a rich, dark bronze color that blends well with brick and stone facades. Many of these coatings are chemically or electrically applied and will last for many years.

It is important to compare air infiltration tests provided by the window manufacturers as a measure of the amount of tempered air within the building that may be lost to the outside on a windy day. Some of the more expensive windows provide a nonconducting thermal break barrier or joint built into the window in an attempt to prevent frost buildup on the interior in extremely cold weather. Good windows should be designed to allow the installation of factory-sealed dual glazing, which can be ³/₄-in. thick or even thicker for extremely large panes of glass. For multistoried structures it may be important to select a window that can hinge or pivot so that the exterior pane of glass can be cleaned from the inside of the building.

Wood windows require more maintenance, primarily painting, but some window manufacturers are providing a factory-applied plastic facing for the exterior portion of the window that does not require painting. Wood windows conserve more energy than metal windows because of the difference in density and conductance between wood and metal. In addition, naturally finished wood windows contribute to the aesthetic beauty of a project. If wood windows are considered, it is important to select one of the better windows on the market, as they are usually made of better materials and hardware and have a more acceptable appearance and longer life cycle.

23.8 DOORS AND FRAMES

Doors and frames for institutions are usually made of wood or hollow metal. Hollow metal doors and frames are fabricated from sheet steel and are strong and durable. The doors and frames are reinforced to fit all types, styles, and sizes of hardware. They are custom fabricated and require a certain amount of lead time for shop drawings, manufacture, and delivery. This is important if the frames are to be in place before the masonry work is to be laid, so that the frame anchors can be built into the masonry joints. Special attention should be given to entrances with heavy traffic to make them sufficiently durable. Wood doors should be solid-core doors to withstand abuse, provide better fire protection and noise control, and serve as a substantial receiver of the various anchoring devices for hardware. It is important that the right mullion be chosen to ensure that it will be able to meet the traffic demands of the installation location.

The selection of high-quality hardware for lock sets, panic devices, closers, and butts is important to provide security, service, and low maintenance costs. Institutions should select a good hardware company and use the same keying system for all buildings, if possible, to reduce the number of keys and master keys. Cheap hardware is quite costly to maintain over a long period of time.

23.9 EXTERIOR INSULATION

Roof and exterior wall insulations are manufactured of many types of materials. They can be classified as loose fill; batts; boards; or poured-in-place, lightweight material. Hollow cells in masonry units are usually filled with a pourable, granular material that is delivered to the job in sacks. It is important to select only materials that will not settle or decay and will not be eaten by termites or rodents.

Insulation boards are commonly used as roof insulation and as the vertical joint between masonry wythes (vertical layers of masonry) in exterior walls. Again, this material should be permanent and not attractive to insects or rodents. Because there are many types of materials on the market, the services of an expert may be required to select the proper board or plank insulation from the available organic, inorganic, plastic, and synthetic materials.

Some roof boards are preformed to install on a flat roof to provide a sloping top surface. This requires special attention to the location and height of roof drains, curbs, scuppers, and flashings. An option to roof board insulation is a poured-in-place, lightweight, concrete-like material that is flexible and can solve many roof slope problems. Care must be exercised to allow sufficient cure time for the wet materials before roofing or vapor barriers are installed.

Batt insulation is quite effective in joist or stud space. It should be permanently anchored to avoid future slipping or sagging. Fiberglass material is common, but other good inorganic materials also exist on the market.

ADDITIONAL RESOURCES

Edward, Allen. "Fundamentals of Building Construction." In *1988 Handbook of Commercial Roofing Systems.* Cleveland: Edgell Communications, Inc., 1988.

TEK Information Series. Herndon, Virginia: National Concrete Masonry Association, 1995. This series of newsletters is distributed by the National Concrete Masonry Association, and information on specific topics can be obtained by contacting the association.

CHAPTER 24

Building Interiors

Grace C. Kelley

With additional contributions by Marneth Ball,
Oklahoma State University

24.1 INTRODUCTION

Interior design as a profession, a specialized branch of architecture, is a relatively new field. Graduates in this field have a thorough education with strong architectural emphasis, and many are finding careers in facilities management. Simultaneous with the development of the interior design profession, and perhaps related to it, has been the growing emphasis on creating and maintaining a higher education environment that is conducive to learning. Until relatively recently, the principal concerns with interiors were that they were kept painted, clean, and adequately lighted and contained serviceable furniture. Choices of interior colors often were left to the occupants, and economy dominated the selection of furniture and interior materials. Currently, however, the value of professional interior design in creating an effective interior environment is widely recognized. Several large institutions have interior designers on their facilities staffs. However, many facilities organizations do not have the resources to include a professionally trained interior designer.

This chapter provides fundamental information that will be useful to the facilities manager. It includes extensive information on interior issues—interior materials and systems, space planning, acoustics, and lighting. It also includes information on standardization of furniture, flammability regulations, and facilities for the handicapped. Some effort is made to provide a sense of the aesthetic issues with which the interior designer is faced.

All information is tailored to the requirements of the institutional environment. Within this environment a design solution must first be functional, and then cost-effective. It must be aesthetically pleasing and commensurate with other factors such as durability, maintainability, and safety.

24.2   BASIC PRINCIPLES OF DESIGN

All interior design schemes endeavor to relate functionally and aesthetically to the existing environment. The more appropriate the design scheme, the clearer the perception that it was created as an integral part of the whole. Design fads that prevail over functional appropriateness generally will be perceived as cluttered, disjointed, or dysfunctional. The designer must distinguish good design from inappropriate design.

Good design does not refer to taste. Taste is a matter of preference. Good design has a basic, lasting quality. Most people can recognize a design scheme that works well, even though they may be unable to fully appreciate why it works.

It is relatively simple to state the basic principles of good design: proportion, balance, rhythm, focus, and harmony. Once these principles are understood, they can be applied almost without effort.

Proportion

Proportion in design is the harmonious relationship of one part to another or to the whole. The early Greeks discovered the secret of good proportion and established rules that have been accepted and followed by designers for centuries. Their standard for good proportion was a rectangle with its sides in a ratio of two parts to three; this is called the *golden rectangle.* The *golden section* involves the division of a form so that the ratio of the smaller proportion to the larger is the same as the larger to the whole. The progression 2, 3, 5, 8, 13, 21, 34, and so on, in which each number is the sum of the two preceding numbers, provides an approximation of this relationship. For example, 2:3 is approximately the same ratio as 3:5 and 5:8 is the same ratio as 8:13. These proportions should be applied when planning the dimensions of a room or selecting a piece of furniture for a particular area. Classic rules of proportion also dictate that the division of a line somewhere between one-half and one-third was most pleasing. This concept was called the *golden mean* and can be applied in interiors when hanging pictures or tying back draperies.

An object is perceived in relation to the area around it. Objects that are too large will crowd a small room, whereas furniture that is too small will seem even smaller in a large room. In addition, a large piece of furniture will seem even larger when surrounded by small furniture.

Form, color, texture, and pattern all influence our perception of scale and proportion. Coarse texture, large patterns, and bold colors will make an object appear larger, whereas smooth textures, small patterns, and light colors will make it appear smaller. Whatever attracts the eye appears larger. These principles can alter the apparent size and proportion of spaces and objects.

Balance

Balance provides a sense of equilibrium and repose, a feeling of the weight of an object. Three types of balance exist: bisymmetric, asymmetric, and radial.

Bisymmetric or *formal balance* uses identical objects arranged similarly on either side of an imaginary line, such as a sofa with identical end tables and matching lamps on both ends. Every arrangement needs some bisymmetry.

Asymmetric or *informal balance* is more subtle. This type of balance requires more thought but remains interesting longer than completely symmetric arrangements. In asymmetric arrangements, objects of different sizes, shapes, and colors may be used in an infinite number of ways. Two small objects may balance one large one, a small shiny object may balance a larger dull one, or a spot of bright color may balance a large neutral area.

Radial balance is an arrangement that radiates from a central point. A round conference table or a dining table with chairs around it is an example of radial balance.

The architectural features of a room, such as doors and windows, should be located in such a way as to give this feeling of balance. A pleasing distribution of highs and lows and large and small features give a room a well-balanced feeling.

Rhythm

In interiors, rhythm is something that allows the eye to move smoothly around the room. It may be achieved by repetition, gradation, opposition, transition, and radiation. By repeating color, pattern, texture, line, or form, a rhythm of a repetitive nature is achieved. Gradation is the succession of the size of objects from large to small, or of colors from dark to light. Opposition occurs wherever lines come together at right angles, or wherever a horizontal line of furniture meets a vertical architectural member. Transition is the rhythm of a curved line that carries the eye over the room, as occurs with an archway. Radial rhythm is a result of lines extending outward from a central axis.

Focus

The focal point is a feature of the room to which the eye is drawn. It creates a feeling of unity and order in the room. A prominent architectural feature commonly serves as a focal point, but if no architectural feature exists, an important piece of furniture can substitute.

Harmony

Harmony involves fitting together parts to form a cohesive whole. Completeness and order are established when the furnishings of a room harmonize in their relationships with other items within the space and with the background. If the room is large or small, furniture should be scaled accordingly. Floor coverings should be selected with a theme or purpose in mind. Fabrics and colors should be appropriate to the style of furniture. Accessories should be appropriate to the style of the other furnishings and should reflect the personality of the space.

24.3 CHARACTERISTICS OF OBJECTS

The basic principles of proportion, balance, rhythm, focus, and harmony are achieved by considering the characteristics of the elements in the interior landscape. Much as one looks at a park-like landscape and perceives the characteristics of trees, shrubs, grass, flowers, sky, water, and pathways, one can look at the interior landscape and perceive the characteristics of desks, chairs, files, carpet, walls, and ceilings. These characteristics include the following elements: texture, pattern, line, form, space, color and light, and color schemes.

Texture

Texture refers to the surface quality conveyed by objects within a space and the dominant texture the architectural background establishes. For example, a room paneled in polished wood or papered in a traditional wall covering will require furniture woods and fabrics with smoother textures than a room paneled with rough-hewn wood or constructed of masonry.

Pattern

Pattern forms the simplest method of surface embellishment. Too much pattern can make a room too busy. Although the total arrangement of the components of a room creates an overall pattern, the more obvious patterns are in carpet, fabric, and wallpaper. These should be appropriate to the general feeling of the room.

Line

Line is expressed by the sense of composition, direction, and whether motion or repose is felt within a space. Line can seemingly alter the proportion of an object or of an entire room. Vertical lines cause the eye to travel upward, causing the area to seem higher. Horizontal lines direct the eye across the area, making it appear wider. Curved lines are graceful and fluid. Diagonal lines give a room a feeling of action; staircases and slanted ceilings are examples of this. Too much line movement gives an unsettled feeling. A proper balance of the various vertical, horizontal, and curved lines achieves harmony.

Form

Form is a major concern in planning interiors. The shape or mass of objects within a space causes a sense of confusion if an excessive variety is used. A lack of variety creates monotony.

Space

Basic rules govern the use of space. Anthropologist and father of proxemics, Edward T. Hall, observed that people have specific, culturally prescribed distances in their daily activities. Designers must be aware of these personal and public space relationships and allow for them in their designs.

Color and Light

Color and light are the most interrelated elements of interior design. Color is a quality of light reflected from an object to the human eye. When light strikes an object, some of it is absorbed, the rest is reflected. The wavelength of the light an object reflects determines its color.

To use color effectively in planning attractive rooms or in selecting furniture, draperies, or floor coverings, it is necessary to consider these factors:

- The relationship of colors to each other, and how light affects the apparent color
- The characteristics of colors and their psychological effects
- Which colors harmonize, and which contrast
- Which combinations are appropriate and practical for the specific project

Color Schemes

Red, yellow, and blue make up the three primary colors or hues. The secondary colors (a combination of two primary colors) include orange, green, and violet. Numerous shades result when one of the primary colors dominates the mixture, as in the case of yellow-green or green-yellow.

Three basic color schemes exist: analogous, complementary, and monochromatic.

- *Analogous.* Colors harmonize with each other when they share a common element. Blue, for example, is a basic element of green, blue-violet, purple, and red-violet.
- *Complementary.* Too many harmonizing elements can become tiresome. It is therefore desirable to introduce contrast by adding a complementary color. For example, the complement of red is green, which is a color containing no red but made up of blue combined with yellow.
- *Monochromatic.* Color schemes that use various shades or values of a single color are called *monochromatic.* Monochromatic schemes can be monotonous unless other elements such as pattern or texture are used.

The color to dominate the room should be the first one selected. This selection is based on the client's preference or the room's size and light conditions. After choosing the dominant color, the related or harmonizing colors are selected. Most rooms should be decorated in an ascending scale from dark to light: rugs or floors should be darkest, walls lighter, and ceilings lightest. Efforts should be made to avoid pure white in favor of off-white shades. Another style of decorating involves a descending scale of values, with colored walls and ceilings of grayed or darkened hues and light-hued floor covering.

Colors should be distributed throughout the room, avoiding the spotty effect of isolated splotches of color concentrated in one area. One way of distributing colors is to upholster at least one piece of furniture in a fabric containing all the colors in the room.

The main areas of a space usually should be the most neutral in value. As areas are reduced in size, the chromatic intensity can be proportionately increased.

In a color scheme, small touches of bright hues are called *accents.* Accessories or an occasional chair can supply the accent color to heighten the effect of a color scheme. Cooler shades such as blue or green, for example, will make small rooms look larger, whereas yellow, orange, and red can make larger rooms more intimate and cozy. Cool tones are quiet and restful, and warm colors are friendly and cheery.

24.4 INTERIOR MATERIALS AND SYSTEMS

Floor Coverings

This section primarily concerns rugs, carpeting, and resilient tiles. However, flooring is not limited to these materials. Terrazzo, wood, ceramic tile, and marble are flooring materials whose initial high cost can be justified under appropriate circumstances. The best approach to flooring design entails considering the specification of flooring in the initial phases of construction and not as an afterthought, limiting it to surface decoration.

Rugs In the 17th and 18th centuries, carpets and other floor coverings were rarely found in the homes of ordinary people; textiles were considered precious and were not used on the floor. The word *rug* referred to a handmade coverlet, and *carpet* referred to a table cover. Woven or hand-knotted rugs did not become common as floor coverings until the early 18th century, when Oriental carpets became popular among the wealthy.

The hand-knotted rug is the best rug or carpet, the Oriental rug being a fine example. However, the demand for Oriental rugs has increased to such an extent that their manufacture has become commercialized. Although Oriental rugs are still made on hand looms, quantity production in factories—with less experienced craftspeople, cheaper materials, and aniline instead of vegetable dyes—has become common.

In the university environment, the designer may have an opportunity to specify such a rug in a special setting, perhaps the office of the president or an important public area. In this case, the designer should depend on the advice of an expert or reputable dealer.

Custom-designed tufted rugs also create a rich effect. For those with a tight budget, manufacturers in Puerto Rico, Hong Kong, and Japan produce fine hand-tufted rugs to specification. In addition, larger mills in the United States also serve as a source of custom-designed products.

A less expensive way to achieve this custom look is to use broadloom carpet and border it with another broadloom, cut to the desired width. Several borders of varying widths and coordinating colors or patterns can be used. After the border is added by attaching it to the field carpet with seaming tape, the seams can be beveled for a sculptured look.

Carpet Carpets and rugs represent beauty, luxury, or status, regardless of the actual function of a space. When considering carpeting, perhaps the most important question is, "How will the space be used?" The following is a list of properties carpets can be expected to provide:

- Acoustical privacy
- Ease of maintenance
- Cost-effectiveness
- Design flexibility
- Feel of luxury

The type and quantity of fiber and the carpet's construction determine the quality. The determining factors in manufacturing are the pitch (number of face yarns per inch), the pile height (the height of the yarn above the backing), the ply of the yarn (number of individual ends of yarn twisted together), and the method of weaving or tufting. Ultimately these factors are reflected in the price, and a reputable manufacturer's price is a good indication of its quality.

To make an intelligent decision when specifying carpet, one must have at least a basic knowledge of carpet fibers and how they perform. The following summarizes the types of natural and man-made fibers used in commercial carpet manufacturing (Figure 24-1).

Natural *carpet fibers* include wool, cotton, and flax. *Wool* carpet has re-emerged in prestigious contract interiors and still represents the standard against which all carpet fibers are measured. The surface of wool scatters optical light, thus improving its appearance by diffusing soil visibility. Because of the scaly character of carpet, crevices in the carpet do not hold surface dirt and dust readily. Below-surface particles release with ease, and resilience is outstanding. The significant characteristics of wool fiber are as follows:

Advantages	Disadvantages
Consumer appeal	Initial cost
Appearance	Stain removal
Feel	Abrasion resistance
Resilience	Styling versatility
Flame resistance	
Soil resistance	
Cleanability	
Solvent resistance	

Economics and technology have resulted in the displacement of wool by man-made fibers. Man-made fibers are in liquid form before solidifying at room temperature. Essentially, they are manufactured by heating the polymer (chemical material) and blowing it through a perforated plate called a *spinnerette*. Fine streams of liquid become solid strands of filament as the material cools in a liquid or air bath.

Nylons are the leading petrochemically derived face fiber materials, followed by acrylics and olefins. Made of chemicals from the polyamide group, nylon is the most widely used carpet fiber. The synthetic yarn is not naturally resilient but is crimped, bent, and twisted to create resilience. Nylon tends to retain its shape because it is nearly 40 percent harder than, for example, olefin. This means that nylon made to the same specifications as olefin and used in the same area will show wear at a lesser rate. However, nylon absorbs and holds limited amounts of moisture, so it is not as stain resistant as olefin, even when solution dyed. (*Solution dyed* means the dye is added while the nylon is still in liquid form, before it is extruded into a fiber.) This capillary characteristic enables nylon to be dyed in many ways, increasing its manufacturing and styling flexibility.

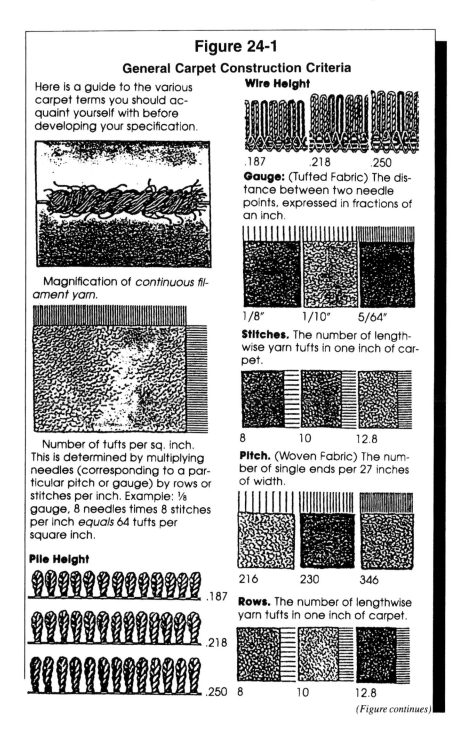

Figure 24-1
General Carpet Construction Criteria

Here is a guide to the various carpet terms you should acquaint yourself with before developing your specification.

Magnification of *continuous filament yarn*.

Number of tufts per sq. inch. This is determined by multiplying needles (corresponding to a particular pitch or gauge) by rows or stitches per inch. Example: ⅛ gauge, 8 needles times 8 stitches per inch *equals* 64 tufts per square inch.

Pile Height

.187
.218
.250

Wire Height

.187   .218   .250

Gauge: (Tufted Fabric) The distance between two needle points, expressed in fractions of an inch.

1/8"   1/10"   5/64"

Stitches. The number of lengthwise yarn tufts in one inch of carpet.

8   10   12.8

Pitch. (Woven Fabric) The number of single ends per 27 inches of width.

216   230   346

Rows. The number of lengthwise yarn tufts in one inch of carpet.

8   10   12.8

(Figure continues)

Pitch to Gauge Conversions

Pitch	108	143.9	172.8	180	189	216	243	252	256	270	346
Needles	4	5.3	6.4	6.6	7	8	9	9.3	9.5	10	12.8
Gauge	1/4	3/16	5/32	9/64		1/8				1/10	5/64

Face Fiber. Specify as follows: "The face yarn of the carpet shall be pile of 100% 'Ultron' advanced-generation nylon *or* its equivalent."

Pile Weight per Square Yard. This is the amount of yarn used in the pile of the carpet, excluding the primary backing. Pile weight is measured in ounces per square yard.

— Face Yarn
— Primary Backing
— Latex
— Secondary Backing

Pile weight + Latex + Backing
   = *Total Weight*

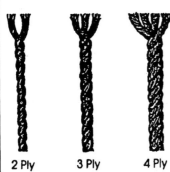

2 Ply   3 Ply   4 Ply

Construction Methods

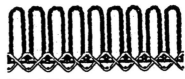

Velvet Weave

Wilton Weave

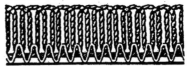

Knitted

Tufted

Fuse-Bonded

Axminster Weave

Nylon's hard, smooth surface reduces the tendency of soil particles to cling to the fiber, so they can be easily removed during vacuuming. The fiber cross section, with its modified triangular shape, breaks up and scatters light rays and tends to hide the soil on the back side of the fiber, making the carpet appear cleaner. The ability of a carpet to retain a good appearance as long as it lasts is the true measure of its effective life. Appearance-retention qualities are inherent in carpet made from nylon.

Nylon fiber exhibits the following advantages and disadvantages:

Advantages	Disadvantages
Good bulk and cover	High static
Crush resistance	
Wearability	
Takes color well	
Good luster range	
Soil resistant	
Good cleaning	
Mildew resistant	

Acrylic fibers are synthetic fibers whose polymer base is composed mostly of acrylonitrile units. The fiber has a high bulk-to-weight ratio; good acid and sunlight resistance; and a bulky, wool-like appearance. It is produced in staple form only to be spun into yarn. High color, long life, and resistance to mildew characterize this fiber. Color is added by stock dyeing in the spinning mill while the acrylic is still in fiber form.

Characteristics of acrylic fibers include the following:

Advantages	Disadvantages
Low static level	Low abrasion resistance
Excellent cover and bulk	Alkali resistant
Wide color range	
Moisture resistant	
Mildew resistant	

Olefin fiber is a synthetic polymer whose base is ethylene, propylene, or other olefin. It is the lightest commercial fiber, has excellent strength, and is resistant to chemicals and abrasion. It is dyed in liquid form prior to being extruded into continuous filament yarns. Thus it is solution dyed. It is highly moisture and stain resistant; however, olefin is a petroleum-based product, and the oily residue in soil will bind more readily to its surface. If spillage is a problem in an area being considered for carpeting, olefin will provide excellent protection against stains. Proportionally more soiling will occur, however, with a resulting increase in maintenance costs. The following advantages and disadvantages characterize olefin:

Advantages	Disadvantages
Color solution dyed	Low resilience
Good cover and bulk	Limited color
Abrasion resistant	
Stain resistant	
Low static level	
Mildew resistant	
Initial cost	

The text that follows describes the major manufacturing processes for carpeting.

Carpet Backing. Carpet specifications must also include the type of backing used, as the backing will seriously affect the quality of the installation. Justifiable arguments exist for specifying both jute and synthetic carpet backings. A choice must be made based on the most appropriate product for the particular installation.

Jute is a natural cellulosic fiber made from certain plants of the linden family that grow in India and Bangladesh. Jute yarns are used for woven carpet backing, yarns, and twine. Woven jute fabrics are used in tufted carpet as primary and secondary carpet backing.

Jute absorbs adhesives well and forms tight mechanical bonds, preventing delamination between carpet and backing. Installers frequently prefer jute backing because it stretches better than synthetics, goes around corners easier, and accommodates irregularities better. It also absorbs moisture, however, and can mold and rot when wet. When wet, jute tends to shrink as it dries.

Synthetics are highly resistant to mold and mildew. They are not damaged by moisture, will not rot, will not contribute to face yarn stain if the carpet is wet, and remain odor free. Synthetic carpet backings reduce static buildup in computer areas. The addition of carbon to the primary backing, combined with face yarn treatment, reduces static by about 80 percent. Together, they eliminate essentially all noticeable and computer-damaging static.

Attached cushion backings are a special type of carpet backing. Made of urethane foam, this backing forms a built-in carpet pad that is glued directly to the floor. One disadvantage is that at replacement time, most of the pad remains glued to the floor. Another type of cushioned backing is vinyl backing.

Tufts of carpet yarns punched into primary backing will pull out unless a "glue" is applied to the back of the carpet. Sometimes this coat is the final step in backing the carpet. In this case, the carpet is said to have a *unitary back.* The coatings, made of latex or polymer, are spread on at room temperature and then baked dry, or spread on hot and allowed to cool.

Both methods lock in each yarn bundle while forming a back ready to be glued down.

Carpeting with unitary backing is made for glue-down installations. Common office areas, libraries, and other heavy traffic areas are ideal locations for glue-down, unitary back carpet. Premium adhesive is essential. Installation of latex unitary tends to be more complicated and therefore more expensive, because installers must follow precise instructions. Latex unitary is nonporous and will resist evaporation. Therefore, it is necessary to let the adhesive tack before the carpet is committed to the adhesive. It is less flexible than conventional backs of jute and must never be folded, creased, bent, or stretched, as it is difficult to flatten. Carpets with unitary backings are extremely durable and provide an exceptionally strong tuft bind. However, the other properties of this thermoplastic backing must be evaluated in relation to the requirements of the specific installation. Just as important, only a crew experienced with this type of backing should be allowed to install it.

Carpet Life Cycle Costing. Low cost is an important consideration, but initial low expense is only one aspect. Although the initial purchase price for carpeting may be higher than for conventional floor coverings such as vinyl tile, long-term cost is demonstrably less, with substantial overall savings from extended wear life at high retention levels for appearance. Total use cost is the best basis for comparison. The use cost concept evaluates the three basic elements of value in relation to cost:

1. How much does it cost to buy and install?
2. How long will it last?
3. What will it cost to maintain?

The installed cost of carpet is, on average, greater than that for noncarpeted floors. However, its combined maintenance costs are so much lower that its total use cost is 45 to 52 percent less than that of noncarpeted floors.

Flammability Testing. It is important to be familiar with potential fire hazards and apply this knowledge in specifying materials and systems. The use of the wrong kinds of interior finishes and furnishings have caused many deadly fires. Often it takes these tragedies to prompt federal or state governments to enact legislation to protect the public.

In the first critical moments of a fire, ignited materials will either contribute to or prevent the fire's spread. Factors to consider when evaluating particular materials for safe use include the amount of heat, smoke, and toxic gases released and interaction with other materials.

Although flammability connotes ease of ignition or flame spread rate, in actual practice it refers to the performance of a product subjected to a specified test. When determining flammability of carpeting, many tests may apply (Figure 24-2):

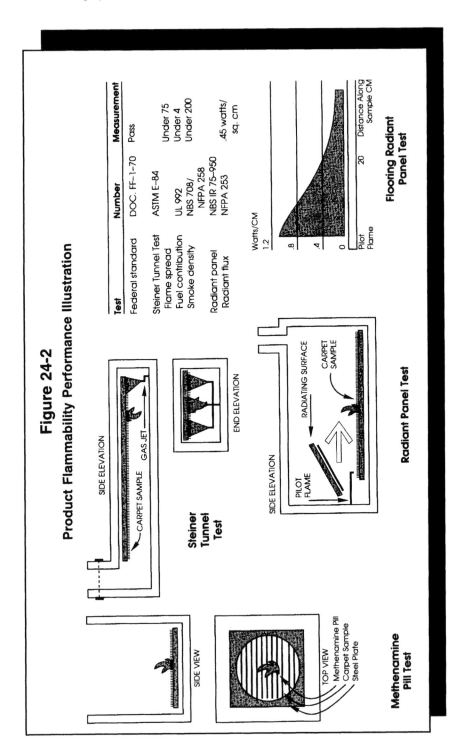

Figure 24-2

Product Flammability Performance Illustration

- *Methenamine Pill Test: DOC FF-1-70.* DOC FF-1-70 applies to carpet, and DOC FF-2-70 applies to rugs. The National Bureau of Standards developed this test, and the Department of Commerce adopted it in 1970. Since 1971 this test has been required by law for all carpet offered for sale in the United States. It is the Federal Trade Commission standard.

- *Steiner Tunnel Test: ASTM E-84, NFPA 255, UL 723.* The Steiner Tunnel Test, developed by Underwriters Laboratories, is required by the Public Health Service and Life Safety Code for contract carpeting installed in health care units participating in Hill-Burton and Medicare programs. This test subjects carpet specimens to a heavy flame source and uses flame spread as a comparative measure. Hill-Burton programs require a flame spread rating of 75 or less.

- *Flooring Radiant Panel Test: ASTM E648; NFPA 253; NBS IR 75-950, NFPA 101.* This test has been approved as a replacement for the E-84 Steiner Tunnel Test and is used by many regulatory agencies; however, some may still require the Steiner Tunnel Test. The Panel Test measures the heat required to sustain combustion in carpets exposed to radiant energy and flame. A pass-fail heat flux level of 0.5 W/cm^2 or higher is generally recommended for corridors and exit ways in hospitals and other such institutions; 0.25 W/cm^2 is recommended for corridors and exit ways in schools, offices, and other buildings.

When the Flooring Radiant Panel Test is used to measure flammability, the National Bureau of Standards Smoke Chamber Test is sometimes used to measure smoke density. Most government regulatory agencies approve the Flooring Radiant Panel Test for nonsprinkled corridors and primary exit ways. The Methenamine Pill Test would then be used for room carpet and sprinkled corridors. The rating system is as follows:

$$\text{Class } 1 = \text{Minimum } 0.45 \ W/cm^2$$
$$\text{Class } 2 = \text{Minimum } 0.22 \ W/cm^2$$

- *Smoke Density Chamber Test: ASTM E662 and NFPA 258.* The Smoke Density Chamber Test is the same as the NFPA 258. It measures solely the smoke density characteristics of carpet under controlled laboratory conditions to determine the specific optical density of smoke within a closed chamber. The rating system includes an index range from 0 to 800 units. Most federal agencies require a smoke density of a maximum of 450 units average of the flaming and nonflaming results.

Carpet Estimating. For planning purposes, a quick estimate of the square yardage required can be figured by calculating the square footage from the floor plan, then dividing by the number of square feet in a square yard (9), and adding 20 percent for waste. This will generally be excessive but will provide a budgetary figure.

Stairs present a special problem. Two methods exist for installing carpet on stairs: the *waterfall* method, which uses one continuous piece of carpet; and the *cap-and-band* method, which covers the stairs in separate pieces.

When estimating carpet, include information about special details such as type and condition of floors and whether shoe mold is to be removed. Note existing carpet to be removed, whether it is glued down or over a pad, and the condition of the flooring beneath the existing carpet.

Most carpet experts use the rule-of-thumb formula of $2^1/2$ linear feet of tackless strip for each square yard of carpet. When drawings or measurements of the areas are available, a closer estimate can be obtained by computing the total linear feet and adding 10 percent for waste.

Most commonly, carpet used commercially is manufactured in 12-ft. widths. Each width of carpet must lie with its pile going in the same direction. Finding the best layout for the carpet will depend on the most economical use of yardage, maximum performance of the carpet, and desired appearance. When planning layout, factors to consider include placement of seams away from high-traffic areas, no right-angle seams through doorways, and minimum of 1-ft.-wide fill pieces.

Carpet Installation. The following issues should be considered in determining whether an installation should be stretched in or glued down:

1. *Size of job.* Generally speaking, jobs with a long run between anchor points will be difficult to properly stretch. This can be aided where separate offices are involved, permitting a 4 x 4 or 2 x 8 board to be used as a base for the power stretches; otherwise, a 25-ft. run would be the maximum.

2. *Layout of job.* Large open areas with furniture or partitions that interfere with proper original stretching (and hinder restretching) should not be stretched in. Likewise, irregularly shaped or angular areas should be avoided. Multiple office space connected by a maze of corridors also presents a distinct stretch and/or restretch problem.

3. *Underlayment.* Underlays that are too soft can cause problems.

4. *Use.* Heavily trafficked areas or areas in which wheeled equipment is routinely used are to be avoided with stretch-in. Padding accentuates traffic patterns and impedes the movement of wheeled equipment.

5. *Style innovations.* If the designer specifies bias cuts, inserts, or other innovations, proper stretch-in is often impossible.

6. *Flammability.* Flame spread performance of any carpet is dramatically and negatively affected by underlayment owing to the insulating effect of the pad, which prevents heat dissipation through the carpet. Where regulations do not specify that the flame spread test must be made over a pad but it is known they cannot be met in such a system, the stretch-in method should not be used.

When stretch-in is appropriate, the following considerations are important:

1. *Amount of stretch.* Each product should be stretched slightly past its stabilization point to avoid growth and buckling as it relaxes. Most installers take responsibility for the first restretch if difficulty in properly stretching occurs initially.
2. *Layout and carpet care.* Proper stretch-in requires proper fastening around the entire perimeter and at all seams. The carpet must be laid in one direction, and the back must not be abused (dragging heavy furniture, exposure to wheeled traffic) prior to or subsequent to installation. Diagonal cuts are to be avoided.
3. *Mechanical requirements.* Power stretchers are recommended in all sizes of contract installations but are essential in larger ones. Large installations also require commercial tackless strips with three rows of pins.

Carpet Wear. Several factors determine the life expectancy of carpeting: abrasion resistance, appearance retention, maintenance, and traffic patterns. Abrasion resistance relates directly to wearability, whereas appearance retention relates to such characteristics as crushing, fading, soiling, and matting. Improper maintenance can accelerate wear as gritty abrasive dirt accumulates, dulling the color and reducing resilience. High traffic will causing crushing and abrasion problems.

Whether a carpet is functionally acceptable is still a subjective matter. Laboratory methods for measuring abrasion resistance employ the Taber Abraser and Rollstuhl tests, among others. These tests are based on the number of cycles required to abrade the pile fiber completely to the primary backing.

The multilobal structure in various deniers of nylon fibers provides exceptional bulk and gives excellent resistance to crushing and matting, even in high-traffic areas. This high degree of resilience minimizes the traffic lanes and deep indentations from heavy furniture.

Tuft bind is also a measure of how well the carpet will wear. The ASTM Test D1335 determines the pounds of force required to pull the tuft out of the back of the carpet. Tuft bind is obtained by proper application of the back coating. A single tuft in a loop pile carpet should withstand a minimum of 8 lbs. of pull force for most end-use applications; a pull force of 4 to 5 lbs. is generally adequate for cut pile carpet. A higher tuft bind is recommended when the possibility of deliberate raveling exists, such as in grade school installations.

Wear is defined as appearance loss, not as fiber loss. The ability of a carpet to retain a good appearance as long as it lasts is the true measure of its effective life. Appearance retention qualities are inherent in properly constructed carpet, installed with an appropriate end use in mind.

Atmospheric contaminants such as ozone and other gases contribute to color fading. The advanced generation nylons have less open physical structure that provides resistance to gaseous penetration and thus aids in color stability.

Solution dyed yarn discourages fading because the color is permanently locked into the fiber. Proper dye selection and fixation will ensure color clarity through resistance to fading, crocking, and repeated cleanings.

Industry test procedures accepted by the American Association of Textile Chemists and Colorists (AATCC) to determine colorfastness include the following:

- *Atmospheric Fading (AATCC 129)*. This test determines the color change of the carpet when exposed to ozone gas in the atmosphere under high humidities. The International Gray Scale ratings are: 5, no change; 4, slightly changed; 3, noticeably changed; 2, considerably changed; and 1, much changed. A shade change rating of at least 3 after three cycles of exposure is desired.
- *Lightfastness (AATCC 16E)*. This method determines the colorfastness of textile materials to light in a xenon arc fade-ometer after material is exposed to 60 "AATCC fading units" and judged visually for change. A gray scale rating of at least 4 is desired.
- *Crocking (AATCC 8)*. This is the degree of color transfer from the carpet to a white cloth rubbed in a standard fashion across the carpet's face. After 20 friction cycles, the color transference is rated visually on the AATCC Chromatic Transference Scale. Results should be 4 or above.
- *Shampooing (AATCC 138)*. This measures the color change caused by severe shampooing. Test specimens are rated visually on the gray scale standard. Results should be below 4.

Pile density is important for the carpet to retain its appearance with wear. Dirt tends to remain on the surface of dense pile and can be vacuumed away. Permanent crushing is less likely with dense pile. A dense surface will feel hard underfoot, but comfort underfoot can be improved with a high-density commercial pad. A cushion can also significantly reduce pile compaction. However, in areas of heavy traffic, especially wheeled traffic including carts and wheelchairs, carpet should be glued directly to the floor.

Preventive maintenance is important. Entry points must have walk-off areas—grids, carpet tile, or other removable and cleanable surfaces. Outside entrances should be covered with outdoor carpet and swept daily. Parking lots should also be swept periodically to prevent foot traffic from carrying dirt into the building.

Multicolored or all-over patterns hide soil better than light or dark solids. Loop-pile carpets are most effective in hiding textural change.

Cut-pile fibers become compressed with wear, whereas loop-pile fibers flex and bounce back.

The maintenance required and the cost of materials are closely related. Nylon, which costs more than olefin, requires less maintenance. On the other hand, olefin is a good choice for areas with less traffic.

Because of heavy traffic in school buildings, even the most durable carpets wear at a rapid rate, especially at the seams. Several manufacturers have developed bonding processes for carpets that are guaranteed against edge or seam ravel, at less than 10 percent weight loss of pile face fiber, and with static resistance effective for the life of the carpet.

An antimicrobial treatment also offered deters bacteria growth, which is a problem in health care facilities, cafeterias, and some classrooms. In offices and classrooms, static can cause problems with computers. Total electric compatibility carpets feature a special conductive backing that manufacturers guarantee will not cause disruption or malfunction of electronic equipment.

Static Buildup. Topically applied static treatments are not permanent. With use, these coatings will wear off. The most effective static dissipator for nylon, olefin, or wool is a carbonized fiber added in the construction of the carpet to absorb and discharge excessive static. With a conductive element added to the yarn bundle, carpets will effectively keep static buildup below the sensitivity level. (For cut pile constructions, a conductive back may be required.) An effective test method to determine the static propensity of carpets is the AATCC-134 Step Test with Neolite soles. This test induces and measures static buildup on carpets by simulating conditions under which the static electricity may become objectionable. Peak charge generation (at 70° F and 20 percent relative humidity), developed by walking or shuffling traffic, is monitored under carefully regulated conditions and related to the threshold of charge above which objectionable shock may be experienced. For areas with delicate electronic equipment, the equipment manufacturer should be consulted for maximum static control levels that can be tolerated.

Alternative Floor Coverings

Alternative floor coverings include modular carpet, resilient flooring, and ceramic or quarry tile.

Modular Carpet Many factors have influenced the popularity of modular carpet, or carpet tile. It is cost-effective, has good quality control, performs well, is easily maintained, and can be used to create unique custom designs. Advantages also include less office disruption during installation, increased flexibility, extended life, and easier access to the subfloor.

The flexibility afforded by carpet modules enables rearrangement in offices without patching the carpeting where power and telephone lines were installed. When workstations are moved in an open plan furniture system, damaged tiles can simply be picked up and replaced.

The life cycle of carpet modules can be extended from two to five years by selectively replacing and rotating tiles subjected to heavy traffic, excessive soiling, or abnormal abuse. In-house employees can easily perform this task.

Carpet tiles are particularly appropriate in installations using flat cable or raised-access floors. With other systems, such as trench ducts, carpet tiles are the most practical floor covering.

Unique designs are possible with carpet tiles because of the variety of colors, patterns, and textures. Contrasting colors can be used as borders or to indicate departmental separation or efficient traffic flow. Many carpet modules are available in coordinating broadloom, so that it is possible to denote a hierarchy of areas with a cut pile broadloom in one area and a coordinating looped pile carpet module in another.

Resilient Flooring A wide selection of resilient flooring materials exists in prices ranging from inexpensive linoleum to the more costly vinyls. Installation is simple and inexpensive; however, some of these materials cannot be installed on grade-level or below-grade floors. Resilient flooring materials on the market include linoleum, asphalt tile, cork, vinyl composition tile, vinyl, and rubber.

Linoleum is a synthetic material manufactured mostly in sheet form. Linoleum has limited resistance to acetone and is unstable to strong alkaline solutions. It was the first synthetic flooring available and has been widely used for heavy duty installations, including flooring for battleships.

Asphalt tile, although still available, has generally been phased out in favor of newer synthetic tiles. It was widely used simply because it was the least expensive flooring. Asphalt tile was used in factories, housing projects, and many other facilities where durability and economy are important. It is brittle and hard underfoot, but practical.

Cork is the only natural material used in resilient flooring. It is resilient and has acoustic properties but is not appropriate for heavy duty use. Cork must be kept waxed to preserve the surface. It can become more attractive with age, developing a rich warm patina from many applications of wax.

Vinyl composition tile (VCT), previously vinyl asbestos tile (VAT), reflects the current concern over asbestos, which has been eliminated from many commonly used products. This type of tile is only slightly more expensive than asphalt but has many advantages, as it is softer underfoot, grease resistant, easily maintained, and available in many colors and patterns. Solid vinyls are the most durable, and many designers prefer their natural appearance over that of imitations made from other materials. A solid color floor will show heel marks and other dirt more readily than a patterned or textured floor. In high-traffic areas, a marbleized appearance or a textured surface is more appropriate.

A current trend in flooring is 100 percent rubber flooring in the form of studded tile. These tiles have raised circular or rounded square patterns and are attractive, resilient, long wearing, and nonslip; they have excellent acoustic properties, resist burns and most chemicals, and minimize breakage. The manufacturer has resolved the problem of dirt accumulation around the raised studs by sloping the sides of each raised stud.

An accessory item used in commercial installations to trim off many resilient floors is vinyl wall base, which is available in cove or toeless base. In finishing resilient flooring, a cove-style base is generally preferred, as it hides the uneven cut edge of the flooring. It is available in the following heights: 1½ in., 2½ in., 3 in., 4 in., 6 in., and 7 in., with 2½ in., 4 in., and 6 in. being the most commonly used. It comes in 4-ft. pieces or 100-ft. rolls and molds to form inside and outside corners. Preformed corners are available. Vinyl wall base is unbreakable and easily cleaned and never needs painting.

Ceramic or Quarry Tile   Tiles are one of the most attractive, durable, and versatile surfacing materials for indoor and outdoor use. They offer tremendous design potential as borders and in patterns. Ceramic tiles come in many colors, from bold shades to subdued hues. For warm colors and earth tones in numerous shapes, textures, shades, and sizes, quarry tile provides a rich, natural look. It exhibits the strength and durability needed in educational environments. In addition, some quarry products are finished with abrasive grain surfaces to improve slip resistance.

Although tiles provide a durable surface in hard-wear areas, they cannot tolerate direct blows, as in the dropping of heavy objects. Cart traffic can crack and break them, especially along unprotected edges. In work areas where people must stand for long periods, rubber mats can cushion the surface and help reduce fatigue.

Trim pieces for tiles come in all colors, with a flat or roll top. Outside corners are precast, and the straight base will not shrink or pull away if properly installed. When selecting tiles, be aware that some imported tiles do not come with a full line of trim pieces.

Wall Coverings

Every wall is a material in itself, and ideally no material need be covered. Designers currently prefer honesty of materials (e.g., brick walls) and will remove many surface layers of old paint and plaster to reach these structural walls in old buildings.

Wallpaper is the material commonly associated with wall coverings for interiors. Many patterns and solids exist in every imaginable color. If the intent is to achieve a particular effect, and if a strong pattern or color is desired, a well-designed wallpaper can be a meaningful asset. Often a strong paper works better on one wall, instead of surrounding the entire space with a dominating pattern.

Most designers note that a well-planned interior, conceived as a total design, does not need the superficial decoration of printed paper; the superfluous pattern and color might actually detract. Plain walls, walls of solid colors or textures, or walls of natural materials are usually more acceptable, especially in the institutional environment.

Attractive wallpapers, however, can serve a specific purpose and have some intrinsic qualities of their own. These include textured papers, often made from natural materials such as silk and grass cloth. Advantages include improved acoustic properties and an atmosphere of interest and warmth. The lamination of linen, burlap, or other textures onto paper backing provides an attractive background in areas where an elegant image is desired. These natural fibers should, however, be treated with a protective sealant to prevent excessive soiling.

Plastic-coated or vinyl wallpapers are useful wall coverings in kitchens and bathrooms. Washable, they stand up better than painted surfaces to steam or grease. The best of these wall coverings are vinyl-coated fabrics rather than coated papers, which are appropriate in all high-traffic areas. Vinyl-coated fabrics come in different weights; the heavier the anticipated wear, the heavier the material should be. The cost of some of the heavy vinyls is quite high but is justifiable, as the material can withstand countless scrubbings.

One special-purpose wall covering provides advance warning in case of fire before there is actually smoke or open flames in the room. Early warning effect wall coverings, when heated to 300°F, emit a colorless, odorless, and harmless vapor that will activate ionization-type smoke detectors, which represent 85 percent of those installed. The 300°F trigger point for these wall coverings is well below the ignition temperatures of most common room materials, including paper, cotton, polyethylene, and polyurethane foam. The early warning effect may be triggered by electrical outlet overloads that heat wall surfaces, electrical fires in walls, fires started in other rooms or in core service areas, and other situations where smoke and fire danger may not be readily detected by smoke alarms.

Depending on the project budget and the designer's imagination, there is almost no limit to the materials that might be used for wall coverings: fabric, leather, wood veneers, wall carpet, or metallic materials. Cork is frequently used for practical purposes, such as tack space or sound-absorbing properties, as well as for its appearance.

Textiles

In judging a fabric for durability, the weave should be examined: The tighter the weave, the longer the fabric will wear. Fabrics in which the colors and pattern are tightly woven in with colored threads will wear better than a printed fabric. To check a fabric, hold it up to the light. The less light showing through, the tighter the weave.

The blending of fibers combines the unique properties of each and can produce a more attractive and durable fabric than a fabric made of one fiber. For example, a fiber that takes color well and is lustrous, but not particularly sturdy, can be woven with one that is duller, but more durable, to produce a vivid and heavy duty fabric. The appearance of a fiber can be altered by its construction or by blending. Textiles can be divided into natural or man-made fibers.

Natural Fibers Natural fibers of animal origin include wool, silk, mohair, felt, and leather. Silk and wool, luxurious and costly natural fibers, offer durability, resilience, and beauty. To clean silk, professionally dry clean. Wool can be spot dry cleaned or washed in a cool, sudsy water solution.

Natural fibers of vegetable origin include cotton, linen, and jute. Cotton has fair resistance to wear and sunlight and a soft feel; it dyes well but must be treated to avoid excessive soiling. Care involves dry cleaning or washing, depending on the other fibers with which it is blended.

Man-Made Fibers Acrylic, nylon, olefin (polypropylene), polyester, and rayon are man-made fibers of chemical origin.

- *Acrylic (Orlon, Creslan, Acrilan, Zefran).* Acrylic has a soft woolly feeling, with fair resistance to sunlight. It has good cleanability characteristics and takes vivid color well. Acrylic is normally used to create plush velvet looks. It wears well and will not bag or stretch after continued seating. It should be cleaned with mild, water-free solvents.
- *Nylon (Antron, Enka, Chemstrand, Caprolan).* Exceptionally rugged and durable, nylon resists signs of wear and tear. A man-made fiber offering the best resistance to soil, it dyes well and will not fade. Professional dry cleaning is recommended.
- *Olefin or Polypropylene (Herculon, Vectra).* Olefin offers high resistance to abrasion and stains. It has a softer feel than nylon, has good resistance to fading when solution dyed, and is very sensitive to heat. In humid climates where mildew can be a problem, it is a good choice. Only water-based cleaning solutions should be used for its care.
- *Polyester (Dacron, Fortrel, Kodel).* Polyester is crisp and strong and fairly resistant to wear and sunlight. It is most like natural cotton in

its appearance and physical properties. Its resistance to heat is low. This fabric accepts color well and is easy to clean using mild, water-free solvents.

- *Rayon (Avril, Enica, Fortisan).* Rayon is composed of regenerated cellulose (a wood by-product). It dyes well, is soft to the touch, and has fair resistance to wear and sunlight. Rayon can be constructed to look like cotton, silk, or wool. It can be dry cleaned or washed, depending on other fiber blends.

There are also fibers of metallic origin. These fibers are made of aluminum, silver, or gold threads, usually in combination with natural or man-made fibers.

Woven Fabrics When any fiber or blend of fibers is woven together, the visual texture and pattern of the fabric are created. Two basic methods of weaving upholstered fabrics exist: flat and pile. Flat weaves include tweeds, twills, and satins. They have no pile, although they may be coarse and nubby because of the uneven size of the yarns. Woven pile fabrics are those in which an extra set of warp or filling yarns is interlaced with the ground warp and filling. In this way loops or cut ends are produced on the surface of the fabric. The base or ground fabric may be either plain or twill weave.

Nonwoven Fabrics Nonwoven fabrics are knitted, flocked, or tufted. Knitting is a method of construction in which yarns are looped and interlocked instead of woven. Flocking involves creating a velvet effect using cut fibers applied electrostatically. Tufting is another method of locking yarns on the surface; the loops can then be cut to create a velvet surface. Another classification of nonwoven fabrics includes those fabrics created through a process of pressing or bonding fibers together with an adhesive.

Window Treatments

Window treatments include many decorative or functional methods for finishing a room. Aside from aesthetics, they have great practical value. They can insulate against winter heat loss and summer heat gain, control glare, provide privacy, absorb noise, and lower maintenance costs. The total environment should be considered when determining whether blinds, shades, draperies, or any other treatment is used. Cost-effectiveness is always an important aspect of window treatments in an institutional environment.

The following items are components of cost-effectiveness. They should be considered when making a decision regarding the least costly window treatment that will satisfy requirements.

- *Initial cost.* Materials, fabrication, and installation costs vary. For example, costs for draperies are considerably higher than those for roller shades.

- *Energy conservation.* This depends on window orientation and could amount to a sizable reduction in air conditioning and heating capacity. The American Society of Heating, Refrigeration, and Air Conditioning Engineers (ASHRAE) has extensive information on the impact of interior shading on heating and cooling loads in its *Handbook of Fundamentals.*
- *Expected service life.* The amount of use and type of maintenance determines service life. Blinds have a service life of approximately 10 years; shades, 3 to 5 years; and draperies, 5 years.
- *Maintenance.* Vacuuming and periodic professional cleaning should be included in regular maintenance costs.

Blinds Flexibility is the key to the popularity of blinds. Many options address every type of window requirement. Blinds that are one color inside and another color outside ensure a unified building exterior appearance while maintaining inside design flexibility. Blinds can be custom made to almost any shape opening: A-frame, bay window, inclined, tapered, cutouts, circular, or arched. Installing two blinds on one headrail enables one to be raised while the other is lowered.

Blinds are available with heat-absorbing or heat-reflecting finishes to help cut energy costs. If heat buildup between the blind and glass is a concern, as it may be with thermal glass, special drop-down brackets allow an additional $1/4$ in. gap at the top of the blind. This permits the heat buildup in the airspace to be vented. Another alternative involves eliminating the top slat from the blind to increase the air gap. Additional air gaps at the sides and bottom of the blind can also be specified.

Vertical blinds combine the flexibility of blinds with the luxury of draperies. They give a dramatic, contemporary look to a room while controlling light and privacy. Vertical blinds are available in many colors and materials, including fabrics, plastics, and aluminum. Replacement fabric vanes allow easy repair; fire retardant vanes are also available.

When specifying blinds, it is best to require that the contractor be responsible for inspection of the site, approval of the mounting surface, installation conditions, and field measurements, rather than simply furnishing measurements to a vendor. Costly mistakes can occur if measurements are not exact.

Types of blinds include painted aluminum, in $1/2$-, 1-, $1^1/2$-, and 2-in. slats; natural wood, in 1- or 2-in. slats; audiovisual, in 2-in. slats; sun controller for the exterior and interior, in 1-, 2-, and $3^1/2$- in. slats; and vertical, in 2- and 3 $1/2$-in. vanes of aluminum, fabric, or plastic.

Shades Shades control light and privacy, maintain interior temperature levels, and accent windows. There are shades for exterior use, pleated shades, and blackout shades for those areas where complete blackout of light is necessary. Shades are available in fiberglass-coated polyester or can be fabricated in the designer's own fabric to coordinate with an interior decorating scheme. Shades present many creative options; provide superior light control; save energy; and are convenient to install, adjust, and remove.

Pleated shades come in many weights: sheer fabrics to filter light, semiopaque fabrics to provide more privacy, and thicker fabrics to keep out the sun. Various types of shades include blackout shades, solar screens, and pleated shades.

Curtains and draperies   Some commonly held misconceptions exist regarding the terminology associated with window treatments. The following definitions will clarify interpretations of these words, which are often used as synonyms.

- *Curtains.* Window coverings fabricated from sheer or lightweight material such as cotton, polyester, rayon, or blends of these fibers are called *curtains.* They are used in less formal interiors. Curtains may be used alone or in combination with overdraperies. They may be lined or unlined and are generally shirred on a rod but may also be pinch pleated. Hung inside a window casing, they should reach to the window sill. Hung on an outside casing, they should hang to the bottom of the apron or to the floor.
- *Draperies.* These are also commonly called *drapes.* They create a more formal mood and are generally of heavier fabrics. They are usually lined, of pleated construction, and weighted at the bottom. Draperies are generally hung on traverse rods that allow them to open and shut. They may be hung to the bottom of the apron or to the floor.
- *Valance.* A separate top or horizontal portion of the drapery treatment is called a *valance.* Valances were originally used to hide drapery hardware but have survived for decorative reasons. Valances are generally 4 to 6 in. deep and made of softly draped, gathered, or pleated fabric.
- *Cornice.* A three-sided box-like top used for a window treatment, with the open side of the box facing the wall, is a *cornice.* This overhanging box is sometimes used, like a valance, to hide drapery hardware. A cornice is generally from 4 to 7 in. deep and fabricated of wood, plaster, or metal. It can be upholstered in a fabric to match the draperies or in another coordinating fabric. A properly designed cornice can add height to short windows.

Although there are no hard-and-fast rules, the following will generally apply:

- Valances should not be used in rooms with extremely low ceilings. A valance with curved lines makes the window appear wider; one with square lines makes it seem narrower. Straight draperies without a valance or cornice make windows seem taller.

- Draperies that are tied back soften a room's severity. A pair of looped-back curtains makes a window seem narrower than a single curtain tied back.
- Tiebacks should be located either above or below the centerline of the window. Usually, the higher the tieback, the taller the window appears. Straight, plain draperies, as a rule, are not looped back.
- French doors should be treated as windows. If located between rooms, they should not be draped. For privacy, they can be fitted with sheer curtains gathered on small rods at the top and bottom.
- Bay windows or two adjacent windows should be treated as one, with a single drapery at the outer edge of each. This gives the effect of one large window and makes them appear much wider. If a valance is used, it should run across the windows.
- Casement windows require special treatment. Those that swing inward are generally fitted with shirred curtains fastened to the top and bottom. Those that swing out are usually fitted with straight-hanging curtains or draperies.
- Draperies and curtains should present a uniform appearance from the street. All windows, at least those on the same level, should have the same kind of curtains, or draperies with white linings.

Maintenance, obstruction of view when closed, and possible glass breakage when used with heat-absorbing glass are some of disadvantages of draperies.

Four major types of drapery construction are used in commercial installations: pinch pleat systems, stack pleat systems, roll pleat systems, and accordion-type pleating systems. Pinch pleat systems are probably the most commonly used. They are constructed with pleater tape and are generally attached to the slide carriers of a standard traverse rod. Installation specifications should include a drawing of the window and wall with bracket locations clearly marked. The drapery workroom will need specific instructions and a sample of the drapery fabric. To avoid delays caused by flaws in the fabric, order extra yardage.

Drapery Flammability Regulations Many codes recognize drapery covering more than 10 percent of the wall area of an interior finish. The most frequently quoted flammability codes are those of New York City, Boston, and California:

1. New York City requires all flame-retardant chemicals to be approved by a board of standards and appeals. After receiving approval from the board, a number is issued. This number is used on a certificate stating that the fabric has been treated with the approved flame retardant.

2. The city of Boston requires that a sample of the treated fabric be tested by the Boston Fire Department. A certificate of flame retardancy and a statement of intended use must be furnished with the sample provided to the fire department.
3. The state of California requires that the state fire marshal approve all flame-retardant chemicals. California law also requires that the flame-retardant treatment be applied in an approved manner by a licensed finishing company.

Fabrics such as fiberglass, wool, and some modacrylics are considered inherently flame resistant. These fabrics do not require flame-retardant treatment. Fabrics containing a blend of Saran, Verel-modacrylic, and rayon are mildew, rot, and vermin proof and will melt rather than support flame. Even though these fabrics are inherently flame retardant, designers should have materials tested to determine whether fabrication or finishing has impaired their natural properties.

Cabinetry

One indication of quality in cabinetry and case goods is drawer construction. The drawers should have concealed dovetail construction at the front and, in better quality cabinetry, also at the back. The drawer bottom should be substantial and grooved into the sides. For added strength, small glue blocks are used. Better grades of cabinets have side and back panels of five-ply veneer. The sides and usually the back should be grooved into the posts.

If a project includes custom-built cabinetry, it is sometimes preferable to separate the cabinetry contract from the general contract. This permits careful selection of a cabinetmaker. Drawings should be prepared covering type and grade of materials, finishes, hardware, and special equipment. Cabinet details are usually drawn on a large scale (e.g., 1 in. equals 12 in.) or even full scale. If the designer is not familiar with detailing, he or she may prepare small-scale drawings and require shop drawings from the cabinetmaker for approval.

Furniture

Furniture styles   Designers should know the history of furniture and be able to recognize the more important periods. The strongest influence on the current interpretation of traditional furnishings originated in the 17th century.

In general, period furniture can be divided into two major categories: formal and informal styles. Formal styles consist of furniture originally designed for the royal courts and for spacious homes of the wealthy. Informal styles include simpler pieces made by local craftspeople using crude tools and local woods. Provincial styles and Early American furniture are examples of informal furniture.

Modern furniture consists of a new design form that breaks away from previous forms. An expression of the 20th century, it uses new materials in new ways. Functionalism is the key and determines form, with an emphasis on line, proportion, color, texture, and finish.

Both traditional and contemporary furniture is used in institutional settings. For the most appropriate selection, the total environment, rather than personal stylistic preferences, should be considered.

Well-designed furniture exists in both contemporary and period styles. In planning furniture requirements, consider the desired image—solid, traditional values or modest, efficient simplicity.

Seating The selection of seating is important. Workers spend many hours each day at a desk, and the right chair promotes efficiency, relaxation, and production. A determination must first be made of the functions the seating must perform. Then it should be evaluated for comfort, durability, cost, ergonomics, appearance, space savings, safety, and availability.

Seating comfort is important, as it can affect the degree of learning in a classroom situation, the acceptance of waiting in a lounge area, and the pleasure of a banquet event. Comfort in office seating increases efficiency and job satisfaction.

Durability ensures the institution's investment. Test the seating under actual conditions, and ask for references of others using the product under similar conditions. If it is a new product, ask to compare test laboratory results with existing lines of seating to determine suitability for particular applications.

In addition to the initial cost, consider the handling and maintenance costs. The cost of arranging and moving furniture to and from storage or rearranging or moving seating for cleaning purposes can become greater than the initial cost.

Ergonomics relates to all elements in the person's work environment, including sound and lighting, layout, carpeting, desks, files, and seating. The following factors should be taken into consideration when evaluating seating: seat height, appearance, space savings, safety, and availability. Occupational Safety and Health Administration (OSHA) guidelines currently require that businesses and public institutions provide seating that offers adjustable seat height, arms, back, and seat pan angle if requested by an employee to reduce the occurrence of various physical ailments caused by long hours spent seated with little range of motion.

Upholstered furniture Purchasing upholstered furniture can be deceptive. Unlike case goods, quality construction in upholstered goods can be difficult to detect. An attractive fabric can hide inferior products and workmanship. Items to be considered when evaluating upholstered furniture are frame construction, springs, padding material, and fillings.

Upholstery Flammability Regulations. The designer must be aware of particular flammability hazards in certain types of installations (e.g., areas where smoking is permitted; places where people sit for extended periods, such as transportation seating; and lounge areas in public buildings). Areas where the lighting level is low and where live-in accommodations include bedding, such as dormitories, are also hazardous.

When planning upholstery for these areas, the fabric should be inherently flame resistant or protected by a flame-retardant treatment. Upholstery fabrics that do not meet the Class 1 requirements of the U.S. Department of Commerce Commercial Standard 191-53 should not be used on upholstered furniture. Self-extinguishing foam cushions are also available from major seating manufacturers at a slightly higher cost than normal polyurethane, but peace of mind is well worth the extra expense.

Surface treatments such as tufting and seams on seat areas should be avoided. Tufting should be limited to vertical surfaces. Cigarettes rank high among ignition sources, and each tuft in a seat cushion provides an area where a cigarette can burn unnoticed. Seams in seats tend to split, exposing the filling to possible ignition. If possible, specify seating with at least a 1-in. gap between the seat and the back.

Office Landscaping. Office landscape designates an informal, open, flexible system of furniture arrangement based on the interrelationships among groups to allow more efficient communication. Partitions and components required to support the work of each area compose the landscape. The partitions are available in various heights to provide different levels of privacy. Generally, they do not reach the ceiling; however, some manufacturers, in an effort to make them more appealing, have provided a complementary line of partitions that reach the ceiling, forming private offices where required.

Because these systems are open, provisions for noise control must be carefully planned. Ceilings should be of acoustic materials, and floors should be carpeted. Acoustic panels should be placed where noise is generated to absorb as much of it as possible. Sound-masking systems can be helpful in lowering the noise level. Draperies and plants are also effective.

Although some form of flexible office furniture system has been available since the late 1940s, this type of furniture became popular during the 1970s and seemed to be the answer to all office layout problems. However, in the late 1970s this trend began to change as management and staff complained about the lack of privacy and problems with noise. Full-height partitions (called by various names, including movable, demountable, or relocatable) became popular. Their popularity seems to be continuing, as these partitions are compatible with and offer the flexibility of the lower-height systems, but provide greater privacy and sound control. Current open offices use both types of systems. More consideration is now given to the employee's need for privacy, as well as to the need for accessibility.

Many important factors should be considered before purchasing a system; these include flexibility and ease of reconfiguration; simplicity of installation; capability for handling electrical requirements for power; and communication and data processing equipment, as needed. It should provide design options such as a variety of panel sizes, components, and finishes.

Furniture Standards. The objectives of a standards program are as follows:

- *Volume purchasing savings.* Vendors will often offer advantageous terms above and beyond the usual quantity discounts when the line is made "standard."
- *Management time savings.* Streamlining the selection and acquisition process eliminates catalog skimming.
- *Reduced competitive behavior among employees.* Consistently applied standards reduce competitive behavior among employees while helping eliminate jealousy and resentment over office size and furniture quantity.
- *Aesthetic consistency.* Furniture standards programs often deal with visual design and attractiveness of office environments.
- *Improved environmental function.* Problems of using furniture and space to the best functional advantage are often addressed through standards, rather than leaving each employee to use trial and error to find the best arrangement for a particular job function.
- *Multiple workstation enhancement.* Function problems involving the arrangement of many employees and such overall considerations as OSHA's fire and safety regulations are more easily and consistently dealt with on a standards level than case by case.

Lighting

Four general categories of light sources are available: natural daylight, incandescent electric light, fluorescent light, and high-intensity discharge lighting.

Daylight constantly changes throughout the day in position, intensity, diffusion, and color. In any design using natural daylight, three conditions should be considered: light directly from the sun combined with reflected light from a clear sky, light from a clear sky only, and light from an overcast sky. Various indirect variables, including local terrain, landscaping, water, fenestration, daylight control systems (shades and louvers), decor, and artificial light, also affect daylight.

Artificial interior lighting is discussed in Chapter 26, which reviews incandescent, fluorescent, and high-intensity discharge lighting.

Acoustics

Sound masking is one effective way to increase sound privacy. By slightly increasing the ambient sound level, sound masking covers distracting noises. Sound masking can be adjusted and fine-tuned to give exactly the right level of acoustic privacy without becoming a nuisance. Masking systems must be programmed to their most desired level in each particular case.

In an open office environment, sound masking is critical. For the closed office, sound masking can result in confidential privacy without resorting to more expensive conventional construction techniques, such as insulating the drywalls or extending the walls through the suspended ceiling to the structural ceiling above. The masking system can also accommodate other audio functions, such as paging and music, utilizing the same speaker system.

24.5 FURNITURE ARRANGEMENT AND SPACE PLANNING

Furniture Arrangement

Standardized furniture arrangements are almost useless, as each client's needs are unique. To provide a furniture arrangement that will be both aesthetically and functionally successful, the designer must analyze the space, determining the flow of traffic and activities. Relationships of scale, mass of furniture to the area, and contrast of elements all must be considered, as follows:

- *Analysis of space.* The designer should first make a thorough analysis of the space. On the first visit to the area, the designer should note dimensions and indicate all architectural features; existing lighting; heating, ventilation, and air conditioning supply; electrical outlets; and existing finishes in the area. Photographs are particularly helpful.
- *Flow of traffic.* The designer must know the intended use to determine furniture requirements and assign activities to zones based on the square footage required for each activity.
- *Scale and mass of furnishings.* Scale and mass of furnishings in an interior are based on the relative proportion of such furnishings to persons, other objects, and the space they occupy. An interior should

appear neither crowded nor underfurnished. If an object were removed from a properly furnished space, the space would appear incomplete.

Large areas allow large-scale furnishings. Small-scale objects generally look insignificant in large spaces. Distribution of mass should be balanced throughout the space. The specific location, as well as the overall space, must be considered. For example, a piece of furniture should relate in proportion to the wall against which it is placed.

- *Contrast of elements.* To avoid monotony, the lines of furnishings and architectural features should vary. A room whose focus is high windows and doors needs horizontal balance. Conversely, a space with strong horizontal architectural accents needs the balance of high pieces of furniture to heighten the vertical line of the space.

Balance in the space and between objects gives a harmonious composition. Too much similarity or lack of contrast causes a space and the objects in it to appear dull and uninteresting. The success or failure of the design depends on how well the space functions and how well it serves the needs of the client.

Basic Guidelines

The following general guidelines offer suggestions for planning the arrangement of furniture:

1. Plan each room with a purpose. Decide what the room will be used for and by whom.
2. Use furniture in keeping with the scale of the room.
3. Provide space for traffic. Doorways should be free, major traffic lanes must be unobstructed. It is sometimes necessary to redirect traffic. This can be accomplished by turning a sofa, a desk, or chairs toward the room and at right angles to the door, with a passageway left for traffic.
4. Arrange furnishings to give the room a sense of equilibrium.
5. Achieve a good balance of high and low, angular and rounded furniture. Where furniture is all or predominantly low, the feeling of height may be created by incorporating shelves, mirrors, and pictures in a grouping.
6. Consider architectural and mechanical features. There should be no interference with the opening of windows, swinging of doors, or heating or air conditioning devices. Lamps should be placed near electrical outlets.
7. Do not overcrowd a room. It is always better for a room to be underfurnished than overfurnished.
8. Generally, large pieces of furniture should be placed parallel to the walls.

9. Avoid pushing large pieces tightly into a corner or close against floor-to-ceiling windows where a passageway should be allowed.
10. Arrange the heaviest furniture grouping along the highest wall in rooms with slanted ceilings.
11. Provide adequate lighting for all activities.

In planning office layout, the following may be helpful (Figure 24-3):

- Office floor space must be conserved, but not at the expense of appearance, production, or comfort.
- Place related departments near each other.
- Each employee, including his or her desk, chair space, and share of the aisle, requires a minimum of 50 to 75 sq. ft. of working space.
- A general conference room where confidential meetings may be held will eliminate the need for many private offices.
- A minimum of 9 ft. x 12 ft. is a standard size for small private offices.
- Standard widths for main circulating aisles vary from 5 ft. to 8 ft. Less important aisles vary from 3 ft. to 5 ft.
- Allow 7 sq. ft. for computers and related items.

Space Planning

Programming   Programming is the determination of the parts and their interrelationships. Precise information is vital to the effective planning and layout of office environments. Considerations to address include the amount, kind, and configuration of space and the relative closeness or relationships of each unit.

Space determination should begin early in the planning process. Decisions must be made as to which location is appropriate for each function. This preliminary figure, however, is only an estimate, determined without detailed analysis. Functional relationship should be established before the division of space gets too involved. Later, particular pieces of equipment may necessitate detailed planning.

A flowchart, sometimes called a *bubble chart* or *space relationship diagram,* should be prepared to indicate all related activities and groups, without concern for actual space required by each. Next, through careful programming, the designer establishes the space required for each activity. Then the area is fitted into the relationship diagram. This preliminary layout, once refined, becomes the final floor plan.

Several ways of establishing space requirements are calculation, conversion, space standards, and rough layout.

The *calculation method* involves a physical inventory prepared for each area. Inventory control lists are usually available for this purpose. Each activity or area is broken down into subareas. An assessment of space required

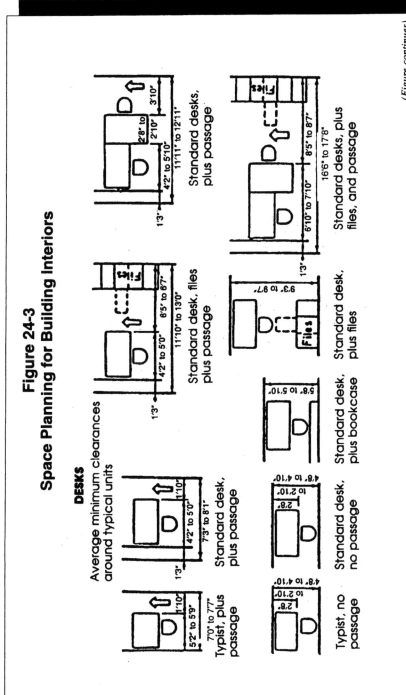

Figure 24-3
Space Planning for Building Interiors

(Figure continues)

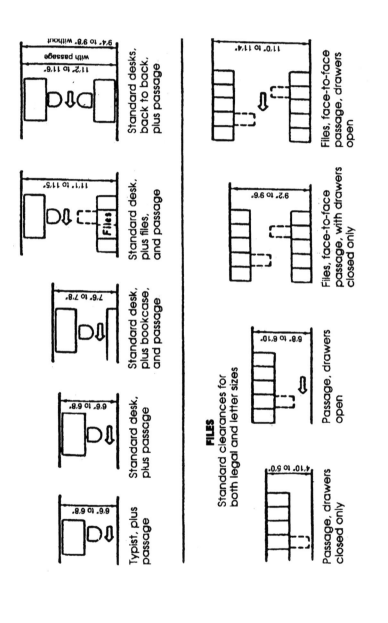

Standard desks, back to back, plus passage

9'4" to 9'8" without
with passage
11'2" to 11'6"

Standard desk, plus files, and passage

11'1" to 11'5"

Standard desk, plus bookcase, and passage

7'6" to 7'8"

Standard desk, plus passage

6'6" to 6'8"

Typist, plus passage

6'6" to 6'8"

Files, face-to-face passage, drawers open

11'0" to 11'4"

Files, face-to-face passage, with drawers closed only

9'2" to 9'6"

FILES
Standard clearances for both legal and letter sizes

Passage, drawers open

6'8" to 6'10"

Passage, drawers closed only

4'10" to 5'0"

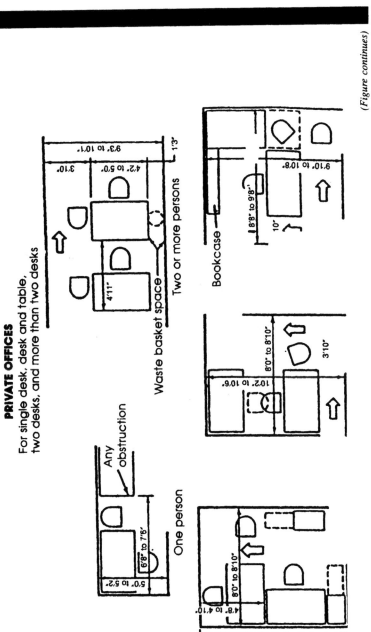

PRIVATE OFFICES

For single desk, desk and table,
two desks, and more than two desks

One person

Any obstruction

Two or more persons

Waste basket space

Bookcase

9'3" to 10'1"

4'2" to 5'0"

3'10"

1'3"

4'11"

6'8" to 7'6"

5'0" to 5'2"

10'8" to 11'8"

4'8" to 4'10"

8'0" to 8'10"

10'2" to 10'6"

8'0" to 8'10"

3'10"

8'8" to 9'8"

9'10" to 10'8"

10'

(Figure continues)

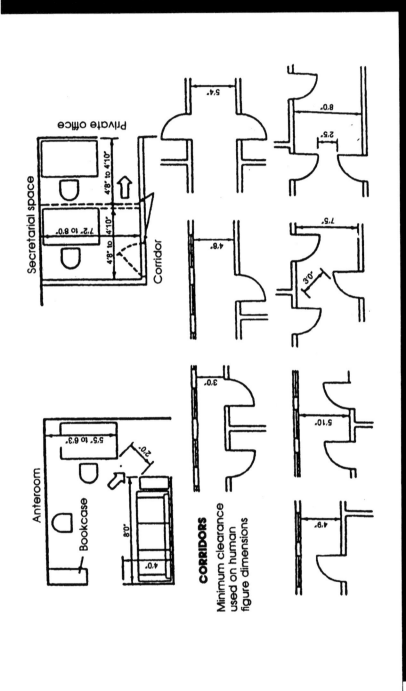

Private office

Secretarial space

4'8" to 4'10"

7'2" to 8'0"

4'8" to 4'10"

Corridor

Anteroom

Bookcase

5'5" to 6'3"

2'0"

8'0"

4'0"

5'4"

8'0"

2'5"

4'8"

7'5"

3'0"

3'0"

5'10"

4'9"

CORRIDORS
Minimum clearance
used on human
figure dimensions

for each subarea is made. This is then multiplied by the number of elements required to do the job, with extra space added for storage and service.

The *conversion method* establishes the space presently occupied and converts it to the proposed amount. Before making the conversion to the new plan, the designer must decide if the occupied space is less than the amount actually needed.

Another method, *established space standards,* is a practical way to determine space requirements any time a given situation arises. Standards must be applied consistently. It is important to record how the standards were established and apply them fairly if they are to have credibility.

The planner or designer should meet first with management and establish goals. An organizational chart should be obtained. Meetings should be scheduled with personnel in various departments. Questions that should be asked are: "How many people are in the group?" "What are their jobs?" "What are the furniture and equipment requirements to support the work performed there?" "Are there any plans for expansion?"

After the planner obtains this information, it must be evaluated for actual needs versus wants of the client. The planner will prepare a written program, listing personnel, furniture, equipment, and services. He or she will then compute the square footage required for these functions. Finally, approval must be obtained on the program.

During this phase a furniture, furnishings, and equipment inventory should be prepared. Three categories of furnishings and equipment must be listed:

1. *As is:* Items that can be moved without any attention.
2. *Refinish:* Items that must be refinished or reupholstered before they are acceptable.
3. *New:* Items that must be purchased.

In purchasing furnishings and equipment, count back from the planned moving date to determine the lead time necessary, always allowing extra time for unexpected delays.

The most precise method of determining space requirements is the *rough layout method.* This method is advisable in certain critical areas involving fixed equipment, large machinery, or multiple workstations. With this method, scale floor plans and templates are used to determine exact location of furniture and equipment.

Most projects are restricted by space limitations. As a rule, they are more restrictive than any other factor except funding. If requirements exceed availability, reductions must be made. Lower priority area reductions are preferable to across-the-board cuts.

The problem of limited space can be solved in the following ways:

1. Authorize overtime.
2. Improve procedures.

3. Organize wasted space.
4. Move storage areas to leased spaces.
5. Divide operations into groups and spread to available buildings.

Preliminary design   At this point, a cost analysis should be made for comparison and justification. During this phase the analysis of the written program and the flow diagram are turned into a floor plan. With this floor plan, a construction estimate can be prepared by multiplying the floor area required by the square foot construction cost. The furniture and equipment budget should be prepared at this point, and then the total preliminary budget can be quoted. This phase of design is sometimes called a *feasibility study.*

Final design   The floor plans are prepared during this phase. The space should be accurately measured, including the size and location of structural elements. Furniture and equipment layouts are drawn, architectural materials are chosen, and finishes are specified. Mechanical engineers are retained, and air handling equipment is located and sized. This plan and an outline specification of materials and equipment permit refinement of the initial estimate.

Presentations   Presentation to the client is generally done in graphic form, arranging floor plans and elevations, rendered perspectives, samples of fabrics and floor coverings, and photographs of furniture on display boards. Computer graphics can aid in the client's understanding. With information on plans and elevations, many computers can create perspective drawings. Some even "walk through" the space, simulating reality for the client, who is usually less able to visualize the space from a two-dimensional drawing. Computer simulations are still a costly venture and are reserved primarily for large projects, where they serve as funding tools.

Working drawings, specifications, and bidding   The general contractor, the facilities department, or both use the final drawings and specifications to properly estimate the job for bid. The drawings should contain all the information necessary to complete the project. Any item not accounted for in these documents may necessitate a change order, with an almost certain increase in costs.

All necessary approvals should be in writing. The specifications should be designed to protect and provide the quantity and quality of work expected. After bids are received and approved, the drawings and specifications, approvals, and costs are incorporated into the contract and actual work begins.

Many public institutions issue a contract (usually on a yearly basis) for furniture, equipment, and carpet. These represent desired products at guaranteed rates of sale and may include installation and delivery. This alleviates ecraft/

the need for bidding every small project, with a considerable savings in time and paperwork.

Supervision   The last phase of a project involves supervision. If the job is a large one, a representative should be on site at all times. At this stage a designer's job involves, at the least, interpreting the documents; approving shop drawings; making decisions regarding alternate submittals; preparing change orders; expediting purchases; coordinating contractors' installations of new furniture and equipment, the physical move into a new space, and "punch lists"; and making pay estimates.

Installation   In most institutions the designer supervises furniture and equipment installations. Preparation for installation is most important and includes preparing installation plans and drawings, arranging for any maintenance work needed, scheduling the moves, and notifying everyone involved. Drawings and written instructions are the best way to communicate. Every drawing should show the North arrow, the scale being used, identification of the building, sheet number, project identification, date, and the name of the person preparing the drawing.

There are three basic phases to an installation: planning, actual installation, and follow-up.

In the first phase, planning and scheduling begin, inventories are prepared, disposition of any existing equipment and furnishings is determined, communications are set up among in-house personnel and contractors, work orders are prepared, departmental personnel are notified, and decisions are made regarding furniture or equipment to be relocated. In coordinating the installation, the designer should use checklists of items likely to be overlooked. These lists should establish what is to be done, when it is needed, and who is to do it.

In the second phase, the actual installation, the designer must be on hand for layout interpretation. As the installation proceeds, the designer makes periodic checks on the status of the work and keeps work crews informed and coordinated.

The third phase, follow-up, requires the designer to inspect the installation to verify that all items are delivered undamaged and the layout conforms to the drawings. If not, the designer must have any discrepancies corrected.

24.6   BARRIER-FREE ACCESSIBILITY

"Barrier-free" does not merely imply ramps and adequate door widths for accessibility. *Removing the Barriers: Accessibility Guidelines and Specifications,* published by APPA: The Association of Higher Education Facilities Officers, provides technical data and illustrations of standards. Aside from the obvious architectural barriers, psychological barriers exist. When persons must enter by a service entrance or ask assistance in opening doors, they feel humiliated and helpless.

There are two major federal acts guaranteeing the right of equal access for the disabled. The first is the Rehabilitation Act of 1973, Section 504, which states that "no otherwise qualified handicapped individual in the United States . . . shall, solely by reason of handicap, be excluded from participation in, be denied the benefits of, or be subjected to discrimination under any program or activity receiving Federal financial assistance." The second is the Americans with Disabilities Act of 1990, which prohibits discrimination on the basis of disability and requires that all new places of public accommodation and commercial facilities be designed and constructed so as to be readily accessible to and usable by persons with disabilities. The act also prohibits discrimination against qualified individuals with disabilities in all aspects of employment.

Colleges and universities are required to make reasonable adjustments to enable disabled students to fulfill academic requirements and to ensure that these students are not excluded from programs because of the absence of necessary modifications. Depending on the nature of the disability, modifications may include changes in the length of time required for completion of degree requirements, substitution of specific courses required, and changes in the way in which specific courses are conducted.

Most colleges and universities have taken the position that all common space should be accessible, with other areas made accessible on an as-needed basis. It is not necessary to modify every structure if programs can be offered in an already accessible space, but areas that might not have been altered for student use may have to be changed to accommodate a qualified individual who becomes employed there.

On a campus-wide basis, curb cuts, ramps, beepers for street intersections, and less hazardous signage and landscaping are necessary. On an individual building basis, restroom modifications, lowered telephones and water fountains, widened doors, ramps, and, where feasible, accessible elevators are needed.

Mobility-impaired individuals are commonly provided housing on the ground level of dormitory buildings. This policy is not well received by many proponents of rights for the disabled, as they believe this is discriminatory and isolates the person from the mainstream of activity.

There are exceptions to the rules and regulations requiring elimination of architectural barriers. The following exceptions are currently allowed by the federal government in accordance with Public Law 90-480, Subpart 101-19.604, "Exceptions," paragraphs a through d, as amended in 1968:

1. The design, construction, or alteration of any portion which need not, because of its intended use, be made accessible to, or usable by, the public or physically disabled people.
2. The alteration of an existing building if the alteration does not involve the installation of, or work on, existing stairs, doors, elevators, toilets, entrances, drinking fountains, floors, telephone locations, curbs, parking areas, or other facilities susceptible to installation or improvements to accommodate physically disabled people.
3. The alteration of an existing building, or of such portions thereof, to which application is not structurally possible.
4. The construction or alteration of a building for which plans and specifications were completed or substantially completed on or before September 2, 1969. However, any building constructed under the National Transportation Act of 1960, the National Capital Transportation Act of 1965, or Title III of the Washington Metropolitan Area Transit Regulation Compact shall be designed, constructed, or altered in accordance with American National Standards Institute standards regardless of design status or bid solicitation as of September 2, 1969.

In many states a state governmental agency (e.g., in Texas, the Texas State Purchasing and General Services Commission's Architectural Barriers Department) will review for compliance and approve plans and specifications submitted prior to bidding and award of contract. In other states a separate agency reviews only college and university construction plans. In still others the college or university is governed by its own board of regents.

Whatever the governing body, after review it will decide to what extent the project shall be made to comply. Plans and specifications will be approved only when the documents reflect compliance with the appropriate accessibility standards and specifications, generally those of the American National Standards Institute, found in *Providing Accessibility and Usability for Physically Handicapped People* (ANSI A117.1-1986).

Upon completion of projects, on-site inspectors determine whether the appropriate standards have been met during construction. All complaints received by the commission or board must be investigated and resolved or referred to the proper authority, with possible legal recourse.

24.7 SUMMARY

This chapter is intended to provide an understanding of the critical issues of building interior design. It serves to assist the facilities manager in developing guidelines for making routine in-house decisions, as well as in communicating better with consultants.

ADDITIONAL RESOURCES

American Society of Heating, Refrigeration, and Air Conditioning Engineers. *Handbook of Fundamentals.* Atlanta: ASHRAE, 1993.

Antes, Victor. "Some Viewpoints on Office and Task Lighting: A Recent Seminar." *Architectural Lighting,* Vol. 1, No. 7, 1987.

Bell, Doreen. "Efficient, Effective Lighting." *School and College Product News,* September 1987.

Coons, Maggie, and Margaret Milner, eds. *Creating an Accessible Campus.* Washington, D.C.: APPA, 1979.

Cotler, Stephen R. *Removing the Barriers: Accessibility Guidelines and Specifications.* Alexandria, Virginia: APPA, 1991.

Friedman, Arnold, John Pile, and Forrest Wilson. *Interior Design—An Introduction to Architectural Interiors.* New York: American Elsevier Publishing Company, 1976.

McMillan, Lorel. "Carpet Backs: The Underside View of Your Carpet Selection." *Facilities Design & Management,* Vol. 5, No. 7, 1986.

Reznikoff, S. C. *Specifications for Commercial Interiors—Professional Liabilities Regulations and Performance Criteria.* New York: Whitney Library Design, 1979.

Whiton, Sherrill. *Interior Design and Decoration.* New York: J. B. Lippincott Company, 1974.

CHAPTER 25

Mechanical Systems

Gary L. Reynolds
The Colorado College

25.1 INTRODUCTION

Building mechanical systems exist to provide an environment that protects the building structure, creates safe and healthy surroundings for the occupants, and allows equipment that is housed within the facility to operate properly.

A key to proper application of heating, ventilation, and air conditioning (HVAC) systems is an understanding of the needs that the systems are designed to address. These needs include, for example, building material response to high humidity or freezing conditions, human thermal comfort, and the impact of environmental conditions on teaching or research equipment. It is also a requirement that the systems be socially responsible through proper application of energy conservation and pollution limitation strategies. This chapter will provide the background that is necessary to ensure a good match between facility and societal needs and the operation and maintenance characteristics of HVAC systems.

This background will include a discussion of the impact of initial design decisions on operating and energy management strategies, along with descriptions of the various heating, cooling, and ventilation systems found in campus buildings; a discussion of their pros and cons; and consideration of appropriate maintenance and operations issues. It will also include descriptions of key components of HVAC systems—pumps and piping, fans and ducts, chillers, cooling towers, etc.—and energy management control strategies as they relate to building HVAC systems as a whole. The chapter will discuss laboratory fume hood systems, including descriptions of various types of fume hoods, system design issues, pros and cons, control strategies, and maintenance and operations issues. It will also include a discussion of codes and standards where appropriate.

25.2 DESIGN CONSIDERATIONS

General Considerations

First and foremost, the mechanical systems of a facility should be designed to accommodate the program of the facility. However, a number of issues surround the mechanical design that should be balanced with the program's needs. These issues include such things as functionality, cost, aesthetics, and energy consumption. Thus, the selection of mechanical systems and the subsequent design of those systems is an optimization process in which the effectiveness of the mechanical systems is judged by the system's ability to balance the various issues.

Building Functions   The mechanical systems should "fit" the building's functional use patterns, or *architectural program,* as it is often referred to during the architectural design phase of the building. Systems, or parts thereof, should operate only as needed. When a space is unoccupied, the system should maintain the minimum conditions required to protect the equipment or materials located therein and permit optimum energy expenditure in returning the space to occupancy conditions.

It is important that special conditions (e.g, fume hoods, computer facilities, clean rooms, animal rooms) be understood early in the design process. Selection, installation design, and operating needs will be dictated by codes, standards, and common practices for these types of special needs and will require a balance between mandates and user desires. Proper selection of HVAC equipment for the architectural program is important so that environmental conditions can be met for these special needs. Certain system designs are not able to provide accurate humidity control or hold tight tolerances, or have other limitations. In many cases these features are not needed, and less complicated systems can be chosen. However, once the system is installed, its intrinsic operating characteristics cannot be easily changed, and the facility will have to live with the resulting levels of environmental control. These system characteristics and their limitations will be discussed in more detail later in the chapter.

The decisions about HVAC system selections are dependent on some of the initial decisions made by the architect. Thus, it is important that the engineering team be on board and have some input during the schematic design phase. It is during this phase that space relationships are worked out and decisions are made about relative locations of functions. Engineering counseling during this part of the design process will more likely result in a facility that can be designed with an energy-efficient, functional HVAC system.

Budget Considerations   Unfortunately, budget restrictions often dictate the selection and design of mechanical systems. Those decisions often are short-term and primarily cost oriented, to the detriment of future operational and maintenance costs. Early budget planning must incorporate planning for appropriate mechanical systems; therefore, an understanding of the numerous options that may be available and appropriate is vital. When the budget will not allow for the desired number of separate systems, a larger system that can be divided in the future into subsystems or zones, and thus offer economical part-load operation, may be possible. Designing for future enhancements can often be done at little or no extra cost when funds are not available for the desired system. When faced with critical choices between what is desired and budget restrictions, life cycle cost analysis can often prove that the desired system will have a short payback period, thereby justifying reevaluation of the budget based on the savings over the life of the building.

Budget planning should never compromise the design of critical medical, research, or similar facilities for which appropriate environmental control is essential.

System Zoning   Zoning of building mechanical systems can be defined as providing for specific areas (zones) that will have individual control of the space environment. This control can be accomplished by having a separate system for each zone, or a large system capable of providing specific areas with separate control. A building can have several separate systems that are able to provide additional zoning to specific areas within the gross area served by each separate system. This is probably the most common arrangement.

Basic minimum zoning is mandatory, but excessive zoning is costly to build and can add unnecessary maintenance expense. Here again, a balance must be achieved.

Basic Zones. The basic building zones are determined by the impact of weather on the building. The primary factor, of course, is the solar effect on each exposure as the sun moves around the building each day and changes elevation with the season. In some parts of the United States, the peak cooling demand for southern exposures may occur in the fall of the year, when northern exposures require heating. Appropriate design can accommodate off-season cooling. The interior zones are not affected by the weather except for the roof, and therefore the interior zones can also be separate.

A typical building might have a perimeter zone on each of the four sides and an interior zone. If it is a classroom or dormitory building with a single double-loaded corridor, the zones can be reduced from eight to two zones. If the building is multistory, the upper floor might contain a separate set of zones because of the roof exposure. It is common for interior zones to have no heat because they are surrounded by spaces at the same temperature. Such zones typically need year-round cooling.

Functional Zones. Most building functions require system zoning beyond that dictated by the weather. These zones are created to serve such diverse areas as offices, classrooms, auditoriums, common areas, and computer rooms. If individual spaces have the same exterior exposure or interior zone function, they can be combined into a larger functional zone. Sometimes, as a result of budget concerns, spaces are grouped together and provided with one environmental control point. This cost-cutting approach should be avoided. For example, offices are often physically zoned together based on a hierarchy of individual zones for managerial or department heads and multioffice zones for staff personnel. If several rooms are placed in a common functional zone and some of the rooms do not have typical heat loss and gain, the controls cannot be located to provide a suitable environment for all the rooms.

Individual spaces within a zone cannot have temperature, humidity, air purity, or hours of operation requirements that are significantly different from the remainder of the zone unless the entire zone is upgraded to that level. Most often spaces have one or more significant differences, so they cannot be controlled from one control point. However, it may be economical to include a few spaces with simple demands as a part of a larger sophisticated zone. An example would be support rooms, such as offices for a research operation or small spaces adjacent to a surgical suite.

Time Zones. The best building mechanical system arrangement, within reasonable limits, is to have separate systems for separate building functions. A common building type on a campus is one that contains offices, classrooms, auditoriums, and, occasionally, food service. These functions can have different hours of operation, including nighttime, weekends, and vacation periods. If separate systems are provided for each of these functions, then only those that are required should operate, and the others should be shut down. Maximum energy conservation is usually achieved when energy-consuming devices do not operate except when each space is occupied. For example, quite often these functions can be divided into three time zones of operation: those that operate from 8:00 am to 5:00 pm (i.e., offices), those that operate from 8:00 am to 10:00 pm (i.e., classrooms), and those that operate 24 hours per day, 7 days per week (i.e., research laboratories). System selection and design should take these time zones into account.

It is not unusual in poor system design to find a 200-horsepower fan and pump system operating to provide night or weekend cooling to a small area that would require only a few horsepower if a separate system had been provided. Although large central systems can be designed to operate at low load to provide environmental control for a few small areas during off-hours, such a design is usually not as economical as separate systems.

In general it will be life cycle cost effective to add a small separate system for that one function.

Architectural Considerations

One of the least understood and most often neglected aspects of building design is the need for adequate space for installation, maintenance, and future modifications of the mechanical systems. Restricted mechanical space results in extra costs to the architect, engineer, and contractor during the construction process, and to the occupants during the life of the building. This situation is most often created by efforts to provide maximum usable space and by aesthetic considerations. Successful building design does not lose sight of the long-term goal: to serve its purpose during its useful lifetime without imposing unnecessary hardships on the occupants. If a mechanical system provides the required environment, is accessible for routine maintenance, and the space occupants are unaware of its existence, it is most successful.

A mechanical system must suit the building functionally. There must be ample space for installation of ducts, pipes, and terminal units, in addition to the other systems involving ceiling space and shafts. There must be space for future modifications that are certain to occur, and for maintenance access to equipment and controls. Mechanical systems operate for many years, and components will fail and must be replaced. This can include fan shafts, chillers, or heat exchanger tubes. It is not unusual to find that a new chiller cannot be installed in the space occupied by the old chiller because of inadequate access for removal and replacement. Some architectural designs may include vertical canyons or interstitial spaces between floors (horizontal canyons) to accommodate some of these needs. Mechanical space can sometimes be reduced in large systems if the air handling unit is field built instead of factory built. This is because factory-built equipment is generally short and wide owing to manufacturing economics and transportation issues.

Site plays an important role. Location and orientation will have an impact on equipment selection, sizing, and placement. Certain sites on a campus may allow connection to central utilities, and others may not. A site that can take advantage of tree shading or allow orientation to reduce solar gain may enable reduced equipment sizes and lower annual energy consumption. Access to equipment, and thus its placement in the facility, can be influenced by the site.

It is important that outside air intakes and exhaust systems be adequately separated. Placement of outside air intakes has become more critical as demands for appropriate air quality are made. Avoid loading docks, plan for wind effects, and consider other facilities that are nearby.

Aesthetics Mechanical systems have an impact on the architecture of the building. Placement of air intake and exhaust grilles, fume hood exhaust stacks, and mechanical room access will all affect a facility. Functional layout will play a role by placing mechanical systems in certain places and relationships with facility spaces. Specialty systems such as boilers, chillers, refrigeration systems, cooling towers, greenhouses, and incinerators will have an impact. Surface materials should be reviewed for durability based on possible exhaust streams that may result from the operation of the mechanical systems.

To minimize the effect of mechanical systems on the aesthetics of the building, it will be important to identify requirements in the early stages of design. Choice of mechanical systems and the layout of the systems can minimize the impact on aesthetics if integrated into the design.

Heating–Cooling Source The source of the heating and cooling media can be within the building or supplied by a central plant. Occasionally, the choice of mechanical system will dictate the source. The following systems generally dictate that the source be at the building:

- Evaporative cooling
- Packaged heat pumps—water, air, or earth source
- Packaged unitary equipment, such as rooftop and through-wall types

Others can generally operate from any source.

Noise and Vibration Noise and vibration from mechanical systems can make an otherwise successful system unacceptable. Noise and vibration control must be a team effort involving the department occupying the space, the architect, the structural engineer, and the mechanical engineer. The occupying department is involved in specifying the location of the mechanical spaces, which should be away from critical spaces. Otherwise, the construction costs will increase because of the types of floor or walls required to attenuate the noise and vibration. The architect must be aware of the equipment's limits in its ability to attenuate the noise it generates. Beyond those limits, the noise and vibration must be contained or reduced via mass—concrete foundations or slabs, or massive walls and ceilings. The design must be reviewed with regard to noise interference with functional use.

Design criteria should be carefully reviewed and understood. Noise criteria levels that are typically chosen for spaces may not meet the specific facilities needs. Selection and sizing of components plays a critical role in the resulting sound levels. It is difficult and expensive to address noise problems once they are built into the system. Fan types and sizes and duct and diffuser sizes play a critical role in noise and vibration levels. You may want to consider increased fan size at slower speeds, noise diffusers, lined duct, noise attenuating duct components, and air diffusers of a different design.

System selection can have an impact. For example, hot water heating systems can provide a low noise environment.

Human Thermal Comfort

Human beings continuously generate heat, which must be dissipated to the surroundings. For people to feel comfortable, they must attain a condition of *thermal neutrality,* which is defined as a condition in which the subject would prefer the surroundings to be neither warmer nor cooler. Considerable research has been done by various authorities to define the conditions under which thermal neutrality can be obtained. These studies have been done across age, sex, and nationality lines and have produced similar results for all groups. The results show that there are six factors that affect human thermal comfort, with a seventh factor coming into play when individuals move out of the comfort zone into a stressful condition.

These six variables can be divided into two types; four environmental and two adaptive variables. The four environmental factors are air temperature, relative humidity, air velocity, and mean radiant temperature. The two adaptive variables are clothing level and activity level.

Air Temperature   Air temperature has a direct effect on heat loss through convective and respiratory losses. Skin temperature is approximately 95°F, and heat loss is directly related to the difference between skin temperature and air temperature. In addition, most surfaces with which a person comes in contact are normally stabilized at room air temperature. Thus, heat loss from the body through conduction with the surfaces with which a person comes in contact are directly proportional to air temperature. During sedentary activity, about 25 percent of a person's heat is lost is by conduction and 50 percent by convection. All other factors being equal, the average person likes the air temperature to be about 76°F.

Relative Humidity   Relative humidity plays a role in heat loss. Heat energy is lost through perspiration and respiration. Relative humidity can actually vary over quite a wide range without creating discomfort. However, there are limits. A relative humidity that is too high will lead to virus spread and bacteria growth, and relative humidity that is too low will cause respiratory and static electricity problems. Generally the relative humidity should be kept within a range of 20 to 60 percent. All other factors being equal, the average person likes the relative humidity to be about 50 percent.

Air Velocity   Air velocity can provide a cooling effect by increasing convective heat losses. However, there are limits. Above about 150 ft. per minute (fpm), air will move papers around on a desk and possibly cause a drafty feeling. If a person is exposed to an air velocity of 150 fpm or higher for eight hours or more, he or she will become dehydrated. Generally speaking, below about 40 fpm, a person cannot feel air movement. All other factors being equal, the average person likes the air velocity to be about 50 fpm.

Mean radiant temperature Mean radiant temperature is probably the most difficult concept to grasp. All objects that are at a temperature above absolute zero (–459°F) are radiating energy from their surfaces. Thus because the exterior of the human body is about 95°F, the human body is radiating energy. All of the surfaces surrounding the human body are also radiating energy, so in reality the human body is radiating to all surfaces, while at the same time all surfaces are radiating energy to the human body. If the human body is warmer than the surrounding surfaces, then the net exchange of energy is from the human body to these surfaces; if the surfaces are warmer than the human body, then the net exchange of energy is to the human body. During sedentary activity, about 25 percent of a human's heat loss is by radiation to cooler surfaces. The amount of heat loss by radiation can change dramatically if nonstandard environmental conditions exist (e.g., a cold window surface). Normally, it can be assumed that most surfaces in a space are at air temperature.

Clothing level Clothing level can vary considerably and has a major impact on comfort level. The unit that is used to measure clothing insulating levels in the heating and cooling industry is the clo. A person wearing a short-sleeve shirt, slacks, socks, and shoes is wearing about 0.6 clo. A woman wearing a light blouse, skirt, panty hose, and shoes is wearing about 0.5 clo. A man wearing a long-sleeve shirt, sport coat and tie, slacks, socks, and shoes is wearing about 1.0 clo. A three-piece suit is equal to about 1.1 clo.

Activity level Activity level is measured in metabolic equivalents, or mets. As you sit and read, you are working at about 1 met. If you lie down, you are at about 0.5 met. If you are taking a test while seated, then you are at about 1.2 met. This can increase to about 3 or 4 mets for athletic activities.

Summary Research shows that when putting all of these factors together, the ideal set of conditions is a temperature of 76°F, relative humidity of 50 percent, air velocity of 50 fpm, mean radiant temperature equal to air temperature, a clothing level of 0.6 clo, and an activity level of 1.0 met. Interestingly, even with these ideal conditions, not everyone will be satisfied. Figure 25-1 shows that with increasing deviation from the ideal conditions, more and more people will be dissatisfied, but even under optimal conditions, 5 percent of the people will still not be satisfied. This dissatisfied group is the result of variations in the metabolism of people.

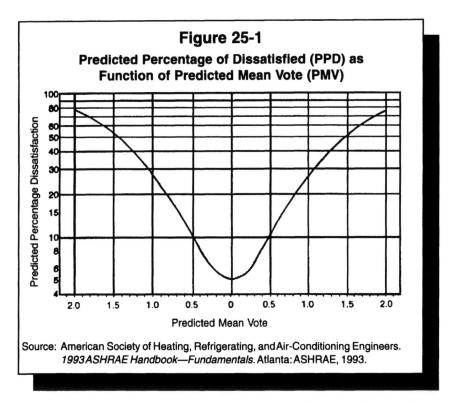

Figure 25-1

Predicted Percentage of Dissatisfied (PPD) as Function of Predicted Mean Vote (PMV)

Source: American Society of Heating, Refrigerating, and Air-Conditioning Engineers. *1993 ASHRAE Handbook—Fundamentals.* Atlanta: ASHRAE, 1993.

Figure 25-2 shows the relationship of these variables to the variable of air temperature. This table can be used to determine equivalency of conditions. For instance, if the relative humidity is raised by 15 percent, then the temperature must be lowered 1°F, as indicated by the negative sign. If the air velocity is raised by 50 fpm, then the air temperature must also be raised 1°F. If the clothing level of a person is increased by 0.08 clo then the air temperature must be lowered by 1°F. It is not uncommon to have a woman dressed at 0.5 clo and a man dressed at 1.0 clo, which is equal to about a 6°F difference in temperature.

The essence of this discussion is that factors other than air temperature and relative humidity determine human thermal comfort. Many times one of the other variables can be adjusted to accomplish the equivalent effect. Are there radiation effects that are causing the sensation of a draft? Can the clothing level be adjusted in the summer? Can the ergonomics of the work situation be changed to lower the met level? These are the type of questions that should be asked. Figure 25-3 shows the thermal comfort envelope for various air temperature and relative humidity ranges based on set conditions of the other variables, as noted at the bottom of the figure. As can be seen in this figure, there is quite a range from summer to winter.

Figure 25-2
Human Thermal Comfort Variable Relationship

Variable	Valid Range	Relationship
Relative humidity (RH)	20% to 80%	1 F = –15% RH
Air velocity (AV)	40 fpm to 160 fpm	1 F = 52.4 fpm
Mean radiant temperature (MRT)	within + or –20°F of air temp.	1 F = –0.7 MRT
Clothing insulation (Clo)	0 clo to 1.5 clo	1 F = –0.08 clo
Activity level (MET)	0.7 met to 3.0 met	1 F = –0.14 met

This variation is due to the adaption of clothing levels from summer to winter. Note that the ideal conditions fall within the comfort envelopes. As mentioned previously, in Figure 25-1, with increasing deviation from ideal conditions, more and more people will be dissatisfied. For example, operating at 68°F and low relative humidity, as some schools try to do in the winter time, will place the building at the extreme edge of the comfort envelope and create a high percentage of people who are dissatisfied.

Finally, up to this point the discussion has centered on human thermal *comfort.* It is possible to create conditions that will cause a person to experience either heat or cold *stress.* As people move outside the comfort level envelope, their bodies expend energy to adapt to the environment. If a person is exposed to these conditions for too long, the body will be unable to maintain body temperature, and the person will experience stress. Thus, a seventh factor, *time,* comes into play when talking about stress. If you are trying to determine whether work conditions are thermally stressful, then appropriate heat stress indices, such as the wet bulb-globe temperature or Belding and Hatch Index, should be used, not just temperature and humidity, as these indices account for all the variables of thermal stress. The National Institute of Occupational Safety and Health (NIOSH) has endorsed the wet bulb-globe temperature measurement for heat stress. The Belding and Hatch Index can be use for either heat or cold stress.

Ventilation

Ventilation is required for two fundamental reasons: to protect the occupants and to protect the facility. Ventilation of attic spaces or crawl spaces are typical examples of ventilation provided to protect the facility. This section will not cover those cases, but will focus on the ventilation requirements for the occupants.

Indoor air quality can be controlled in three fundamental ways: removing the source of pollution, providing sufficient ventilation air, or masking

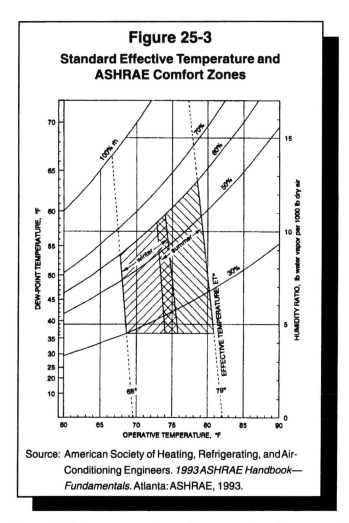

Figure 25-3

Standard Effective Temperature and ASHRAE Comfort Zones

Source: American Society of Heating, Refrigerating, and Air-Conditioning Engineers. *1993 ASHRAE Handbook—Fundamentals.* Atlanta: ASHRAE, 1993.

the odor. Removal of the pollutant is usually a cost-effective means of dealing with an indoor air quality problem but is not always practical. Similarly, masking (often used in residences) is not always a practical strategy. Thus, most educational facilities focus on ventilation for indoor air quality control. Thus, this section of the chapter will focus on proper design of ventilation systems.

Ventilation Requirements   There are two issues when dealing with providing proper ventilation. The first is to ensure that the correct amount of clean air is provided; the second is to ensure that the clean air gets to the occupants. Before proceeding further, it is important to understand the terms used in discussing ventilation. Figure 25-4 shows the various airstreams in an air handling system. When we discuss proper ventilation rates, we are talking about outdoor air or makeup air. There is a common misconception

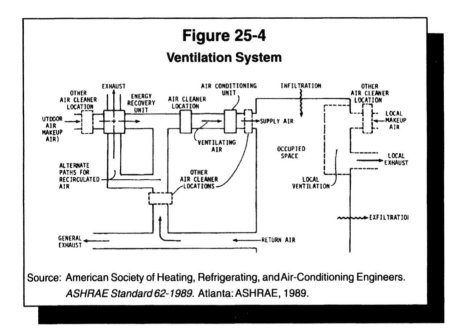

Figure 25-4

Ventilation System

Source: American Society of Heating, Refrigerating, and Air-Conditioning Engineers. *ASHRAE Standard 62-1989.* Atlanta: ASHRAE, 1989.

that the ventilation rate is based on supply air, but this is incorrect. Note that *supply air* consists of two fundamental streams of air: *outdoor air* and *recirculated air.* These two airstreams are mixed together to create *mixed air.* The *mixed air* then goes through an air filter and becomes *ventilation air,* which is than heated or cooled to become supply air. Figure 25-5 shows examples of ventilation rates of outdoor air that are recommended in Standard 62 of the American Society of Heating, Refrigeration, and Air Conditioning Engineers (ASHRAE).

There are many circumstances that require special attention, such as operating rooms, animal housing facilities, and laboratories. Specific standards and regulations govern these types of spaces and should be consulted during the design phase. Early consideration should be given to these spaces to ensure that they have access to outdoor air and that there is room to accommodate the ventilation systems these types of spaces need.

Thus, it is up to the design engineer to ensure that the proper amounts of outdoor air are delivered to the space. It is also up to the design engineer to ensure that the outdoor air meets the needs of clean air. Thus, proper placement of outdoor air intakes relative to building exhaust systems, vehicular traffic, or other pollution sources is important. In addition, some air handling systems, such as variable air volume systems, do not always provide the proper amounts of outdoor air under certain operating conditions. A careful analysis of the full range of operating conditions must be undertaken to ensure that the system always delivers the proper amounts of outdoor air for ventilation.

Figure 25-5
Outdoor Air Requirements
from ASHRAE Standard 62-1989
(cfm/person)

Space Type	cfm/person
Office	20
Reception Area	15
Lobbies	15
Classroom	15
Laboratory	20
Conference Room	20
Assembly Room	15
Music Room	15
Library	15
Spectator Areas	15
Locker Rooms	0.5 cfm/sq. ft.
Smoking Lounges	60
Dining Room	20

Ventilation Delivery   The second issue in proper ventilation system design is to ensure that supply air (which contains the outdoor air for ventilation) reaches the occupants. Many poorly designed systems introduce the supply air at the ceiling level and direct it straight at the return air grille. Thus, the supply air essentially bypasses the space and never reaches the occupants. The ratio of supply air that reaches the occupied area of the space, divided by the total amount of supply air introduced into the space, is called the *ventilation efficiency.* Manufacturers of supply air diffusers and grilles have specifically engineered the throw and the drop of the supply air devices to ensure good ventilation efficiencies (Figure 25-6). Engineering studies have been done to show that the proper selection of a grille or diffuser can provide the right throw and drop to ensure that supply air is introduced with enough velocity that it is forced down into the occupied area of a space, but not so much velocity that it creates a drafty condition. The complete explanation of this design approach is outside the scope of this chapter, but the facilities administrator should be aware that proper design of the air supply system in a space is crucial to providing proper indoor air quality.

Certain heating and cooling systems, such as variable air volume systems, can have an effect on the proper operation of grilles and diffusers. In a variable air volume system, the amount of air exiting the grille or diffuser varies, and with it, the throw and drop. One way to overcome

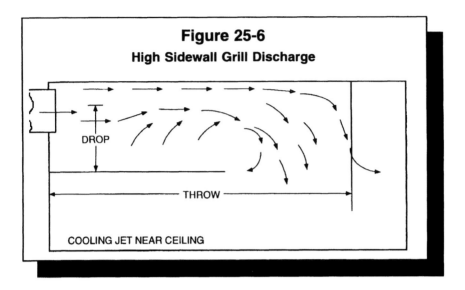

Figure 25-6

High Sidewall Grill Discharge

DROP

THROW

COOLING JET NEAR CEILING

this limitation is to use an induction or fan-powered variable air volume box as shown in Figure 25-7. This type of box uses local air from the room for recirculation, with the supply air coming from the central air handler. Thus, as the supply air from the central air handler is reduced, the local recirculated air makes up the difference, ensuring that the same amount of air continues to be supplied through the grille or diffuser, thus ensuring that the carefully designed throw and drop are not affected.

Indoor air quality Indoor air quality problems are divided into two basic categories: building-related illness and sick building syndrome. Building-related illness is the result of actual viruses or bacteria that have grown in a building's systems. Legionnaire's disease is an example of a building-related illness. Proper maintenance and cleaning procedures will ensure that building-related illness is not a problem. Sick building syndrome is a problem that is poorly understood. Symptoms include headaches, eye irritation, and reports of general malaise by the occupants. Industrial hygienist evaluations usually do not turn up any one thing that can be seen as the culprit. The general consensus is that as we have tightened up buildings for energy conservation reasons, not paid attention to air handling system designs, and reduced maintenance because of budget cuts, the compounded result is sick building syndrome. Also, there is some evidence that although levels of individual pollutants may be below industrial hygienist guidelines, the synergistic affect of several pollutants together may cause problems.

In recognition of the growing concern for indoor air quality issues, ASHRAE modified its Standard 62 and republished it in 1989. In most cases the recommendations for ventilation were tripled or quadrupled. For

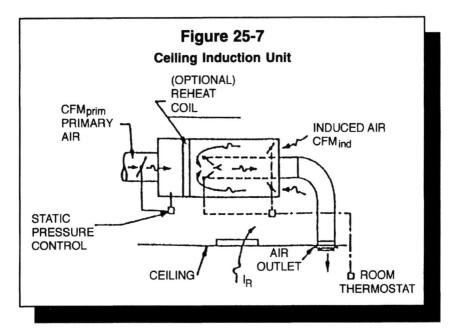

Figure 25-7

Ceiling Induction Unit

example, the previous recommendation for outdoor air was 5 cu. ft. per minute (cfm) per person in an office. The new recommendation is 20 cfm per person.

Another area that is causing problems is the introduction of new pollution sources within the space. Quite often this happens after the building has been in operation. Products that give off formaldehyde, such as furniture and carpet, or printers and copiers that give off various chemicals are introduced into spaces, with no attention paid to ventilation requirements. If there are point source pollutants in a space, these can be addressed locally by providing a local exhaust system or by replacement with alternate products. These alternatives should be considered, as they can often be the most cost-effective solutions in the long run.

To ensure that a facility will not suffer from indoor air quality problems, it is important that the air handling system be designed, installed, and maintained properly. Proper attention to the details of control strategies over the full range of operation of the air handling system to ensure high ventilation efficiencies, with the proper quantities of outdoor air and proper maintenance of the system, is the key to success.

Filters Filters can be an important part of a system that provides quality air to a space. They can also play an important role in reducing energy consumption by allowing recirculation of greater amounts of air to the limits as required by standards. For example, changes in medical facility ventilation requirements and improved filters have permitted the recycling of air in surgical suites and similar areas, resulting in increased air quality control and greatly reduced energy requirements.

The choice of filters available for buildings is almost endless. They come under two general classifications: disposable and permanent. Within each type, various degrees of filtration efficiency are available. A filter should be only as efficient as required, as the cost of filter installation and maintenance increases with efficiency. A 25-percent efficient filter means that 75 percent of the dirt passes through the filter. All high-efficiency filters should be downstream from lower cost roughing filters, whose prefiltering action will greatly extend the life of high-efficiency filters.

Disposable Filters. Disposable filters are available in rectangular form and in blanket form that can be cut to size and spread over a suitable mounting surface. A variety of thicknesses and media are available. The media can be fiberglass or a proprietary-type plastic material, treated or untreated. Treated filters generally contain viscous material for impingement-type filtration. Roll-type filters with disposable blanket media are operated by advancing the filter media manually or automatically, based on lapsed time or pressure differential across the filter.

A bag-type filter provides a relatively high level of efficiency. Such filters are gaining in popularity and have relatively low air resistance and greatly extended periods of time between filter changes.

High-Efficiency Particulate Air Filters. A high-efficiency particulate air filter must be used in certain medical and research operations. These filters generally require special airtight mounting frames, roughing filters, and careful monitoring of maintenance. This is generally true of all high-efficiency filters but is most critical for high-efficiency particulate air filters.

Chemical-Type Filters. Chemical-type filters use citrus derivatives or similar compounds to absorb the odors contained in the airstream. They are available in many different sizes and types and are energy efficient in that the exhaust air can be recirculated.

Carbon Filters. Carbon filters have a long history of use for odor removal. They have been used on submarines for many years. In the past, the low cost of energy made their use uneconomical, but with current energy prices, carbon filters can often be used effectively for the removal of odors and permit the recirculation of air. Carbon filters require periodic regeneration of the carbon and eventual replacement.

Electrostatic Filters. Electrostatic filters work on the principle of electrically charging particles as they pass through the filter, which causes the charged particles to be attracted to an oppositely charged dust collector plate. Electrostatic filters can achieve efficiencies of greater than 95 percent.

Special Space Ventilation Considerations

As noted earlier, the major ventilation standard currently used is ASHRAE Standard 62, which sets forth the ventilation requirements for buildings based on occupancies and functions within the building. In addition, there are other standards that must be met, depending on the nature of the facility. Medical facilities must comply with U.S. Public Health Standards. In the case of contaminants that adversely affect human health, the organization must meet Public Health Service industrial hygiene standards.

Toilets and Locker Rooms The buildup of objectionable odors is one of the major motivating factors for adequate ventilation of these spaces. Often the odors themselves are not harmful but merely objectionable. The nature of the odor source should be considered when determining the method of odor removal and operation of the ventilation system. Some odors are generated only during a specific function and are short-lived. Others are pervasive and continue beyond the function involved. Locker rooms and other athletic facilities are examples of the latter.

Moisture generation, an additional contaminant, usually ceases when the use of the space terminates. These factors may necessitate different operating levels of ventilation at different times. The design of a ventilation system for a locker room should consider the moisture problem, and suitable corrosion-resistant materials should be used for ductwork, air grilles and diffusers, and interior space finishes.

Food Service Food service ventilation involves applicable health and safety codes, including fire protection. Range hood exhaust systems can require large volumes of exhaust air, imposing the need for makeup air and the consequent high operating costs. Careful design of kitchen range hoods by knowledgeable people can greatly reduce the ventilation exhaust and makeup air requirement. The specific location of cooking equipment below the hood can also reduce ventilation requirements by grouping similar items in a smaller area.

High-temperature cooking operations involving open flames, frying, and similar grease-related operations should be grouped together if at all possible. These sources of contamination require relatively high rates of ventilation compared with ovens or vegetable steamers. The latter require moisture and heat ventilation but do not require ventilation for contamination.

The removal of fumes from grease-related operations involves a potential fire hazard. As a result, range hoods must meet applicable fire codes and include an automatic fire suppression system. The construction of the range hood and the exhaust ductwork must meet health and fire codes, which require internal cleaning of the ductwork and the ability of the ductwork to withstand the higher temperatures generated by a grease fire. Filters are not 100 percent efficient, and grease will migrate throughout the system and can contaminate the surfaces of the building near the point of discharge. Generally this is an unsightly condition and is not particularly hazardous, but it increases maintenance. Roof-mounted upblast fans are preferred for exhaust of grease-laden air. Building food service exhaust systems that are properly designed in accordance with codes will result in a more easily maintained and safer exhaust system. The exhaust of air over dishwashers, for example, is done primarily to reduce the moisture. The ductwork for this operation should be noncorrosive, and the points of pickup should be as close to the source as possible to reduce the amount of ventilation required.

Makeup air can be provided through a variety of types of equipment depending on the season of the year and the climate. In arid climates, it can be provided by evaporative coolers to furnish makeup air that is conditioned. In other climates, direct mechanical refrigeration is usually required. In the winter, direct-fired gas makeup heaters are the most energy efficient means available if this equipment is compatible with the facilities operations. Makeup air can be introduced directly at the hood with or without tempering. Under certain conditions, no tempering of the makeup air is required if the hood and makeup air ductwork system are properly designed.

Maintenance Operations   The types of ventilation required for maintenance operations vary greatly. Some maintenance operations are quite simple, consisting mostly of the use of power-driven bench tools, soldering, light welding, and occasional manual painting. These types of operations need general room ventilation of the type that is ordinarily found in maintenance buildings with operable windows and doors and the typical wall exhaust fan. Ventilation for maintenance operations, however, should be designed to conform to code and, in cases of hazardous operations, should be specifically designed to meet the standards set forth in the American Conference of Governmental Industrial Hygienists' *Manual of Recommended Practice,* "Industrial Ventilation,"[1] and the applicable National Fire Protection Association (NFPA) standard.

Shops   Maintenance shops on the typical college campus are usually fairly large and have diverse operations. The ventilation should consist of specific local ventilation for individual tasks and general ventilation for the entire area for removal of fumes that escape from localized systems or that are generated from operations conducted in inappropriate locations in the shops. The use of solvents and cleaning fluids presents a health and an explosion hazard that is often not recognized. These operations, if they are extensive or highly concentrated, cannot be handled by general ventilation and should be performed in a location designed for the use of the fluids.

Painting   Painting is one of the most common yet most critical types of operation found in maintenance that requires ventilation. Maintenance workers paint repeatedly throughout their years of employment and, as a result, are subject to the cumulative effect of different hazardous components. Paint exhaust systems should be designed for the size of the objects to be painted and the types of paints used. Small bench-top booths are appropriate for minor spray-painting operations. Exhaust systems for such booths can be relatively simple, with a fan discharging the air captured by the hood directly outside.

As the operation becomes larger, such as where furniture or even vehicles are painted, the exhaust system arrangement should provide an individual room for the operations. The exhaust air path through the paint room should be directed toward one end at an appropriate velocity, usually 100 ft. per minute, and where required, the exhaust air should be filtered prior to discharge. Filtration can consist of dry filters, water-wash filters, or a combination. The fresh uncontaminated air should pass over the operator first and then the paint operation.

Some painting operations also involve paint-drying areas. These areas should also receive appropriate ventilation, or the time spent in the space should be limited, with the room flushed with air between operations. Appropriate design of the electrical system is required in the painting and drying areas to avoid explosion hazards.

Maintenance Garages   Garages for vehicle maintenance should be provided with two types of ventilation:

1. General ventilation should be provided throughout the entire area to dilute the contaminants to an acceptable threshold level for the space occupants.
2. In cases where internal combustion engines are being operated for maintenance and testing purposes, tailpipe exhaust systems should be mandatory.

Woodworking Operations   The ventilation for many woodworking operations can be quite simple, such as a portable vacuum unit at the power tool. For larger operations such as a carpenter shop, a central exhaust system can be installed. Such a system should be designed in conformance with the *Industrial Ventilation Manual*[2] and should have a central dust collector for separating the wood chips and sawdust from the exhaust air before discharge. The exhaust from the collector can be filtered with an appropriate grade of filter and a major portion of the air recirculated back to the space to provide significant energy savings in makeup air heating.

Grinding Operations   Ventilation of grinding operations can be accomplished relatively simply by having small ducts with flexible connections and appropriate hoods attached to the power grinding operation if it is bench-mounted. Hand grinding of large objects should be done in an appropriate exhaust booth. It is possible to recirculate air from a grinding operation if it is filtered. However, the filters become contaminated and loaded quite readily, so this is not normally cost-effective.

Parking Garages   Parking garages present a special problem because of the widely varying rate of contamination within the space. Parking garages that are used throughout the day in facilities with large numbers of visitors have a relatively constant rate of contamination. Facilities for office workers and staff have peak contamination on arrival and departure, with far less between those times.

Many garage ventilation codes require a specific and continuous rate of ventilation on the order of 1 cu. ft. per square foot of enclosed parking garage throughout the period of occupancy. This rate is generally unnecessary for many hours of the day and has a major impact on energy requirements. It is preferable to design garage ventilation systems based on carbon monoxide monitoring devices located throughout the facility. These devices can vary the rate of ventilation depending on the concentration of contaminants. In general, enclosed parking garages should have a relatively small ventilation fan that runs continuously, with larger fans for primary ventilation. Detection of contaminants can sequence additional fans to increase the ventilation. Carbon monoxide sensors do not necessarily respond to diesel fumes. Depending on the climate, tempering of the makeup air may be required if there is potential for freeze-up of piping within the garage.

General Issues Regarding Ventilation

Ventilation System Maintenance and Safety   Exhaust systems for maintenance operations are important, because the very nature of the maintenance operation presents a hazard to the occupants of the space and to those performing the work. In many cases, the exhaust product not only is hazardous, but also presents a fire or explosion hazard, and the system design should account for both of these. However, the best exhaust system will be ineffective with inappropriate maintenance and operation. Filters must be changed, and all electrical wiring and devices should be maintained in an explosion-proof manner where required.

The location of the exhaust discharge must be carefully selected relative to the surrounding structures and building openings to avoid introducing the contaminants into an occupied space.

Makeup Air and Energy Conservation   A variety of methods provide makeup air to ventilation requirements. These include the simple introduction of unconditioned air to the space for dilution. Direct-fired gas heaters are the most efficient to use in the winter. Evaporative coolers operate most efficiently in summer and in arid climates.

Careful planning and appropriate strategies can greatly reduce the investment in initial cost and fuel for makeup air systems. Wherever air must be conditioned, energy conservation measures should be introduced. Generally, they are cost-effective on a life cycle basis if the air quantities are large. In many cases, makeup air for exhaust operations can be supplied from adjacent spaces or systems, thereby avoiding excessive energy costs.

A widely used and successful strategy is the economizer cycle. This control strategy allows the use of 100 percent outside air during times when it is beneficial and controls ventilation to some required minimum when the use of outside air would cause high energy consumption.

Moisture Control Ventilation systems are used for the control of moisture at swimming pools and other facilities involving high rates of moisture generation. It is a mistake, however, to rely solely on ventilation for protection of the envelope of the building. An enclosed swimming pool area or similar facility must be designed with adequate vapor barriers or constructed of materials not subject to deterioration by moisture. Such materials include concrete and various types of metals, either noncorrosive or appropriately clad. A building in a cold climate that has paths with a high rate of heat transfer directly to the outside should be monitored carefully for potential building damage. It is not unusual to see the entire veneer of a building that encloses a swimming pool become detached from the basic envelope structure because of moisture migration. Ice can form within walls to the point where the pressure of the ice expansion causes the wall components to separate.

Moisture in the swimming pool area can be kept at reasonable levels with ventilation and, in many cases, with appropriate application of heat pump systems. Ventilation can range from simple dilution in mild climates to energy-intensive systems that mandate serious consideration of energy recovery.

Operating Considerations Although ventilation is necessary, it is a constant source of potential energy consumption. Heat recovery systems are desirable but are not 100 percent efficient, and the loss is continuous during system operation. Monitoring of ventilation systems and education of responsible parties constitute a difficult and ongoing process. If a facility can be metered, the parties involved can be made aware of and possibly held responsible for the impact they have on the overall operating costs of a building. Metering of individual utilities at points of use is a desirable arrangement. Ventilation operation success or failure can be pinpointed by such metering in buildings with high ventilation requirements.

Equipment Needs

Although the main reason for environmental control of indoor spaces is the comfort and safety of human occupants, we often provide environmental control for equipment that is in spaces. Research equipment and computers often dictate what the environmental conditions should be. Manufacturers provide this information and often include installation guidelines to ensure proper connection to utilities.

It is important that equipment needs be considered early in the facility design. Such needs should be considered in zoning choices and in location for access to an outside wall if necessary. Special utility considerations and temperature and humidity needs may require special local heating and cooling equipment to ensure proper space conditions. In some cases equipment requires not only accurately maintained environmental conditions, but also consistent conditions. These types of requirements will have budget as well as design and installation implications. Standard commercial technology quite often is not up to being able to maintain exotic equipment needs, and these should be identified early in the design phase so that they can be accommodated.

However, many manufacturers are designing new equipment that no longer requires special environmental conditions. In the past computers often needed special conditions, but currently many computer systems do not. It is important that the manufacturers' recommendations be obtained to ensure that overdesign and excessive energy consumption do not result from a misunderstanding of equipment needs.

25.3 BUILDING CENTRAL PLANT SYSTEMS

Heating Source Components

The heat source for a building can be derived from numerous options. The option selected is usually determined by the budget for the building, energy costs, history of previous success, and the institution's preferences. If a facility has no need for steam, hot water might be the obvious choice. In a facility housing medical operations, food service, or laboratories where steam is required, the most economical source of heat could be steam.

Types of Boilers Types of boilers include steam, hot water, cast iron sectional, modular, and electric boilers.

Steam Boilers. Steam boilers require careful monitoring of the condition of the water and generally require water treatment. Water treatment is particularly required if there is any loss of water because of leaks or steam consumption for humidification. Water treatment is also required to provide protection for the condensate return system, which is subject to corrosion because of the presence of air. Chemicals that provide a film inside the condensate return pipes are generally added to the boiler water for this purpose. Steam boiler systems must be "blown down" periodically to remove the mineral deposits left by the evaporative process.

Fire tube boilers are most commonly used in steam heating systems for average buildings. Cast iron sectional boilers are also used, although not as often. The steam from the boiler can be used in a heat exchanger (converter) to heat the water required for a hot water heating system.

Hot Water Boilers. Hot water boilers are found in all the same models and configurations as steam boilers. They are completely filled with water and fired in the conventional manner, with water temperature being the controlling factor. The controls on the boiler are also similar except for the need for a water level control on the steam boiler. Both types of boilers need low water cutoffs and an automatic means of providing makeup water for safety.

Hot water boilers are popular in this country because in many cases codes require that a boiler operator be in constant attendance for a steam boiler but not for hot water boilers. In many hot water heating systems, it is desirable to vary the temperature of the hot water to meet the actual load in the building. This increases occupant comfort and energy efficiency. It is not advisable, however, to vary the water temperature by varying the temperature of the water in the boiler. Most boiler manufacturers caution against operating the boiler at too low a water temperature, as this causes condensation of the products of combustion in the boiler and the flue, and subsequent corrosion. It is recommended that water temperature variation be accomplished by blending valves, with the boiler operating at a constant temperature. The blending valves mix cooler return water with hot boiler water to provide the desired supply temperature.

The advent of pulse-type combustion burners has led to the development of high-efficiency hot water boilers. Reduction of flue gas temperatures by extraction of heat is one of the main sources of added heat in a high-efficiency boiler. This reduction of flue gas temperatures leads to limitations of the temperature of the hot water generated, in effect requiring a low-temperature hot water system. Obtaining equally high efficiencies with a steam boiler is impossible because of the need for higher temperatures to boil the water to form steam.

Cast Iron Sectional Boilers. An older form of boiler is a cast iron sectional type. The design consists of a number of cast iron sections bolted together and connected by openings in both the top and the bottom of each section. Steel sleeves form a watertight seal between sections, allowing free circulation of water or steam throughout the boiler. Cast iron sectional boilers can be used for steam or hot water. They are highly resistant to corrosion but susceptible to cracking under thermal shock by introduction of the hot water heating system's relatively cool return water against the end plates of the boiler. This temperature stress can cause distortion of the boiler shell and significant damage to the boiler. Careful design can avoid this problem. A steam boiler operates at a relatively constant temperature throughout all its internal flow circuits and usually never experiences thermal shock.

The cast iron sectional boilers require a relatively small amount of floor space, and initial costs are comparable to those of a fire tube boiler. Cast iron sectional boilers have been successfully used for many years. They are often employed to replace boilers in relatively inaccessible boiler rooms, as they can be field assembled for a relatively moderate additional cost.

Modular Boilers. Many boiler systems within buildings were grossly oversized prior to the energy crisis of the 1970s. Typically, the heat loss calculations were made, and 25 percent was added for pickup and 10 to 15 percent for piping losses. Then two boilers were installed, each capable of providing 75 percent of the inflated figure. Often, only one boiler was required, even under the most severe weather conditions. The net result was a typical seasonal operating fuel efficiency of 30 to 40 percent.

Heating equipment efficiency is rated at full-load conditions. At part-load conditions, the efficiency generally is significantly reduced. The use of numerous small boilers, each operating at full-load conditions in a sequential fashion, offers a vastly improved seasonal operating efficiency. Although these systems were available prior to the energy crisis, in many cases their higher initial cost could not be justified in view of cheap energy prices. Currently, however, they are finding wide application in both steam and hot water systems. Proprietary and nonproprietary designs are available.

Some systems give the appearance of a series of domestic hot water heaters connected to pipe manifolds and individual heater pumps. Typically, on a call for heating, one boiler is placed in operation and functions until the load increases beyond its capacity. At that point, subsequent boilers are brought into operation, maintaining essentially a full-load operation on each.

Where oversized boilers have been taken out and replaced with a series of modular boilers, energy consumption reductions of up to 30 percent have been obtained. Multiple boilers of relatively small capacity are used with individual pumps. Fewer and larger boilers are used in much the same manner. The larger boilers, however, tend to be typical fire tube boilers used for steam or hot water heating systems, but of smaller sizes.

Modular boiler systems can also provide domestic hot water efficiently. This is particularly important in installations where domestic hot water consumption is relatively high.

Often a number of the boilers in the modular boiler installation never come into operation. In effect, the package is oversized. This is not necessarily undesirable, as the extra boilers provide standby capacity and the ability to add to the load in the future as the building heating requirements expand, and they do not affect the operating efficiency. The extra boilers indicate that a true heating load is difficult to predict because of diversity and other factors. In a new building, the tight fit of the building components reduces the load that the designer anticipates will develop as the building ages.

It is interesting to note that most central heating plants with multiple boilers operate on a modular basis. The plants often have a summer boiler sized to match the summer load.

Electric Boilers. As previously discussed, electricity has a number of highly desirable features for use as a source of heating energy. Its undesirable features are its cost and its impact on nonrenewable energy sources. In many buildings it might be considered wise to use electric boilers for the heating source rather than electric heating units distributed throughout the building. This facilitates switching the types of boilers as energy source economics and availability change.

A typical air conditioning and heating system for a building using forced water flow is essentially independent of the fuel source for heating the water. Under certain applications (e.g., cogeneration), it is possible to install a supplemental electric boiler to improve the overall energy efficiency of the plant. Electric boilers are typically 200- to 480-V, three-phase units. Capacity control is obtained by modulating the number of heating elements made active to match the load. The heating elements in an electric boiler are also subject to deterioration as a result of mineral deposits. This can be avoided by selection of appropriately enclosed or sheathed heating elements. Electric boilers are available for steam or hot water.

Heat Pumps The heat source for a building can consist of one or more heat pumps. In many applications this can provide an extremely flexible system. Small modular heat pumps can be relatively economical and are low in initial cost. However, size limits, maintenance in multiple locations, and noise may be a consideration. Normally these types of installations are found in smaller facilities with just one or two zones.

Cooling Source Components

Although humans have survived the evolutionary process from prehistoric to relatively recent times without air conditioning, most Americans expect their working and living accommodations to be comfortable. In addition, many of the processes involved in modern activities (e.g., computer rooms and research and medical facilities) require air conditioning for successful operation. Most of us are familiar with old-fashioned methods of cooling, but one that should be noted here was the use of ice. One ton of air conditioning is the energy equivalent of melting one ton of ice in a 24-hour period, equivalent to 12,000 British thermal units (Btus) per hour. This has become a convenient method of measuring air conditioning. Current sources of cooling involve mechanical devices of various degrees of simplicity or complexity of design, operation, and maintenance, and varied energy sources.

Although low initial cost of a system is still an important consideration, energy costs have become a major factor in the selection of air conditioning primary equipment. Larger electrically driven mechanical cooling equipment should be provided with a combination demand and Watt-hour meter, as well as an elapsed time meter. In the case of heat source machines, condensate or fuel supply can be metered. The following is a discussion of the more common types of equipment that provide a source of cooling for air conditioning systems in colleges, universities, and institutional facilities.

Refrigeration systems typically use chlorofluorocarbons (CFCs) and hydrochlorofluorocarbons (HCFCs) as the working media. International agreements are phasing out the production of these gases. The CFCs that are affected are commonly used in facilities equipment and include CFC-11, CFC-12, CFC-113, CFC-114, and R-500. CFC-12 is often used in refrigerators, freezers, refrigerated water coolers, and automobile and truck air conditioners. R-502 is used in low-temperature equipment such as low-temperature centrifuges and freezers. HCFCs are being phased out at a slower pace but ultimately will be obsolete. The most common HCFC is HCFC-22, which is used in air conditioners, heat pumps, smaller building chillers, and other small refrigeration equipment. The design and installation of new equipment should consider the effect of the phase-out of these gases.

Direct Expansion Systems   The direct expansion system is one of the oldest and most popular refrigeration systems. Figure 25-8 illustrates a typical compressor–direct expansion refrigeration cycle. The refrigeration effect in a compression cycle is caused by the expansion of the liquid refrigerant into the gaseous state.

The liquid refrigerant is vaporized through an expansion value to a lower pressure into a cooling coil, causing cooling. The compressor draws the refrigerant gas out of the coil and raises it to a high pressure, which also raises the temperature. The high-pressure gas is cooled in a condenser and becomes liquid that is stored in a receiver tank to repeat the process.

Reciprocating Compressor Chillers   Reciprocating compressor chillers are piston-type machines and are typically sold as package components complete with all necessary operating and safety controls. They can be connected to a remote air-cooled condenser or be provided with a water-cooled condenser, with water being supplied from a cooling tower or other source. Reciprocating compressor chillers are available in sizes up to approximately 200 tons. The larger sizes of package-type units are often equipped with multiple compressors. Most are electrically driven. The units may be hermetic, semihermetic, or open. Capacity modulation for reciprocating compressor chillers is provided by using multiple compressors or a combination of multiple compressors and step unloading of compressor cylinders.

Reciprocating compressor chillers tend to be noisy, and their locations should be carefully chosen. They have a higher energy consumption than centrifugal compressors for a net cooling effect, although improved design has significantly reduced the energy consumption of the reciprocating type. All compression-type equipment can be modified for use as a heat pump. Reciprocating compressor chillers can be connected to a direct expansion system serving one or more cooling coils. They can also be provided with

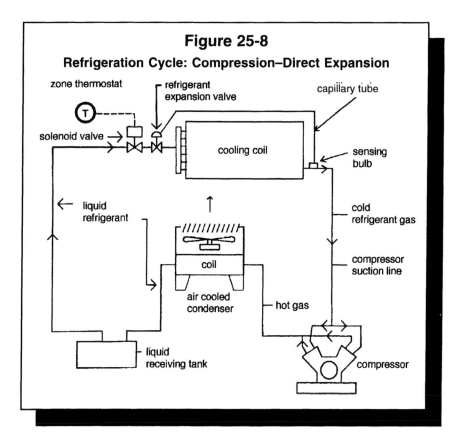

Figure 25-8

Refrigeration Cycle: Compression–Direct Expansion

an evaporator shell to provide chilled water for an air conditioning system. Well-designed and well-maintained reciprocating equipment can have a life expectancy of 20 to 25 years.

The dimensions of modern equipment are such that most equipment will pass through a normal door. Solid-state controls are available with current reciprocating equipment, monitoring operating conditions and providing safety protection and diagnostics of abnormal conditions. The controls can also be designed to be interconnected with energy management/monitoring systems. These types of controls should be seriously considered on any 20-ton or larger unit. Typically, an air-cooled compressor could require 1.2 to 1.4 kW per ton, compared with 0.9 to 1.1 kW per ton for a water-cooled type.

Reciprocating compressor chillers can be driven by gas engines fueled with natural gas. With the potential for gas engine exhaust heat reclaim, the operation of a gas engine drive can be economical.

Screw-Type Chillers   Screw-type chillers are a type of compressor chiller. Rotary helical screws, as illustrated in Figure 25-9, provide the compression.

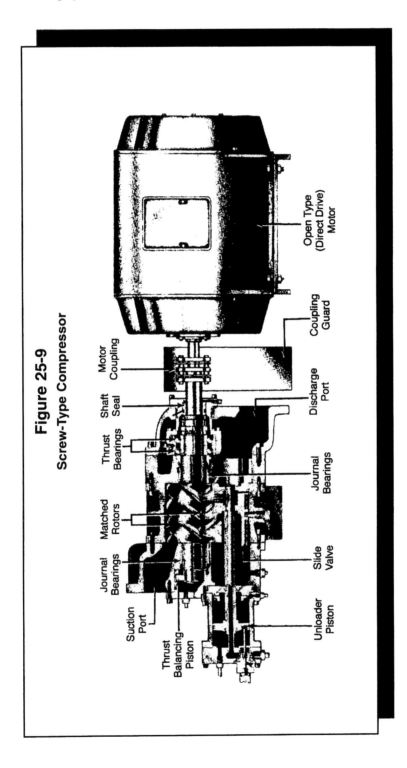

Figure 25-9
Screw-Type Compressor

Capacity control is obtained by varying the distance between the helical screws or the portion of the helical screw that is exposed to the refrigeration circuit. Screw-type chillers can operate with either direct expansion or chilled water systems and may be air or water cooled. They can be exceptionally quiet in operation and have fewer moving parts than do reciprocating or rotary-type compressors. Screw-type machines are available in capacities ranging from 40 to approximately 800 tons. More major manufacturers are offering screw-type chillers that are competitive with reciprocating and centrifugal chillers in the 80- to 150-ton range.

Centrifugal Chillers   The centrifugal unit is also a compression cycle chiller, using the compression effect of the centrifugal force of an impeller rotating at high speed. The tip speed of the impeller determines the compression of the refrigerant gas expelled. The design of the impeller provides the low suction pressure necessary to evacuate the refrigeration chamber and the higher discharge pressure required for the cycle. Centrifugal chillers have a higher initial cost than reciprocating chillers and are available in sizes ranging from 100 to several thousand tons. They generally are water cooled, although air-cooled centrifugal chillers are available.

Centrifugal chillers are designed with single-, dual-, or multiple-stage compressor units. They can be hermetic, semihermetic, or open drive. Centrifugal chillers provide chilled water to an air conditioning system; they are rarely used to provide direct expansion refrigeration. A centrifugal chiller typically has a long life and is easily maintained. The water-cooled condenser can receive water from a cooling tower, a process source, a well, or a heating system, in which the chiller can function as a heat source (heat pump).

Centrifugal chillers are relatively quiet in operation and are available with low energy requirements per ton capacity. A reading of 0.64 kW per ton at full load is not unusual. Capacity control of a centrifugal chiller is achieved by controlling suction inlet vanes or by varying the speed of the impeller.

Drives for centrifugal chillers may consist of electric drives, gas engines, or turbine drives.

Electric Drives. The majority of centrifugal chillers currently in use are driven by electric motors. The electric motors can be either hermetically sealed within the refrigeration cycle or externally (open) mounted. The choice is usually based on personal preference. Some manufacturers do not offer externally mounted motors.

Gas Engines. Gas engines have been used successfully to drive centrifugal compressors. A gas engine is adaptable to a combination system where a standby generator can be placed on the opposite end of the drive shaft of the gas engine. Under this arrangement, the gas engine offers standby power at a relatively nominal cost. The engine cooling system can provide heat for reclaim, including steam for heating or absorption cooling. The gas engine increases maintenance costs over those of a centrifugal chiller with an electric drive.

Turbine Drives. A steam turbine drive on a centrifugal chiller can be quite economical, particularly if the steam is from a waste heat source or if the turbine exhaust is used as the steam supply for low-pressure absorption-type chillers. With an inlet pressure of approximately 200 lbs. and with an exhaust pressure of approximately 12 lbs. to the absorption machine, which then condenses the steam, a steam rate of 9 to 10 lbs. of steam per ton of refrigeration can be achieved. In plants with a heavy investment in existing absorption machines and the availability of high-pressure steam, this can be a retrofit option. Steam turbine installations of this type are called *"piggy-back" operations,* and the combined chilled water system can operate down to 10 to 15 percent of the total capacity. The drawback is that, under light loads with the absorption machines providing the only cooling, the steam rate returns to the typical 19 to 20 lbs. per ton. Steam turbine installations of this type have been provided in capacities of several thousand tons in a single unit.

Absorption Chillers   Absorption chillers may be single-effect or double-effect chillers.

Single-Effect Chillers. The absorption chiller became popular during the period of low energy costs. The coefficient of performance, a measure of the ability to convert energy into cooling, is approximately 1 for a low-pressure-type absorption machine. This compares to a much better coefficient of performance, 3 or better, for an electric-driven refrigeration cycle. At a coefficient of 1, it takes approximately the same amount of energy input to produce the same cooling effect output. With a coefficient of 3, for every Btu energy input, the net cooling effect is 3 Btu.

The low-pressure absorption machines have fallen into disfavor because of the high operating cost but are making a comeback through combining the chillers with energy reclaim cycles such as cogeneration, or where waste heat is available for operation of the chillers. Absorption refrigeration is available in a wide range of capacities, from household refrigerators to machines of 1,800-ton capacity.

Absorption machines have sensitive operation conditions. They use lithium bromide salts in liquid form for the refrigeration cycle; the salts are corrosive in the presence of air. The cycle requires a high vacuum, making it difficult to prevent air leakage into the system.

Water, the refrigerant, is sprayed into the high-vacuum chamber. Under these high-vacuum conditions, water boils at a temperature of approximately 32°F. The evaporation of the water in the chiller chamber cools the tubes containing the circulating chilled water.

The lithium bromide salts form the "pumping" part of the cycle, absorbing the moisture that is evaporated. The moisture-laden salt liquid collects in the bottom of the container and is pumped to a section where heat is introduced to boil off the moisture. The moisture is recondensed into water ready to repeat the cycle. High energy costs have led to the development of two-stage, high-pressure absorption machines that have energy consumption requirements of approximately one-half that of the low-pressure-type unit. They can be competitive against electric drive chillers under certain utility rate schedules. Absorption machines are relatively quiet in operation. They can use gas, steam, or hot water as an energy source and generally require water for condensing. A low-pressure absorption machine is sometimes referred to as a *single-effect unit* because of the nature of its cycle as compared to a high-pressure machine.

Double-Effect Chillers. Double-effect absorption machines were developed to improve operating efficiency. These machines typically have a double-effect cycle, and today's modern design can use high-pressure steam or high-temperature hot water or be direct fired. Owing to modifications of cycle design and improved efficiency, the units can operate at an energy rate of 9 to 10 lbs. of steam per ton or 11 to 12 cu. ft. of gas per ton. These efficiencies reduce the energy requirements by 50 percent when compared with a single-effect absorption type chiller. Double-effect absorption chillers operate in much the same fashion relative to the use of lithium bromide and water for the refrigeration cycle as the single-effect units. Double-effect absorption chillers are air and water cooled and available in sizes from 5 to 1,500 tons. Generally, the chillers are water cooled, but air-cooled units are available in smaller sizes.

Evaporative Coolers   In arid climates it is possible to accomplish acceptable cooling for many operations through the evaporation of water. In a typical evaporative cooler used for a single-zone application, water is sprayed into a chamber with air passing through it, or water flows through an absorbent pad placed in an airstream. As the moisture evaporates, the evaporation process extracts the latent heat of vaporization from the air, thus cooling the air. This process also raises the relative humidity. In a dry climate the resultant humidity is still at an acceptable level for human comfort.

Evaporative coolers have been incorporated into more sophisticated air conditioning systems where mechanical refrigeration and evaporative cooling are combined. The evaporative cooler can be used for the cooling of fresh air or, through a heat transfer process, to supplement the mechanical refrigeration.

Evaporative cooling offers a significantly low initial cost when used as the only source of cooling. When used with mechanical refrigeration, it reduces the capital investment and lowers energy requirements. Whether cooling occurs by water introduced over an evaporative media similar to a filter or by an air spray, evaporative cooling provides a degree of air cleaning. Evaporative cooling presents major maintenance and potential health problems because of the buildup of mineral deposits and mildew within the evaporative cooler.

Well-Water Cooling   The use of well water for cooling has a long history in the United States. Early uses consisted of cooling for movie theaters, department stores, and places of assembly. Many of the early applications were unsatisfactory because the well-water temperature, normally between 55°F and 60°F, was too high to remove moisture from the air in the cooling coils. It is possible to use well water for cooling a building for the major portion of the cooling hours per year in climates with design conditions as high as 95°F dry bulb and 78°F wet bulb, which would be typical for many areas of the United States. When used in combination with mechanical refrigeration, it has been found in certain instances that the mechanical refrigeration was used only 20 to 40 hours per year.

Well-water cooling can be used without adverse impact on the water quality and availability from the source. It has been used successfully where the water contained iron on the order of 15 to 20 parts per million. If the circuit is completely enclosed, the mineral precipitation is usually minor and easily cleaned. It is not unusual for well-water systems to be in use for 20 years, with little evidence of deposits in the coils and pipes. The well-water coils should have removable heads for inspection and cleaning of the tubes.

The water can be returned to the ground uncontaminated and at a slightly elevated temperature, on the order of 5°F or less. Well water cooling can be achieved at one-third to one-fifth the energy requirements of a typical electric installation. In addition to use with a mechanical system, well water can precool the fresh air in public assembly areas or food service operations before being used as condenser water. After use as condenser water, it can be used for lawn irrigation or similar purposes, or as makeup water for another operation. Well water can be used in precooling coils to provide a major portion of the cooling of the return air, with the mechanical refrigeration being brought on for dehumidification or additional cooling.

A well can be used to replace a cooling tower. In many parts of the country, a well-water system is far cheaper than a conventional cooling tower system and avoids the use of chemicals. A well is available at any time of the year, which is not always possible with open cooling tower systems. Wells require little space and have exceptionally low maintenance requirements.

Wells have been used as a heat sink for heat pump installations. In certain applications, the supply and discharge wells can have their functions reversed to take advantage of the thermal conditions available. This also provides automatic flushing of the well screen. The use of well water for air conditioning, when properly designed, has been approved by public health and water resource authorities in a number of states. The same wells have been approved as a source of water for fire protection when the design meets the necessary requirements. The supply well and discharge (recharge) wells must have significant separation; otherwise, short-circuiting will occur, and the water temperature will drastically increase. A minimum separation of 300 ft. is common, with greater distance preferred.

Heat Rejection Systems All mechanical refrigeration systems need a means of heat rejection. This heat rejection generally takes place in a condenser. The condenser transfers the heat from the refrigeration cycle to the heat sink, either water or air. Condensing can be provided by air-cooled refrigerant condensers, cooling towers, evaporative condensers, water from wells or bodies of water, or by atmospheric cooling.

Air-Cooled Condensers. Air-cooled condensers find applications in systems up to 150 tons. Many reasons exist for using air-cooled condensers: low initial cost, low maintenance cost, no chemical treatment required, and ready availability for operation at any time of the year. Air-cooled condensers can be an integral part of the equipment package, as in unitary equipment, or they can be located remote from the compressor. During times of high outdoor temperatures, air-cooled condensers cause the refrigerant cycle to operate at higher pressures, with resultant increases in energy consumption. The remainder of the year, at lower ambient temperatures, the condenser provides sufficient cooling surfaces for more economical operating costs.

Cooling Towers. Most large refrigeration systems use water as the cooling medium. A cooling tower provides the point of heat rejection into the atmosphere by evaporation of the water. Cooling towers offer the advantage of lower energy costs in the refrigeration cycle, the ability to be located quite remotely from the refrigeration equipment, and the ability to closely control the condensing pressure and temperature of the refrigeration circuit.

Cooling towers have the disadvantage of requiring water for makeup, blow-down for removal of solids resulting from the evaporation process, and chemical treatment. Cooling-tower condensing systems require more maintenance than air-cooled condensers because of the water impact on the tower components, the condenser water piping, and the condenser water tubes. The refrigeration condenser tubes require periodic cleaning to maintain system capacity and efficiency. Cooling towers are available in factory-assembled packages in sizes ranging from 10 to 700 tons, and in field-built models 200 tons and above. Towers are available in induced draft, forced draft, single-flow, and double-flow configurations. The choice of the type of tower is based primarily on economics and the application. A cooling-tower system requires pumps to circulate the water from the tower to the refrigerant equipment.

Corrosion and deterioration of the components can severely limit the life of certain types of cooling towers. Towers should be specified to meet the appropriate Cooling Tower Institute and federal standards.

One of the disadvantages of a cooling-tower system is that the water may freeze in cold climates. With appropriate design, cooling-tower systems can be made to operate year-round, using bypasses that direct the warm water from the chiller directly into the cooling tower basin. In climates where freezing can occur, any well-designed condenser water system should include a cooling-tower basin or sump where the water can be stored without danger of freeze-up. Vertical turbine pumps can be used in an installation of this type. Such a system would be readily available for cooling at any sudden onset of warm temperatures in the winter. This is particularly vital in installations such as hospitals or research centers.

Evaporative Condensers. Evaporative condensers can be used on closed refrigeration circuits and for cooling condenser water in heat pump circuits where exposure of the heat pump condenser water to the atmosphere is undesirable. The evaporative condenser circulates water through spray nozzles in a circulating airstream. The net effect is to cool the refrigerant tube bundle in the evaporative condenser. These condensers are easily adaptable to year-round operation by connecting the discharge and the intake openings with a return or bypass duct. Dampers can be installed that modulate the proportions of fresh air and return air to maintain the desired temperature within the condenser section. In cold weather they can be operated dry, without circulation of the water for the spray system.

Well Water. Well water, as previously discussed, is an appropriate source of condenser water. All of the previously listed advantages and auxiliary uses of well water apply. A well water system can be considerably lower in initial cost than a condenser water system and eliminates the problem posed by the unsightly appearance or size and weight of the typical piece of condensing equipment and the need for chemical treatment.

Atmospheric Cooling. Today's modern buildings and the functions carried out therein often create the need for carefully controlled year-round air conditioning. Computer operations are the most common examples of such functions. In northern climates, outside air temperatures during a great portion of the year are sufficiently cold to provide the immediate cooling for such spaces. This is often referred to as *free cooling,* similar in many respects to cooling with outside air on an economizer cycle. The typical computer room operation cannot use outside air for direct cooling because of the stringent moisture control requirements.

Figure 25-10 illustrates a method of obtaining free cooling in cold weather by the use of a cooling tower; a water circulation circuit with or without antifreeze, depending on the weather; and plate-type heat exchangers. Plate-type heat exchangers have the advantages of being efficient and subject to little or no damage that is due to freeze-up, and having the ability to increase capacity at a relatively moderate cost by simply adding more plates. In a typical northern midwest climate, these systems can provide cooling whenever the outside air temperature is 45°F or lower.

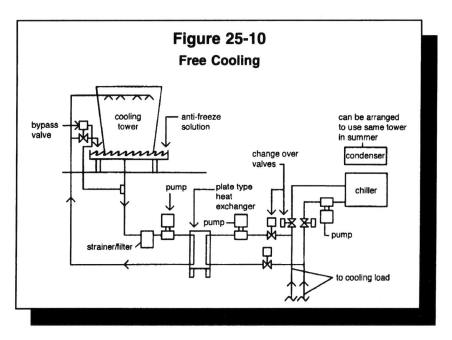

Figure 25-10
Free Cooling

In such applications they can be designed with sufficient capacity to provide up to 100 percent of the cooling requirements as the temperature drops. This can occur for a number of hours of the year in the case of facilities such as computer rooms, which often operate 24 hours a day. The plate-type heat exchanger transfers the cooling from the cooling tower circuit to the chilled water circuit for the air conditioning system. Freeze protection must be provided for the chilled water, as the anti-freeze solution can be below freezing.

Shell-and-tube heat exchangers are subject to severe damage in the event of a freeze-up. Their use is not normally recommended.

In more moderate climates, the antifreeze can be eliminated and the condenser water circulated either directly through the chilled water circuit after appropriate filtration or through a plate-type heat exchanger. In all cases it is recommended that duplex basket-type filters be provided in the circuit ahead of the plate-type cooler where the liquid has been exposed to the atmosphere. This type of filter can be cleaned while the system remains in operation.

Central Plant All buildings can be served by a central chiller plant, discussed in Chapter 45. It is emphasized here that the interface of a building chilled water system with a central plant system must be appropriate if the chilled water distribution system is to be successful. Systems can be designed with all the pumps in the central plant. In most systems, however, each building has its own pump, and herein lies the potential for problems. The pump, in effect, becomes a secondary pump for the chilled water system. It must be capable of circulating the water through the building without imposing excessive back pressures on the central distribution system and should be provided with valves and controls to maintain the required temperature differential flow within the building. A small temperature differential indicates excessive flow that can potentially rob other buildings on the system.

The maintenance of a relatively high temperature difference between the entering and leaving water in a chilled water system design for new buildings can reduce the overall pumping requirements for a campus system or for an individual building. Many central chilled water distribution systems have experienced trouble or have failed because of a lack of appreciation of the interface needs between the building and the distribution system. Therefore, the designer of any new system must respond to the needs of the central distribution system in the proposed design of the new building system.

25.4 MECHANICAL DISTRIBUTION SYSTEMS

Piping System Components

The design of piping systems for heating and cooling of buildings has evolved into five or six major systems, a few minor and little-used systems, and combinations of the various systems depending on the types of building and the HVAC systems installed. Prior to the development of air conditioning, piping systems were used for heat only and were limited to hot water heating or steam heating. Many times the choice of piping systems was dictated strictly by economics. On other occasions the actual physical construction of the building determined whether water or steam was used for heating.

Steam tends to be more restrictive than water systems relative to placement of the pipes in the building, because the pipes must be pitched to drain condensate that accumulates in the supply line and is discharged in the return line from the heating equipment. This water must be continuously drained. Water piping, on the other hand, has the unique ability to pass over or under obstructions and to be placed horizontally without adversely affecting the operation of the system.

There are a wide variety of piping systems applicable to heating and cooling systems within buildings. The final selection generally is a function of whether the system is heating, cooling, or both, and the construction budget for the facility. Excessively elaborate piping systems are not necessarily the best choice, as they offer more opportunities for improper operation, which results in higher energy consumption and the very discomfort that they are purported to avoid. This is particularly true where interconnections exist between heating and cooling systems.

Hot and Chilled Water Piping The choice of the type of piping system can greatly affect the energy required for pumping and the volumes of water to be pumped for system operation. A well-designed system recognizes the impact of the design on the pumping and is able to take advantage of the best features of each system without imposing undue energy requirements for pumping. The type of control valve can also influence the pumping requirements. When three-way valves are used so water bypasses the coil when full flow is not required, constant volume of pumping is required. The pumping system must always pump a quantity of water equal to the sum of the peak loads in the entire system. With a two-way valve that throttles the flow of water, the pump will only pump the amount of water required for the load.

One-Pipe System. The success of a single-pipe system depends on the use of monoflow fittings, specially designed T-fittings to divert the water from the main into the radiator and then back into the same main. As the water passes through the radiator, it is cooled and reintroduced into the main, which lowers the supply water temperature to the next radiator. Each successive radiator must have its size increased because of the lower supply water temperature. One-pipe systems are not recommended for cooling.

Two-Pipe System. A two-pipe system is the most common for heating and cooling. It has a supply main and a return main, with pipe sizes that vary based on the water flow within each portion of the system. Two-pipe systems can be divided into two categories: direct return and reverse return (Figures 25-11 and 25-12). In the direct return system, the first connection to the supply main is the last connection to the return and therefore has the shortest travel. The water can short-circuit through the first heating or cooling device, affecting the flow through the other devices unless the system is carefully balanced, and the balance can be lost if the balancing valve is closed and not opened to the previous setting. The system can also be affected by a change in one of the heating and cooling devices, which would cause a pressure loss differently from the previously installed device, when the system was initially balanced. Direct return systems are used quite often because of the economics of installation but are not recommended if a reverse return arrangement can be installed. A balancing valve installed in such a system should have a permanent notation of its setting if it has to be closed for maintenance.

Reverse return systems result in equal water travel for all the heating or cooling devices. The first connection to the supply main is also the first connection to the return main. The water traveling through any device essentially has an equal length path; therefore, all have an equal path, and the system tends to be self-balancing.

It is not unusual to have combinations of reverse and direct return. In some high-rise buildings, the horizontal main around the perimeter of the building can be installed in reverse return, and the vertical risers to spaces

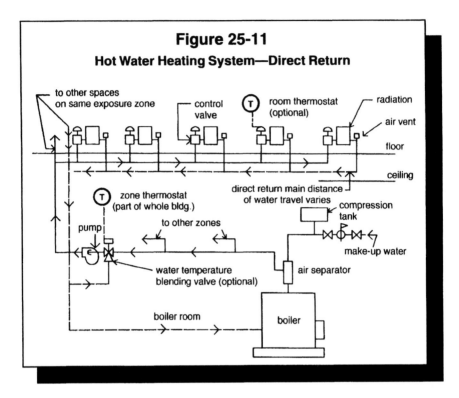

Figure 25-11
Hot Water Heating System—Direct Return

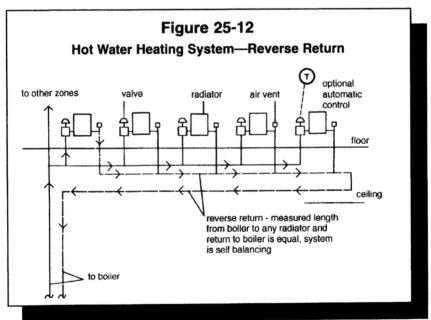

Figure 25-12
Hot Water Heating System—Reverse Return

can be direct return. This is not particularly desirable, but the building construction may not permit a reverse return to be installed economically or practically for the riser portion of the system.

Three-Pipe System. Three-pipe systems were developed in the 1950s, when energy was inexpensive. They have a chilled water supply main, a hot water supply main, and a common return. Both the heating and the cooling devices discharge the water into the common return main, mixing hot and cool water. Through a complicated system of controls, the system directs the flow of water back to one of the supply mains. The system can provide heating and cooling simultaneously; however, the control of water temperature is not accurate, and energy is wasted. This system should not be used.

Four-Pipe System. A four-pipe system is simply the use of two two-pipe systems, one for heating and one for cooling. The need for such a system is based on the need to have cooling and heating available simultaneously. Each heating and cooling device has four pipes connected, a supply and a return for both the hot and chilled water. The equipment can have a heating coil and a cooling coil, or a single coil with an arrangement of valves that permit either cool or hot water to flow to the coil.

One problem with this system is that often valves do not seat properly, causing leakage, and significant amounts of energy can be wasted if the four pipes are connected to a common coil wherein the heating and cooling water become mixed. In four-pipe systems it is advisable to have additional control valves to isolate a heating or cooling zone. Leaking coil valves would not have as much impact, because the zone supply is cut off. A good compromise in many buildings is to have a combination of four-pipe and two-pipe systems.

A four-pipe system is more costly and should be investigated carefully. Some buildings require chilled water year-round because of computer operations or similar functions. This dictates that the chilled water system be separate from the hot water heating system.

Primary/Secondary Systems   Piping systems in themselves generally are operated with a single pumping source. The pumping source may be one or more pumps operating in parallel. These are referred to as *primary pumping systems.* Under certain conditions, it is advantageous to have what is called *primary/secondary pumping systems.* The main advantage of this system can be found in the ability to provide better control over the flow of water and the pressures required in the various sections of the system. The primary system essentially becomes a reservoir of either cold or hot water, circulating the water throughout a loop to which is connected all of the secondary systems. The secondary pumping system can serve a single air handling unit, a group of air handling units, or an entire building. Primary and secondary pumping systems can be used to great advantage in a central chilled water plant. Some professionals, however, would not choose a primary/secondary distribution system because of the imbalance that can be caused by the individual pump pressure added to the secondary system. In an improperly designed or operated primary/secondary system, it is possible for the return water pressure in the primary return main to become higher than the supply water pressure and actually reverse the flow of water through a building and back into the primary supply main, or restrict the water's ability to have adequate flow.

Steam Piping Systems   Steam is rarely used for heating modern buildings. Water systems can be designed to handle both chilled and hot water, and a water piping system is easier to install. Steam systems can be divided into low-pressure and high-pressure systems, which are similar in their operation. Low-pressure systems can be further divided into gravity return, pumped return, and vacuum return systems.

One-Pipe Gravity System. In a one-pipe gravity return system, the supply main also functions as a return. The supply main rises from the boiler to the point of entry into the heating system and from there pitches downward back to the boiler. The radiators are connected by a single pipe to the main. The steam flow is upward in the connection to the radiation; the condensate flows out the bottom of the radiator opening, down into the same pipe, and back into the main. One-pipe systems exist in some older homes that have been acquired by universities and other institutions.

Two-Pipe Gravity System. A two-pipe gravity system is similar to the one-pipe system. It is more flexible in that the return and the supply are separate, so less water hammer noise potential exists in the supply main. The supply main can be run in a different location than the return.

Two-Pipe Pumped Return System. A two-pipe pumped condensate return system is similar to the two-pipe gravity system except that the water is returned to the boiler by a pump. The radiation devices can be below the water level of the boiler or in any location. The condensate is always drained to a condensate receiver connected to a pump. This receiver is placed at the low point in the system, which can be well below the location of the boiler and quite remote. This is the most common type of steam heating system currently in use.

Vacuum Return System. In the vacuum return system, the condensate return has a specially designed pump capable of creating a vacuum on the return line. By varying the pressure in the system, it is possible to have steam at temperatures considerably below the traditional 212°F. The steam pressure can be modulated; this is turn modulates the temperature in much the same manner as the water temperature of a heating system.

A vacuum system requires a tight piping system. If the system has leaks, the pumping becomes excessive and the energy costs increase. These systems are rare today.

Piping Materials   The most widely used materials for piping systems are copper and steel. Steel pipe up to 2 in. in diameter is generally threaded; pipe of larger diameters is welded. Copper pipe is fabricated with soldered systems throughout, except at valves and connections to equipment. Generally, the selection of materials is based on the initial cost of the installation. It is not unusual to find copper used for pipe up to 2 in. in diameter, with steel used for larger pipe. Whenever dissimilar metals such as copper and steel are connected, they require proper isolation connections to avoid electrolytic corrosion between different materials in piping systems or between the piping system and connected equipment.

Plastic piping systems are widely used in the renovation of buildings because of the ease of handling and fabrication. Standards have been established for all types of plastic piping to ensure appropriate strength and quality for the intended use. Plastic systems offer the advantage of relatively low fabrication costs. If hot water is to be involved, selection of material and the support are extremely important. Problems will develop if the plastic pipe is not supported in strict accordance with the manufacturer's recommendations for the operating temperature.

Pumps   Pumps are used to move liquid through the piping systems. Pumps are classified into two types; positive displacement or nonpositive displacement. The positive displacement pump normally is a reciprocating type that creates lift and pressure through the positive displacement of liquid from the piston chamber. The only limits on liquid movement are in the structural integrity of the plumbing system. Another type of positive displacement pump is the rotary type, which uses cams or lobes that rotate together to force liquid movement.

Nonpositive displacement pumps are usually of the centrifugal type and are the most commonly used type of pump in HVAC applications. In these types of pumps the liquid is moved through the centrifugal force that is created by a rotating spiral-shaped element called the *impeller*. These types of pumps do not develop unlimited pressure and thus must be properly sized to ensure that proper pressures can be obtained at the desired flow rates. This type of pump is simple in design, has low initial cost, is easy to maintain, and is fairly quiet and vibration free. Centrifugal pumps are normally driven by a constant-speed electric motor and may be in-line, close-coupled, or base-mounted.

In-line pumps are normally used with fractional horsepower motors up to 1 hp, and close-coupled pumps are normally used with $1/4$ to 40 hp motors. Base-mounted pumps are also used with smaller motors, up to the largest needed.

Valves   Valves are needed for isolation of a system for maintenance or for emergencies. They must be placed at every piece of equipment. Selection of valve types is also important. Gate valves can be used for isolation because their initial cost is low and they perform acceptably. Gate valves should not be used for throttling because of poor control and rapid deterioration. Globe valves should be used for throttling but not for isolation because of their high cost.

Butterfly valves are an acceptable alternative to gate valves and in certain circumstances can be used for modulating control. Ball valves in smaller sizes are gaining wide acceptance for isolation and balancing. Plug valves have been used for shutoff and balancing for many years. All hydronic systems require balancing for proper operation. Balancing valves should be designed for the intended purpose and should be capable of permanent marking or stopping at the point of balance. Valves used for balancing flow should have a mechanical stop-open position so that the valve can be closed for maintenance work and reopened to the balanced condition.

Valves also have opening and closing characteristics that affect the operation of the system. Some valves allow a lot of flow in the first part of the opening movement, with lesser flow in the latter part of the opening movement. Other valves have the opposite characteristic, whereas still others have more "linear" characteristics. The valve characteristic must be properly matched to the device it is trying to control.

Finally, the pressure drop through the valve, compared to the system it is controlling, must be properly designed. Over- or undersized valves will cause operating problems and lack of control of the system.

Duct System Components

Duct systems are simple in appearance yet can have a significant impact on the initial cost and the operating costs of a facility. High-quality ductwork generally costs more but is a good investment. Poor-quality ductwork generally results in high operating costs because of leakage. Ductwork systems are divided into high-pressure/high-velocity and low-pressure/low-velocity systems.

The generally accepted standards for ductwork construction have been established by the Sheet Metal and Air Conditioning Contractors National Association (SMACNA) and ASHRAE. Currently, the SMACNA standards have more than one classification based on the system air pressure. Duct construction standards do not specifically address the subject of air distribution design. Instead, they set standards for the quality of construction of the duct system relative to structural integrity and potential leakage. The lower the air velocity and pressure, the more economically the air distribution system will operate. A low-velocity system requires more space for installation and possibly higher initial cost but, because of lower operating costs, is the most economical on a life cycle cost basis. The traditional high-velocity system is generally unacceptable by current energy conservation standards.

High-Velocity/High-Pressure Systems   The velocity and pressures involved in a high-velocity/high-pressure system can be as high as 5,000 ft. per minute and 8 to 10 in. of water pressure, compared to 1,000 to 2,000 ft. per minute and 3 to 4 in. for a low-velocity/low-pressure system. The horsepower requirements for a high-pressure system are approximately 300 percent greater than those for the low-velocity system. The duct system requires relatively high-quality construction to eliminate air leakage and to limit noise levels. A majority of the ductwork is spiral, machine fabricated, and essentially airtight. Fittings are of relatively high quality and are welded or stamped and formed. Connection between duct and fittings involves the use of high-quality tapes and sealants. A high-velocity ductwork system can reduce the cross-sectional area of the required duct by as much as 60 percent and thus can fit into smaller areas. Some savings may result by reducing building size, but this is generally offset by the higher operating costs.

Low-Velocity/Low-Pressure Systems   These systems generally consist of rectangular ductwork. Round ductwork is increasing in popularity because of its low cost, airtightness, ease of installation, and efficiency. The structural requirements for a low-velocity/low-pressure system are far less than those for a high-velocity system, and it tends to have far greater leakage if improperly constructed. All low-velocity ductwork specifications should be written specifically with regard to leakage. Leakage as high as 25 percent is not uncommon in poorly built systems, but low-pressure proprietary duct construction systems are available that limit leakage to as little as 1 percent. Low-velocity systems generate less noise than high-velocity systems. However, improper fittings, construction, and design can result in unacceptable noise levels.

Duct Material   Currently the most widely used material for ductwork is galvanized sheet metal. The gauges and construction methods for galvanized sheet metal have been well established by ASHRAE and SMACNA.

Fiberglass Ductwork. Proprietary systems involving fiberglass boards can be formed into acceptable ductwork systems. These systems should be limited to areas where the duct is not exposed to damage from impact. The system must be carefully designed relative to the air pressures involved.

Some of the early fiberglass duct systems experienced failure of the materials and adhesives used to form connections between duct sections and fittings. This problem has been recognized by the industry and apparently solved, although the history of success of the materials used in a fiberglass system should be checked. A fiberglass system offers relatively moderate initial cost, low fan horsepower requirements, and quiet operation.

Flexible Ductwork. Flexible ductwork, insulated or noninsulated, has been available for many years. Although it offers lower initial cost, its main advantage is that it can be fitted into tight spaces. The potential for increasing fan horsepower as a result of high resistance or improper installation must be recognized. A sharp turn in a flexible duct can result in excessive pressure loss and reduced flow to the space, as well as excessive noise. Flexible ducts should be limited in length to no more than 4 ft., and should be installed in a relatively straight route.

Flexible Connectors. Flexible ductwork must meet the requirements of NFPA standards for ductwork. A flexible connector does not meet the ductwork standard but meets the NFPA requirements for the final connection between an air distribution system and an air outlet device or piece of equipment. The length of these connectors is strictly limited by code. An attempt to use this material for ductwork can result in rejection of the system and expensive changes, in addition to reducing the life safety of the installation.

Lined Ductwork. Ductwork has been lined for many years to reduce noise levels and to provide insulation. Generally, lined ductwork is cheaper than ductwork with insulation applied to the outer surface. Lined ductwork imposes a modest fan horsepower penalty. This can be overcome by increasing the duct size, although this increases the initial cost. Lined ductwork is particularly applicable in mechanical spaces and other areas where exposed insulation would be subject to damage. It cannot be used in facilities or operations where the liner fibers can be entrained in the air delivered to the space, such as in critical medical facilities.

Special Ductwork. Many systems require special ductwork because of the nature of the material conveyed. This can include moisture; corrosive fumes; and air laden with grease, dirt, and dust. Corrosion-resistant ductwork is used in removal of moisture or chemical fumes from laboratory hoods and maintenance cleaning operations. Stainless steel solves many problems, but it can fail rapidly in certain atmospheres. Polyvinyl chloride and similar materials are used for laboratories. Aluminum can be used in systems where moisture is removed and chlorine content is not excessive. Any system involving the transportation of fumes and dirty or corrosive atmospheres should have the material specifications carefully verified against codes and industry standards. Food service exhaust ducts must comply with NFPA recommendations and local codes.

Return Air Plenums. Return air ductwork is frequently eliminated and the ceiling space used for return air. Life safety codes specify the required materials of construction for the plenum. Ceiling return air plenums and return air shafts operate successfully and greatly reduce the cost of ductwork. A return air plenum that is not appropriately constructed can greatly increase energy costs because of infiltration of outside air or air from noncooled or heated spaces. It must be recognized that the plenum is under a slight negative pressure and therefore must meet the construction standard for airtightness of a ductwork system. All too often the exterior fascia of a building serves as the outer boundary of the plenum and is poorly constructed for such a service. These designs should be carefully checked and the construction monitored.

Fans Fans are used to move air through ductwork systems. They work similarly to centrifugal pumps in that a rotating element is used to impart energy to the air to create air movement and pressure. Fans and compressors are really just the same device, except that fans usually operate at lower pressures. For example, even fans that operate in high pressure/high velocity systems rarely operate above 10 to 12 in. water gauge (wg). The following industry standards have been established for fans based on the pressure they produce:

- Class I: Maximum total pressure of 3 ³/₄ in. wg.
- Class II: Maximum total pressure of 6 ³/₄ in. wg.
- Class III: Maximum total pressure of 12¹/₄ in. wg.

Fans can be divided into two types: centrifugal or radial flow, and axial flow. Centrifugal fans transmit their energy to the air by rotating an impeller that moves the air in a radial fashion from the axial while axial flow fans move air parallel to the axial of the fan.

Centrifugal fans can move air over a wide range of volumes and pressures, and certain impeller designs have been created to accomplish different tasks. Impellers may be of the straight, forward curve, backward curve, or airfoil-type design. Straight-blade designs tend to be used where dense material is to be moved (e.g., in dust collection systems). Forward curve fans tend to be used in small compact locations such as in furnaces. Backward curve and airfoil-type fans are used in larger HVAC installations; both have similar characteristics, but the airfoil fan is a little more expensive.

Propeller fans are a type of axial flow fan and are typically used in a through-the-wall type application to move a lot of air at a very low pressure. Most often these types of fans are not connected to ductwork, as they have poor pressure capability.

Tube axial fans move the air parallel to the axis of the fan through a tube-type housing. These are commonly used on large air volume applications requiring modest pressure. They are fairly inexpensive and can often be mounted right in the round ductwork system. They typically are used in applications where pressures do not exceed 2 ¹/₂ in. wg.

Vane axial fans are like tube axial fans except that vanes are added to help direct the flow of the air inside the fan. This increases the efficiency of the fan and allows it to operate at much higher pressures. These fans typically can be used in applications where pressures can be as high as 6 in. wg.

Dampers Dampers are to ductwork what valves are to piping systems. There are essentially two types of dampers: parallel blade dampers and opposed blade dampers. In parallel blade dampers, the blades of the damper are like a venetian blind and open and close in parallel. The parallel blade damper is best used in shutoff-type situations where the opening is to be fully open or fully closed. Opposed blade dampers are similar in appearance to parallel blade dampers when fully open, but during closing, every other blade turns in the opposite direction, so that the blades close by pinching off the air between opposing sets of blades. The flow characteristics of opposed blade dampers are more suitable for modulating-type situations and should be used where modulating control is needed.

Another type of damper is the fire-rated damper. These dampers are installed where ductwork penetrates walls and floors and are designed to retain the fire rating of the wall or floor. These types of dampers are typically spring loaded, with a fusible link that will close the damper if there is a fire. In some cases fire dampers are connected to the fire alarm system and are activated based on specified fire alarm system criteria.

Other Components   There are many other components that are used in ductwork systems to enhance the operating characteristics of the system. Turning vanes, sound attenuators, transition sections, bell mouth entrances, and transition takeoffs are examples of these components. Every good ductwork design will incorporate these items where appropriate to improve energy conservation, sound, and balancing capability.

25.5   HEATING, VENTILATION, AND AIR CONDITIONING SYSTEMS

The following discussion covers the basic types of air conditioning systems used in buildings. The entire mechanical system for a building may include several of the systems outlined below, creating an infinite variety of combinations within a building. These variations are often dictated by the differences in requirements for temperature and humidity control for the different functions housed.

All-Air Systems

All-air systems are those having the heating and cooling equipment, including coils, fans, and filters, located in a central point such as a mechanical room, with the conditioned air transported to the spaces by a ductwork system. The heating and cooling media required are connected to the central air unit and are not distributed throughout the spaces. Some of the advantages of all-air systems are as follows:

- Centralized location of equipment consolidates maintenance and operations
- The ability to cool with outside air by incorporating a fresh air or economizer cycle provides free cooling to all spaces during mild weather and to interior spaces in the winter
- A wide choice of zones
- A convenient means of humidity control

- Return air or exhaust fans are often incorporated into air systems to provide improved control of air circulation and building pressurization. On large systems, return air fans are a distinct advantage in preventing doors being blown open or being hard to open. They can also help control "stack effect" in high-rise buildings.

Single-Zone Systems   The single-zone system is the most fundamental type of air conditioning system. It operates successfully only if all the spaces included in the zone have similar exposures to exterior weather conditions and similar space occupancies and operations. A single-zone system contains all the elements necessary to provide the environmental conditions for the space, including cooling and heating coils, filters, fans, and controls. Figure 25-13 shows the basic elements of a single-zone system. The fan draws air through the coils, and therefore, the system is called a *"drawthrough" system.*

A building may have multiple single-zone systems. One of the advantages of this type of arrangement is that if one system fails, the other zones are unaffected. A single-zone system also has the advantage of being

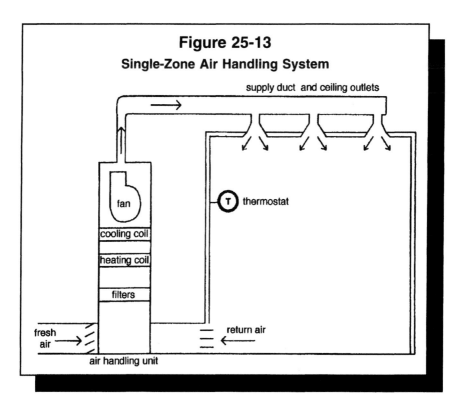

Figure 25-13
Single-Zone Air Handling System

supply duct and ceiling outlets

fan

cooling coil

heating coil

filters

thermostat

return air

fresh air

air handling unit

designed specifically for the zone served, without compromising environmental conditions in other areas of the building. The systems are usually quite simple and easily maintained. They are generally limited to relatively small areas, except in cases of unusually large interior zones. This limitation can necessitate an excessive number of systems in a large building, thereby increasing costs, maintenance, and the requirements for mechanical space. The cooling medium can be chilled water or direct expansion refrigeration from a compressor located within the unit or exterior to it. The compressor/refrigeration cycle can either be water cooled or air cooled. The heating media can be steam, hot water, recovered heat, gas, oil, or electric. The choice of cooling media results in various degrees of operation and maintenance difficulties. A single-zone unit can be arranged to cool with outside air.

Multizone Systems In the air conditioning industry, the phrase *multizone system* refers to an air handling unit that is specifically designed to provide multiple areas throughout a building with individual space temperature control simultaneously. The basic unit has the usual air conditioning components found in a single-zone system, including the heating and cooling coils, filters, fresh and return air dampers, and controls. Figure 25-14 illustrates a typical multizone unit.

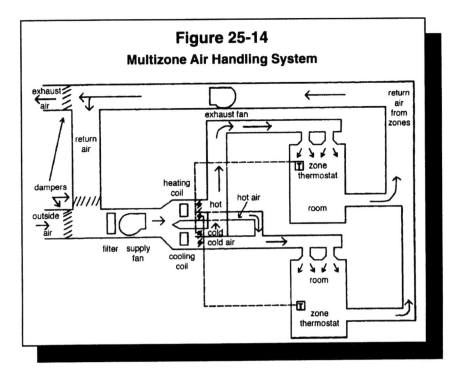

Figure 25-14

Multizone Air Handling System

The multizone unit differs from the single-zone unit in that it is a blow-through unit, meaning that it blows rather than draws the air through the coils. The reason for this arrangement is the way the unit operates. The unit has a cooling coil and a heating coil, stacked one on top of the other. The air blown through the coil then can enter what is referred to as a *cold deck* if it goes through the cooling coil or a *hot deck* if it goes through the heating coil. The "deck" is actually a plenum or discharge space downstream of the coils. This can be thought of as a reservoir of hot or cold air. The hot or cold air can be mixed through a set of dampers into a single duct that is extended to each zone.

A thermostat in the zone controls the hot and cold deck dampers to mix the air to the right temperature to meet the space needs, whether heating or cooling. In the continuous cooling season, no heat is applied to the heating coil, which becomes a bypass section. Room air that has been returned to the unit is thus passed around the cooling coil and mixed with the cold air that has passed through the cooling coil to maintain the required delivery temperature. At full-load conditions in a zone, the bypass or hot deck damper would be closed, with the cooling damper fully open. As the load drops off because of a change in solar conditions or space activities, the cooling damper modulates toward the closed position, and the bypass damper opens to raise the delivery temperature to avoid overcooling the space. In the winter the operation is reversed. A multizone system is also capable of cooling with outside air during mild seasons of the year.

A multizone unit has the advantage of being able to serve many zones at a relatively low initial cost. A disadvantage is that the zone ducts are fixed, and if spaces are remodeled or loads are increased within the zone, major ductwork changes are required.

A multizone unit has certain inherent inefficiencies, particularly in the heating season. Because the unit is serving many zones, some of which may require year-round cooling, there must always be a supply of cold air. The cooling air temperature source is achieved by mixing cold outside air with warmer return air; all the air is then blown through the coils at the colder temperature required for winter cooling. Typically this would be 55°F. Although this air from the cold deck is able to cool the rooms needing it, the air entering the heating coil is below the temperature of the rooms needing heating. Heating the air to match the room temperature wastes energy unless the heating coil itself is using waste heat from possibly another cycle in the building, such as the heat pump. One method of minimizing this disadvantage is the use of controls to detect exactly how cool the air needs to be at any given time and thereby keep the mixed air temperature as warm as possible.

In summary, a multizone unit is an air handling system that can mix warm and cold air to appropriate simultaneous temperatures for different zones and deliver it through separate ducts to maintain required temperatures.

The unit is traditionally a low-velocity, low-pressure system requiring a moderate amount of horsepower for the fan.

Reheat Systems   A reheat system is a variation of an all-air system. A heating coil is installed directly in the duct serving each zone. The central fan system supplies air to all of the zones at a constant temperature, adequate for cooling any space in the zone. Those spaces not requiring the low-temperature air have a space thermostat that will actuate the heating coil to raise the air temperature to the point required to prevent overcooling. The heating coil can be steam, hot water, recovered heat, or electric. In addition to controlling space temperature, reheat systems are used where close humidity control of the space is required. A space humidity controller cools the air to the point where condensation occurs and the air is sufficiently dried. The air is then reheated to the appropriate delivery temperature. Sophisticated reheat systems for humidity control are generally one-zone systems serving such areas as computer rooms.

Reheat has the advantage of providing excellent space temperature and humidity control. The major disadvantage is its relatively high energy costs. All of the air must be cooled to satisfy the warmest space within the building, while the remainder of the air must be reheated to an acceptable delivery temperature. Reheat systems generally require careful maintenance and calibration of controls and monitoring to avoid even greater energy consumption. Installation of the coils and controls above ceilings can lead to problems where space is at a premium. Figure 25-15 shows a typical reheat system.

Operating economies can be achieved by designing a reheat system with the air supply zoned into spaces with similar cooling loads and functional uses, thereby minimizing the amount of reheat. A reheat system generally is undesirable except when the heating is with recovered energy. Most energy codes limit the amount of reheat that can be applied to a building, except for requirements such as computer rooms, hospital operating rooms, and similar applications. The reheat system is relatively inflexible, as coils, pipes, and ducts all have to be modified for building modifications.

Dual-Duct Systems   Dual-duct (sometimes called double-duct) air conditioning systems found great favor in the period between World War II and 1973, when a change in design philosophy resulted from the oil embargo. A dual-duct, constant-volume system is flexible in accommodating building modifications and easily achieves balanced airflow and maintains desired temperatures in the individual zones. Historically, dual-duct systems have been high velocity, high pressure, requiring unusually high fan horsepower, contributing to their inefficiency. In recent years, dual-duct systems have been medium pressure or low pressure, thereby greatly reducing the horsepower requirements.

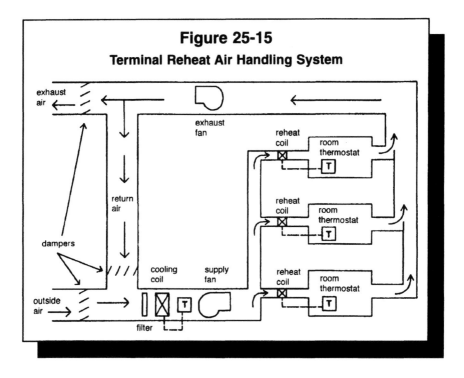

Figure 25-15

Terminal Reheat Air Handling System

A dual-duct system is basically arranged in the same manner as a multizone system. The major difference is that instead of single ducts being extended from the air handling unit to individual zones, the hot deck and the cold deck are each connected to separate ducts, and the two ducts are extended to each zone. The temperatures in the individual ducts are controlled similarly to those in a multizone unit. In some configurations there can be separate fans for the hot deck and the cold deck. The dual-duct system has the same inherent inefficiency as the multizone system in the mixed air cycle during the winter, relative to the heating of the mixed air from approximately 55°F back up to room temperature.

Figure 25-16 shows a typical dual-duct system. Mixing of the air for the space takes place in mixing boxes at the entrance to the zones. The mixing box has dampers controlled by a room thermostat. An automatic device within the dampering system maintains a constant volume of supply air. The box is self-balancing, provided the duct system is capable of delivering air to the box at the required inlet pressure. The mixed air is discharged from the box and enters a low-velocity duct system with diffusers to introduce air into the zone spaces. Large zones may require several mixing boxes.

Figure 25-17 shows a modified dual-duct air handling unit wherein the return air delivered to the hot deck through the coil will not be mixed

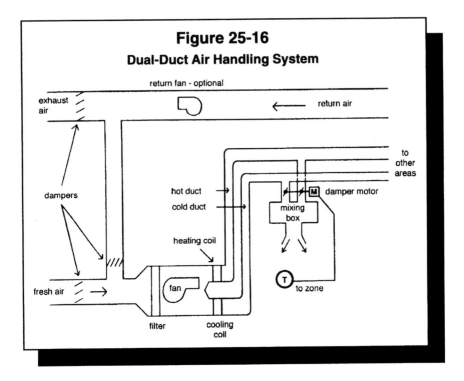

Figure 25-16

Dual-Duct Air Handling System

with outside air, thus eliminating the inefficiency of the standard dual-duct or multizone systems. This arrangement can be very energy efficient in the heating season as it has the capability of transferring heated air from interior spaces to other portions of the building.

A dual-duct system has the disadvantage of requiring an extra duct that claims space above the ceiling and, as evidenced from the drawing, requires extra depth for hot and cold ducts to pass over the main trunk ducts. This can be minimized with careful planning by using a structural system's voids.

A dual-duct system cooling source can consist of chilled water from a remote chiller or direct expansion from a compressor located near the unit. Heating can be furnished by steam, hot water, reclaimed energy, or electricity.

To provide greater economies of operation, dual-duct systems can be modified to provide variable volume to the spaces served, rather than constant volume. This method of operation allows the fan to operate only as required for the sum of the individual loads, whereas a constant-volume system always delivers the total air that is required by each room at its peak load condition. The sum of the peak loads for each space can be 20 to 30 percent higher than the peak load on the system, thereby wasting fan horsepower by circulating more air than needed. Eliminating the hot duct distribution system to the interior zone boxes can lower initial

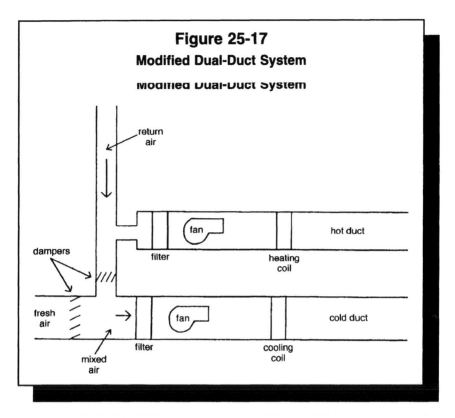

Figure 25-17

Modified Dual-Duct System

Moamea Dual-Duct System

and operating costs. The box then is simply a variable volume box. A dual-duct system is relatively high in initial cost compared with a single-zone or multizone system. However, it is a flexible system with regard to modifications and generally provides good space conditions.

Variable Air Volume Systems   There are two ways to control the temperature of a space. One is to deliver a constant volume of air and vary the air temperature. The second is to deliver air at a constant temperature and vary the quantity. This latter method is called a *variable air volume system.* Variable air volume technology was available for many years prior to the oil embargo of 1973, but was little used because of its more critical design requirements for successful operation and the availability of cheap energy. As a result, the technology was not highly developed. With energy cost increases, the system has become more popular, and the technology has improved significantly.

The basic variable air volume system is a single-duct, cooling-only system. Heat is supplied to the space at the zone level. This can consist of fin tube radiation on the walls or a heating coil in the air duct supplying the zone. Figure 25-18 illustrates the basic variable volume system. The

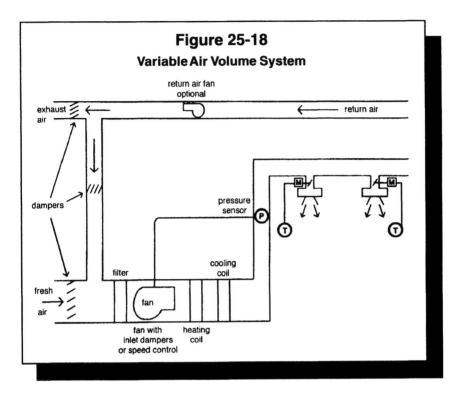

Figure 25-18

Variable Air Volume System

air handling unit is a draw-through type with a fan, cooling coil, optional heating coil, filters, fresh air, and return air dampers. The unit is capable of cooling with outside air during the winter and mild seasons of the year.

The terminal box serving a zone is similar to the box on a double-duct system; however, it does not have a hot duct or a constant-volume control device. A damper controlled by a thermostat varies the amount of cold air entering the box for distribution to the space. The terminal unit can be one of many types, ranging from a simple damper in a duct to a sophisticated box with sound attenuation lining and adjustable minimum and maximum volumes. It can also be an induction type if the air supply is at a sufficiently high pressure. An induction type induces air from the ceiling plenum to mix with the cold air, providing a larger air supply to the space while varying the amount of air that is delivered by the fan system.

A major objection to the variable volume system is that at low loads the air quantities are often greatly reduced, causing a lack of air circulation. In addition, as discussed earlier, the reduced airflow dictates a critical selection of air outlets. Under low airflow, typical air outlets simply "dump" air into the room. Variable air volume systems also tend to have acoustic problems because of the variation in air volume.

Fan-powered terminals can overcome variable air volume flow within the space. This unit is similar to the variable air volume box, as it also has a thermostatically controlled inlet damper that varies the amount of cold air coming into the box from the central system in proportion to the room requirements. However, there is a fan within the box that mixes return air, either directly from the room or from the ceiling plenum, and circulates air at a constant rate to the space.

Control techniques are also available to maintain minimum airflow without overcooling the space. The controls can vary the supply air temperature, requiring more air for cooling as the temperature of the supply air is raised or with the installation of a reheat coil. Zoning of the supply air ducts can also offer greater energy savings by separating zones on the basis of the weather's impact.

In cold climates, the air handling unit should have a heating coil and reverse acting space thermostats to provide for rapid warm-up after a period of setback to cooler temperatures.

Although a variable volume system is fundamentally a cooling-only system, heating coils can be placed at the terminals to form a reheat system; this is, in effect, a variation of a multizone system.

Variable air volume systems can be designed for a set minimum airflow or to terminate all airflow to a room on a call for no cooling. The variable airflow characteristic of the system requires careful design to maintain proper ventilation as required by code and the needs of the occupants.

A variable volume system is flexible relative to future space changes. Generally, its costs are similar to or less than those of a double-duct system, depending on the method of providing heat.

Heat Recovery

The art of heat recovery is not new and has been practiced in major power plants and industrial operations for many years. In subarctic climates, the practice has even involved residential ventilation systems for many years. Heat recovery is not 100 percent efficient, although some systems can achieve energy recovery levels as high as 80 percent. For many building systems designed prior to the energy crisis, heat recovery may be the only means for reducing high energy costs. This is particularly true in buildings that require 100 percent ventilation air, such as medical and research facilities.

Air-to-Air Heat Recovery Figure 25-19 illustrates a typical air-to-air heat recovery system. These systems can achieve efficiencies up to 65 percent. They require that the exhaust and supply airstreams be adjacent to each other at the point of heat recovery. These systems have limited use in highly contaminated airstreams because of the danger of cross-contamination, although there are good prefiltering, purge units available, and seals eliminate this

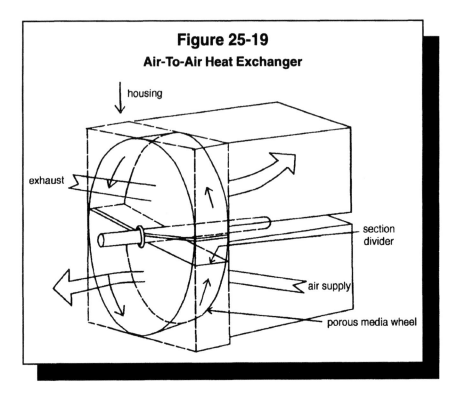

Figure 25-19

Air-To-Air Heat Exchanger

housing

exhaust

section divider

air supply

porous media wheel

danger in most cases. Effectively designed and maintained filtration systems are required for optimum performance. Air-to-air systems are relatively easy to clean. Properly designed and installed, they can be very successful.

Heat Pipe/Coil Equipment   Figure 25-20 illustrates a heat pipe type of air-to-air recovery. The design is based on a refrigeration cycle. The tubes in the finned coil contain a refrigerant and a secondary internal tube with a wick. The opposite ends of the coil are exposed to makeup air and exhaust air. The high-temperature air causes the refrigerant to vaporize and flow to the cooler end, where it condenses and gives up the heat. The liquid refrigerant migrates back to the warm end of the coil via the wick, where it exits the tube, and the process is repeated. Tilting of the coil can control its capacity. Heat recovery devices of this sort can achieve recovery efficiencies as high as 80 percent. These are generally easily maintained systems. They also require that the makeup air and exhaust airstream be adjacent.

Runaround Systems   Figure 25-21 illustrates a runaround system for heat recovery. These systems are applicable where the makeup air and the exhaust airstreams are not adjacent. They are particularly applicable in retrofit

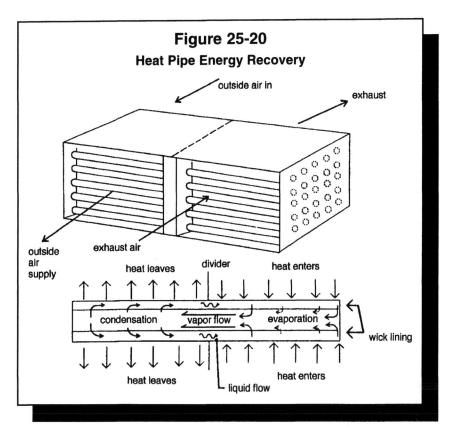

Figure 25-20
Heat Pipe Energy Recovery

applications in existing buildings. The system simply consists of two coils, one in the exhaust airstream and the other in the makeup airstream. A piping system, complete with pumps, circulates a heat transfer medium between the two coils. Control can be achieved by varying the medium flow or simply cycling the pumps. In cold climates, antifreeze must be added to the system, although it reduces the efficiency and increases pumping costs. Such systems can achieve efficiencies as high as 45 percent. Runaround systems have been applied to energy transfer within buildings involving other than exhaust and makeup air systems.

A simple method of energy recovery involving a typical water-type coil used for air conditioning can be applied to a laundry or similar operation requiring a large amount of domestic hot water. The coil is placed in the hot exhaust airstream from the dryers or irons.

Plate-Type Heat Exchangers In some operations it is necessary to transfer heat from one water or air system to the other while keeping the streams separate. Plate-type heat exchangers meet this requirement. They are efficient, easily installed, and easily maintained. However, large sizes

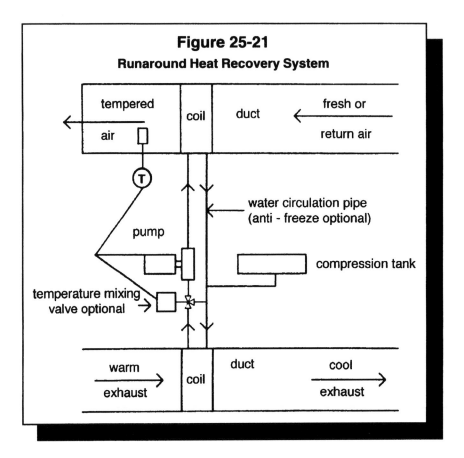

Figure 25-21
Runaround Heat Recovery System

are required to avoid high pressure drops. These systems can achieve heat transfer efficiencies as high as 70 percent.

Laboratory Fume Hoods

This section addresses the selection, application, and maintenance of laboratory fume hoods. Fume hoods should be considered a specialized part of the total ventilation system designed to protect laboratory users, people who work on or near laboratories, and the environment.

Chemical and biological laboratories are the main users of fume hoods. Fume hoods are used to exhaust toxic gases, contaminants, and noxious odors to the outdoors and to avoid aspiration of such substances into the building's HVAC system.

The Scientific Apparatus Makers Association defines a laboratory fume hood as a "ventilated enclosed work space intended to capture, contain and exhaust fumes, vapors and particulate matter generated inside the enclosure. It consists basically of side, back, and top enclosure panels, a work surface or counter-top, an access opening called the face, a sash and an exhaust plenum equipped with a baffle system for the regulation of air flow distribution."[3] The ASHRAE *Handbook of HVAC Applications*[4] is a good source for fume hood ventilation design.

Fume Hood Location   The volume of air exhausted by fume hoods affects building HVAC systems by increasing heating and cooling loads, changing air balances, and introducing noise and drafts. The design of the HVAC system and the location of the fume hood within the laboratory are important to the success of an installation. Stable and consistent airflow into the face of a fume hood may not be possible if the fume hood is too near a doorway or window, in a high traffic area, or near HVAC grilles or diffusers. Actual airflow into the face may be observed by using a smoke bomb; however, any high-efficiency particulate air filters in the exhaust system should first be removed to avoid plugging them with smoke particles.

Face Velocity   The velocity of air entering the fume hood face is the key factor in its safe performance. The volume of air is adjusted to maintain the required velocity. The recommended face velocity may range from about 100 to 150 linear feet per minute (fpm). Scientific Apparatus Makers Association Standard LF 10-1980 lists hoods from type A to type C and suggests 125 to 150 fpm for certain critical operations.[3] Caplan and Knutson, in ASHRAE research project 70, found 60 to 100 fpm to be adequate.[5] Some OSHA regulations require 150 fpm. The old philosophy "more is better" does not apply to fume hood face velocities. On the contrary, there are data showing that face velocities around 100 fpm may be more effective in providing safety than higher velocities. There is a growing consensus among safety experts that 100 to 125 fpm is the safest range for even the most critical materials, such as carcinogens, acids, and radioisotopes.

Velocity readings should be taken with the sash in the normal operating position or at maximum opening.

Measurement of the feet per minute can be accomplished by dividing the face opening into a grid of approximately 12-in. squares and measuring the air velocities at the center of each square using an accurate anemometer (Figure 25-22). The readings should not vary by more than 10 percent within the desired range. A minimum six-point traverse is necessary for a 4-ft. face opening, eight for a 6-ft. hood.

Fume hoods should be tested on a predetermined schedule according to Scientific Apparatus Makers Association Standard LF-10-1980. Face velocities should be checked at least once a year and after blower or duct changes, repairs, or HVAC modifications. For quick reference, the date and feet per minute can be posted on the hood. New installations should be

Figure 25-22
Measuring Fume Hood Air Velocity

Walk-In Hood

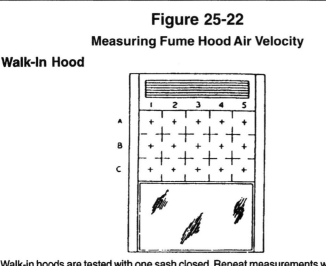

Walk-in hoods are tested with one sash closed. Repeat measurements with bottom open, top closed.

Bench Mounted Hood with Horizontal Sliding Sashes

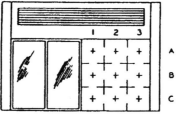

Open face to maximum. Repeat measurements with open area at left, right, and center. Take readings at the center of each sq. ft. of area.

Bench Mounted Hood

Combination Hood (Tested with lower door closed)

Auxiliary Air Hood (Tested with auxiliary air off)

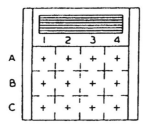

Divide hood face into grid. Readings should be taken at the center of each sq. ft. of area.

tested before use; equally important, and most often overlooked, is testing after remodel work.

Fume Hood Construction A large variety of hoods are available from various manufacturers. The materials used and the method of construction should be carefully tailored to the intended use. The materials with which the fume hood is made should be resistant to the chemicals to be used, lined with impervious surfaces to prevent contamination, and assembled to form a smooth continuity that encourages airflow and prevents accumulation of chemical residuals in cracks and corners that will be difficult to decontaminate.

There are many optional features available. Hoods may be provided with gas, air, steam, electrical, and other utility outlets. Sinks, drains, cup sinks, and even steam tables are commonly included. Explosion-proof or external lights are typically provided. Special accessories might include air bypass openings (Figure 25-23), airflow alarms, and special linings and exhaust filters.

There are several options available in fume hood sash design. The sash is an operable glass for shielding and observation purposes; it may be vertically or horizontally movable. The sash should be positioned to provide the user with maximum protection, but can be moved to enlarge the opening while setting up apparatus in the hood. It can be fitted with an alarm to prevent improper use.

Baffles in the rear plane of the hood arrange the airflow for removal of vapors and particulates by the exhaust system. The three-slot baffle is standard with top, middle, and bottom locations (Figure 25-24). ASHRAE procedure 110-1985 indicates that there is an optimum slot sizing. The bottom and center slots are usually fixed in size. The top baffle is adjustable from closed to a 2-in. opening. The center baffle has a slot opening of 1 to 2 in. The bottom baffle has a slot opening of 2 to 3 in. An airfoil is used to direct air across the bottom of the fume hood to sweep heavier fumes into the exhaust and prevent air turbulence at the working surface (Figure 25-25).

Exhaust System The fume hood exhaust system may consist of a single exhaust blower for each hood or several hoods together on one exhaust blower. Materials used for the duct and blower are determined by the use of the fume hood. Blowers and ducts are available in galvanized or stainless steel, coated with special materials such as epoxy or plastic, or they may be made completely of polyvinyl chloride. Resin or plastic ducts may require a metal sheath to meet fire code requirements. Transite ducts are found in many installations.

For the safety of personnel who might be on the roof, and to ensure that exhausted air is dispersed into the atmosphere, the blower exhaust should be vertical and should extend above the roof or any windbreak, wall, or parapet. To avoid reentry, the exhaust must be well away from and downwind of air

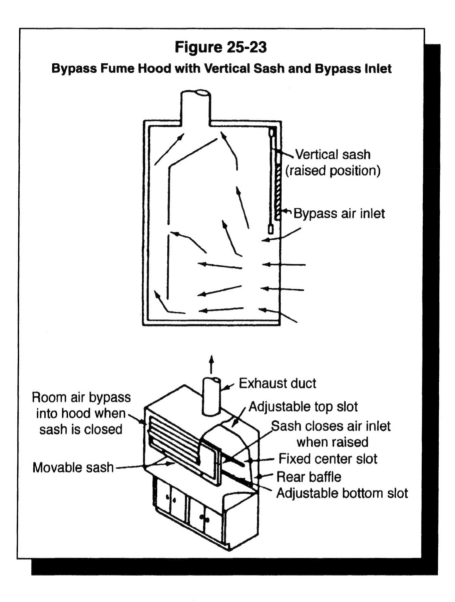

Figure 25-23

Bypass Fume Hood with Vertical Sash and Bypass Inlet

Vertical sash (raised position)

Bypass air inlet

Exhaust duct

Room air bypass into hood when sash is closed

Adjustable top slot

Sash closes air inlet when raised

Fixed center slot

Movable sash

Rear baffle

Adjustable bottom slot

intakes for the building's HVAC system. The ASHRAE *Handbook of Funda-mentals*[6] provides specific guidance on the proper design of exhaust stacks.

Blowers should be mounted at roof level. Auxiliary or in-line fans should not be installed in the hood or duct to ensure a negative pressure in the duct in the event of blower or duct failure.

Quite often multiple hoods are connected to one exhaust system. This is done for several reasons. First, it lowers the overall cost by requiring less

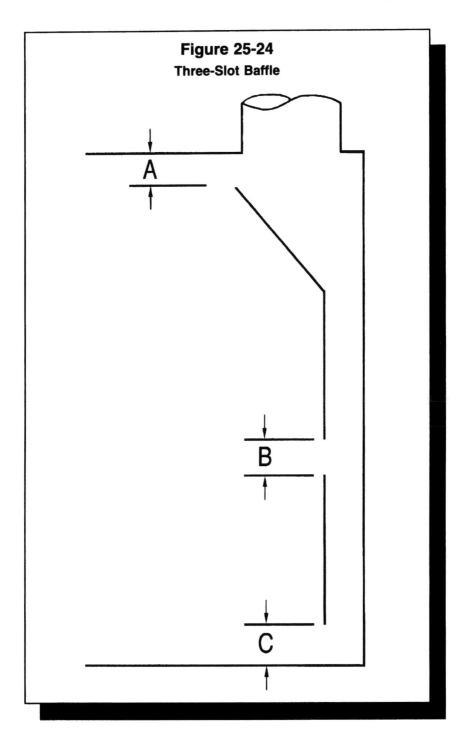

Figure 25-24
Three-Slot Baffle

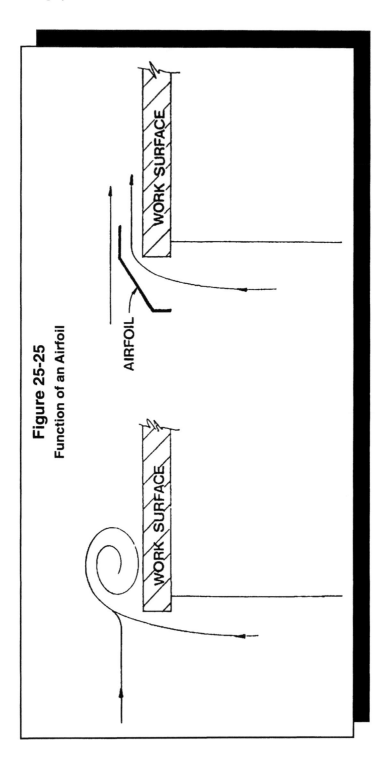

Figure 25-25
Function of an Airfoil

ductwork. It can improve the aesthetics of a facility by reducing the number of exhaust stacks that stick up on the roof. However, there are several issues that should be addressed to ensure that the system will operate properly.

Figure 25-26 is an example of a system that addresses the issues of proper operation of multiple hoods on one exhaust system. In this design a number of fume hoods are connected to a common plenum. The term *plenum* has a special engineering meaning. In this case it means that the duct is so large that there is essentially no pressure drop from one end of the plenum to the other. Thus the negative pressure created in the plenum by the two rooftop exhaust fans is essentially the same for any given fume hood, regardless of where it dumps into the plenum. This is an important design characteristic to ensure that there will be no backflow and to maintain balance in the airflow from each fume hood. The two exhaust fans on the roof are designed to be able to handle 75 percent of the load, with the second fan coming on when the load becomes greater than 75 percent. Also, the fans can be duty cycled for lead/lag to distribute wear evenly on both exhaust fans.

Each fume hood could have its own exhaust, and this will ensure that each exhaust fan can be balanced independent of the other fume hoods. An exhaust fan for each fume hood is not essential, but if the budget allows, this is a good option. The fume hoods will be of the variable volume

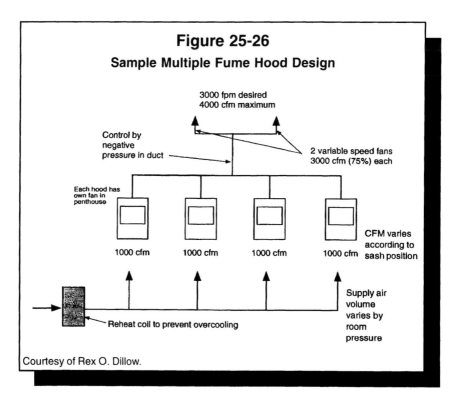

Figure 25-26
Sample Multiple Fume Hood Design

3000 fpm desired
4000 cfm maximum

Control by negative pressure in duct

2 variable speed fans
3000 cfm (75%) each

Each hood has own fan in penthouse

1000 cfm   1000 cfm   1000 cfm   1000 cfm

CFM varies according to sash position

Reheat coil to prevent overcooling

Supply air volume varies by room pressure

Courtesy of Rex O. Dillow.

type so that as the sash is closed, air discharge is reduced, and face velocities are maintained at safe levels. There are various systems on the market that control fume hood exhaust amounts based on various measurements.

Finally, since the volume being exhausted from the room will vary, the amount of makeup air to the room must also vary. This is frequently done by measuring the space pressure relative to the exterior hallway. A set point of positive or negative pressure is established, and the variable air volume system responds accordingly. Because there may be times when the amount of exhaust exceeds the need for supply air for cooling, a reheat coil may be necessary to avoid overcooling. Overcooling is usually only a problem if there are more than several fume hoods in one space.

Maintenance Procedures should be written for both routine maintenance and emergency repairs, and maintenance and laboratory personnel should be trained in safety procedures.

Maintenance workers should not attempt repairs or filter changes with the exhaust system running. Protective clothing and respiratory protection should be used when working on fume hoods or exhaust systems, even when they are shut down. A good safety procedure includes stopping all activity in the fume hood and having laboratory personnel remove or cap all containers in the hood before starting repairs or maintenance. During a system failure, laboratory personnel should immediately stop work and cap or remove chemicals before requesting maintenance. All parts to be worked on should be tested for contamination and, if necessary, decontaminated by qualified technicians.

In the case of multiple hoods on an exhaust system, all hoods must be secured before shutting down the exhaust blower or working on the system. To avoid confusion, each exhaust blower should be marked with all hood locations it serves, and the hoods should be marked with the blower. During repairs, each hood should be posted with an "Out of Service" sign. In addition, all users should be given advance notification of maintenance shutdowns.

Preventive maintenance is important to reliable performance and reduces potential safety liability problems. Planned maintenance should include inspection of the fume hood and its utility services, lubrication of the motor and fan, inspection and adjustment of belts and sheaves, inspection of ducts and fans for deterioration, face velocity verification, and air balancing between ganged hoods.

Chemical Fume Hoods The chemical fume hood is probably the most common type and is used in process and research laboratories for low- to moderate-hazard processes. This type may have a vertical or horizontal sliding sash and is most commonly used where there are known materials and procedures. An indicating manometer will provide the user with visual indication that the hood is functioning.

Radioisotope Fume Hoods   Fume hoods used for radioisotopes in process and research laboratories are usually provided with special shielding and high-efficiency particulate air exhaust filters. An indicating manometer should be installed across the filter to indicate pressure drop. An alarm should be installed with the manometer to indicate excessive pressure drops and inform the operator of the hood of an unsafe condition. This type of hood must be dismantled for decontamination and should be equipped with flanged neoprene gasket joints with disconnectable fasteners. This type of fume hood should not be installed in an exhaust system shared with other fume hoods.

Perchloric Acid Fume Hoods   A perchloric acid fume hood is used in process and research laboratories. Perchloric acids must not be used in other types of hoods because of the explosive characteristics of perchloric salts, which may accumulate on inner hood, duct and blower surfaces, and joints. These salts are water soluble and require a wash-down of the complete exhaust system, from the stack down to the baffles in the rear of the fume hood. A trough and drain system are provided in the hood to collect the wash-down water. To prevent buildup of the unstable salts, wash-downs must be regularly scheduled according to the amount of perchloric acid in use. Explosion may result if the salts are subjected to impact or heat. A thorough wash-down is essential before performing any maintenance.

This kind of hood requires ducts of smooth, impervious, and cleanable materials resistant to acid attack. Nonmetallic material, polyvinyl chloride, or stainless steel with high chromium and nickel content, not less than no. 316, are recommended. Because perchloric acid is an extremely active oxidizing agent, organic materials cannot be used in the exhaust system in such places as joint gaskets or coatings. Perchloric acid hoods should not be installed in an exhaust system shared with other hoods because of the potential for organic materials to mix with the acid in the airstream. All joints should be glued or welded and finished smooth. Sprinkler heads should be installed inside the exhaust system at all changes of direction and connected to a water supply to provide the necessary wash-down. High-efficiency particulate air filters should not be installed in perchloric hood exhaust systems.

Canopy Hoods   Canopy or range hoods are not fume hoods, and although they are useful for exhausting heat and smoke from specific areas, they are not appropriate for chemical and research purposes.

Biological Safety Cabinets   Biological safety cabinets are also referred to as *safety cabinets* or *ventilated safety cabinets*. There are four classes of cabinets and two sizes: 4 ft. and 6 ft.

Class I is a partial containment cabinet. The front opening is fixed, allowing room air to pass, preventing microbial aerosols from escaping into the laboratory. The 10 percent exhaust system is suitable for flammable substances, animal autopsies, and low biological agents. Its main function is to protect people, not materials. It may be filtered by a high-efficiency particulate air filter on the exhaust system.

Class II-A has a fixed opening. This design provides for recirculation of about 70 percent of its total air. Flammables, toxic agents, or radioactive material should not be used in this cabinet. The exhaust and supply are filtered by high-efficiency particulate air filters. Exhaust may be expelled into the laboratory, but it is best to hook it up to building exhaust systems. The cabinet protects people and research materials.

Class II-B has a sliding vertical sash and maintains an inward airflow. It is designed to exhaust 70 percent through the work area and recirculate 30 percent. Use this cabinet with low-level volatile materials and trace levels of chemical carcinogens in tissue cultures. Exhaust is expelled from the building.

Class III is a specially designed unit for use with high-risk biological and chemical materials. This unit has a gas-tight, negative-pressure containment system, so laboratory workers and agents have a complete physical barrier between them. This cabinet has the highest protection factor for the laboratory worker. Highly infectious and radioactive materials are used in this cabinet. A sealed front with rubber sleeves is where all work is performed. High-efficiency particulate air filters are used on the supply and exhaust system, with 100 percent of the air exhausted from the building.

Laminar Flow Designs   Laminar flow hoods (horizontal cross-flow or vertical downflow) are for product protection, not for laboratory worker safety, and should not be used in a biomedical laboratory without adequate risk assessment. The design of the hood is such that it keeps contaminants out of the hood to ensure purity of the experiment but does not ensure that pollutants from the experiment do not get to the worker. Thus, laminar flow hoods are not recommended for biological, chemical, or radioisotope worker protection.

Effect on HVAC System   Current increased concern for laboratory safety has caused HVAC systems to become more sophisticated. Airflows are balanced to maintain minimum room air changes and static pressures relative to adjacent spaces. The air exhausted by fume hoods becomes an integral part of the balanced equation. At one time it was common practice to equip fume hoods with a switch to shut down the exhaust fan during periods of nonuse. However, shutdowns or changes in the volume of air exhausted will disrupt the room's air balance unless the hood utilizes a separate air source or the HVAC system is a variable air volume type that is designed to compensate automatically. The use of multiple fume hoods on a common exhaust system also precludes the use of local switches for user safety.

Energy Conservation   Fume hoods equipped with auxiliary air supply systems are offered by most manufacturers and are designed to introduce a curtain of air at the face opening instead of exhausting conditioned room air. If unconditioned outside air is used, the user and the research materials in the hood will be subjected to prevailing outside temperatures. The use of heated or cooled air in an auxiliary supply system may still result in energy conservation, because the hood and its supply system may be shut down without affecting the room. This type of system is useful in retrofit installations when supply air to a room is not adequate to support the demand of an added fume hood exhaust.

Another way to reduce the unnecessary venting of conditioned air is to reduce the exhaust fan speed to a lower rate and close the sash at night or when the hood is unused. A switch operated when the sash is in the closed position can control the fan speed. Periodic checks of the laboratory air balance will help ensure efficient energy use and reduce complaints about room environments.

Hydronic Systems

Air-Water Systems   Air-water systems are those in which both air and water are distributed to the spaces throughout the building for heating or cooling. In all-air systems, air is the primary means for temperature control at the space level. In an air-water system, both air and water are available at the space level to provide temperature control.

Fan Coil Systems. Fan coil systems were one of the earliest air-water systems to be used to provide temperature control and air circulation for individual spaces. The unit consists of a cabinet containing an air circulation fan, a coil to be used for either heating or cooling, water connections, and a control valve with a thermostatic control device that is remotely mounted or self-contained in the cabinet (Figure 25-27). The fan coil unit can be floor-mounted against the wall, generally below windows, and surface-mounted on the ceiling or above the ceiling, concealed and connected to ducts and air outlets. Ventilation air can be supplied to the fan coil unit through the exterior wall in the case of a floor-mounted unit, through a separate ventilation air duct system directly connected to the floor-mounted unit, or to the concealed units above a ceiling. In cold climates, the wall opening is a source of problems, such as freezing of the coil, dirt, and insects.

Fan coil units, when properly selected, are quiet, easily maintained, and simple to operate. They cannot cool with outside air, but when they are installed around the perimeter of a building with operable windows, cooling with outside air is not needed. Fan coil units provide individual temperature control to a given space, but the water supplied to them has to be zoned relative to the impact of the weather on the building (northern versus southern exposure, for example) and the functional zone requirements.

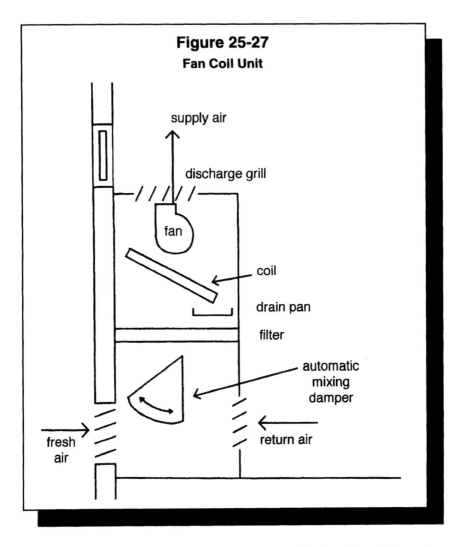

Figure 25-27
Fan Coil Unit

supply air

discharge grill

fan

coil

drain pan

filter

automatic mixing damper

fresh air

return air

The piping system can be a two-pipe system with the pipe dedicated to heating or cooling, which would not permit simultaneous heating and cooling to different areas unless zoned. A four-pipe system can be installed, providing hot and cold water to each fan coil unit at all required times.

Fan coil units are not satisfactory in areas where clean air is important. The condensation on the cooling coil forms a breeding ground for bacteria. Fan coil units and similar equipment using cooling coils should not be used in such areas because of the inability to filter the supply air.

Fan coil systems are relatively energy efficient and are usually equipped with a three-speed motor. Selecting the units based on the middle fan speed— hence, volume—is desirable, because it provides for quieter operation and a cushion for future load increases (for a small added cost). These systems are generally limited to use on the perimeter of the building.

Fin Tube. Fin tube heating and cooling, perhaps the most common type currently used, consists of a copper or a steel pipe with fins along the heating portion of the element. These fins can be made of copper, steel, or aluminum, and the pipe size, fin size, and spacing determine the heating capacity. Fin tubes can be provided singly or stacked in multiple lengths. Radiator-type units provide heat via radiation and convective transfer of heat by the airflow over the unit.

Fin tubes can be built into the perimeter finish of the building by providing air space at the floor and an outlet above the fins for circulation. Generally, this process is convective only, as the cover acts as an insulator. Fin tubes can also be provided with a variety of metal covers designed to meet the economic and aesthetic considerations of the installation. Metal covers tend to become warm and provide some radiant heat.

Fin tube control is achieved by operation of a manual valve, operation of an automatic valve under the control of a space thermostat, or control of the water temperature supplied to the unit. Damper control of the fin tube is not recommended, as the cover will act as a hot radiator when the damper is closed. Fin tube radiation is often used in conjunction with cooling-only variable volume systems. Fin tube radiation has the virtue of providing a blanket of warm air at the perimeter of the building, greatly increasing human comfort. Steam or hot water can be the heat source.

Convectors. A convector is similar to a fin tube, although the heating element can consist of a small coil similar to that in a fan coil unit. A convector coil can be built into the perimeter of the building in much the same fashion as a fin tube or contained in metal cabinets. Styles of metal cabinets are generally determined by economics and aesthetic considerations. Convectors are often used where space will not permit the installation of long sections of fin tube (e.g., in stairways, entryways, and other relatively small spaces). Convectors can use steam or hot water as a heat source.

Unit Ventilators. Unit ventilators, introduced before World War II, are a variation of the fan coil system. The major differences are that unit ventilators generally are larger and can circulate more air and therefore can heat and cool larger spaces. They can also cool with outside air. Figure 25-28 illustrates a typical unit ventilator. The space temperature controls are typically designed to automatically go from heating with no fresh air until the space is warm, to heating with minimum fresh air as required by code when the room temperature has been achieved, and then to modulation of the heating valve to the "off" position as the space temperature becomes satisfied. As the space temperature continues to rise, the fresh air damper opens and provides a cooling effect until the damper is 100 percent open.

Unit ventilators are often installed with fin tube radiation or auxiliary air ducts (connected to the unit air supply) under the windows that provide

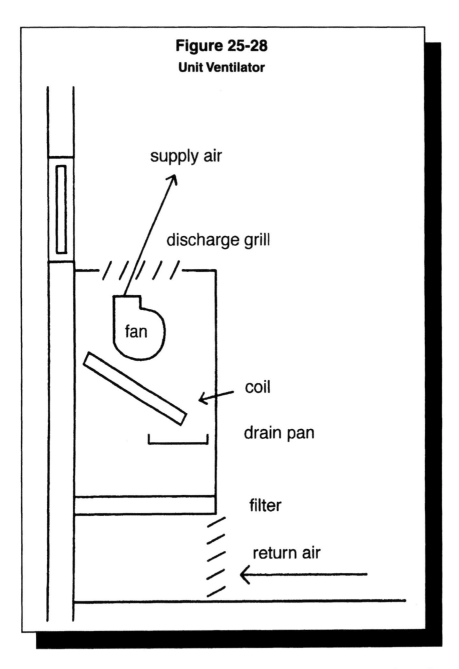

Figure 25-28
Unit Ventilator

supply air

discharge grill

fan

coil

drain pan

filter

return air

an air curtain of warm air over the windows. The fin tube can keep the space warm during unoccupied hours without the fan. The original units functioned for heating only. They subsequently were designed to provide cooling using the same coil but with a combination heating–cooling thermostat. A drain pan is required to collect the condensate from the coil.

The fans of unit ventilators are designed for quiet operation, making them ideal for libraries and classrooms. The fan horsepower required is quite small, and as in the case of the fan coil system, the overall horsepower required for air circulation for either cooling or heating is often substantially less than that of an all-air system. Unit ventilators also have the advantage of relatively simple maintenance of the fans, valves, and coils. When the economizer cycle is used, the controls are more sophisticated and require more maintenance for satisfactory operation.

Because of their size, unit ventilators occupy more floor space than fan coil units. The louvers on the exterior of the building can also have a negative effect on the architectural design. The units are generally limited to use at the perimeter of the building. They can also be installed on or above the ceiling. The wall openings can create problems similar to those experienced by fan coil units if the damper and filters are not well maintained.

Induction Systems. Induction systems were introduced shortly after World War II and achieved great popularity as a solution for quality air conditioning for existing high-rise office buildings. Traditionally these are high-velocity systems using small, round ducts to each unit, with the units placed around the perimeter below windows. Generally, the ducts are installed vertically, parallel to the exterior columns of the building. Figure 25-29 shows the typical unit.

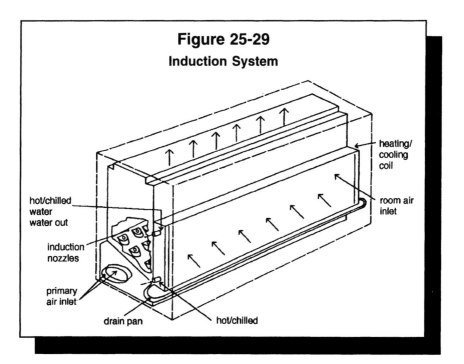

Figure 25-29

Induction System

The induction unit has a plenum to which the high-pressure air duct is connected. Discharge nozzles on the outlet of the plenum provide high-velocity jets of system air, inducing room air to provide a total air discharge quantity equal to approximately four times the amount of high-pressure air introduced. This is similar to the process in which a high-pressure jet of water in a pool circulates large quantities of water. The unit has a coil that can be used for heating and cooling. The induction process draws the room air through the coil and mixes it with the high-pressure air jets. In the cooling mode, when chilled water is circulated through the coil, the room air drawn through the coil is cooled and mixed with the high-pressure air supplied to the unit. In the heating season, hot water is circulated through the coil, and heat is provided in much the same manner. The temperature of the primary air is centrally controlled on an outside air temperature schedule.

During unoccupied hours in the heating season, hot water can be circulated through the coils with the high-pressure primary air fan turned off, in which case the unit acts as a convector. Under extreme cold conditions, the fan may have to be operated periodically. Such a cycle provides significant energy savings.

The minimum amount of primary air is normally determined by ventilation requirements. The air supplied to the induction unit is from a central air handling unit with cooling coils, heating coils, filters, and controls similar to that in a typical all-air system. The main air handling unit can operate with up to 100 percent fresh air. However, because of the small ratio of primary compared to total air, typically 1:4, induction systems cannot cool individual spaces with outside air alone.

In most systems, the primary air is sufficiently dehumidified at the main air handling unit to avoid condensation on the cooling coils in individual units. However, a drain system for condensate in the induction units is advisable. The temperature of the water supplied to the induction unit coil is critical if drainage is not provided. Induction units have the advantage of being able to circulate the total space air requirements without a fan installed within the unit. The horsepower required for moving the air in the room comes from the main fan system.

Induction units have the advantage of providing individual space control, little or no cross-contamination among rooms (depending on design), relatively quiet operation, and minimum maintenance. If the nozzles are too small, they will clog with dirt, and a major maintenance problem will occur. The units have filters that require attention. The induction system has the disadvantage of a high fan horsepower requirement. The system is not highly flexible, but the multiple units generally provide enough space rearrangement flexibility to cover most contingencies. Air-water induction units generally are limited to the spaces around the perimeter of the building, with a separate air conditioning system serving the interior. They are widely used in offices and dormitories, with an induction unit in each room and a separate system serving common use areas. The units are also available in low-pressure designs to save energy.

Hot Water Heating   Hot water heating systems have many advantages, including relatively small pipe sizes, flexibility relative to routing throughout a building, the ability to vary the water temperature through a wide range, ease of zoning, and quiet operation. Their greatest advantage, of course, is the ability to be combined with a chilled water system, providing heating and air conditioning from the same piping system and often the same air conditioning and heating device, such as a fan coil unit and air handling unit.

Hot water heating systems experience little or no corrosion if properly operated and maintained. Entrained air in a hot water heating system causes corrosion and circulation problems. The initial water introduced into the system can attack the pipe for a brief period of time; however, that action soon stabilizes itself. If the water is not replaced, drained off, or lost through leakage, corrosion essentially stops. A mistaken notion on the part of many heating system operators is that chemical treatment should be added periodically to a hot water heating system. This is only true if that judgment is based on careful analysis. The introduction of a small amount of new water will cause only a mild amount of corrosion. It is not unusual to open up gravity hot water heating systems that have been operating for 65 years in residences and find essentially no corrosion of the piping. Excessive chemical treatment of hot water heating systems can destroy radiation and zone valves. Facilities personnel should question regular chemical treatment for a closed water system, be it chilled water or hot water.

Hot water heating systems can be zoned in much the same fashion as a steam system; however, they have the additional advantage of being able to have the temperature varied within each zone. In a high-rise dormitory building with individual room radiators, it is entirely possible to zone the entire face of the building based on exposure and vary the water temperature to maintain uniform temperatures in all spaces in the zone. Rooms with manual radiation control can provide uniform space temperatures throughout the day as the sun strikes the face of the building in the morning and the far side in the afternoon. Uniform temperatures are also maintained throughout the night. A manual radiation valve within each space permits the occupant to select his or her own specific temperature adjustment within the range of the radiator capacity and the water supply temperature provided.

Hot water heating systems, because of their reset capability, have lower heat losses and greater energy economies. Typically, hot water heating systems are designed for a 20°F temperature differential between the temperature of the water leaving the heating source and that of the water at return. This differential is seldom achieved because it is based on peak heating conditions. Systems can be designed for large temperature differentials if they are for heating only. Differentials of 40°F and 60°F can be used, which reduces the pipe size and pumping costs considerably. This approach requires careful sizing of the radiation devices because of the large temperature drop experienced. In long pieces of fin tube radiation, the far end can become excessively cold. Where a hot water heating distribution system also serves as a chilled water system, in most cases the pipe sizing is determined by the chilled water flow and the temperature drop.

Figure 25-11 shows the layout of a typical hot water heating system with perimeter radiation. The system is shown with zones for the various exposures of the building, individual zone pumps, and mixing valves. Hot water heating systems should be balanced when they are started up, and the balancing valves should be of a type that can be closed and reopened to the balance point by a marking or a mechanical stop device. As discussed earlier, balancing of systems can be achieved with a reverse return piping system, such as that shown in Figure 25-12. With this type of installation, the actual distance through any circuit is equal to the distance through any other circuit. The system, then, is essentially self-balancing. Although this system may require more pipe, it is an easily maintained system relative to flows throughout the various parts of the building.

A properly sized hot water heating system will operate with relatively low pumping costs. It can be provided with multiple pumps or one large individual pump with a standby pump and single- or multiple-zone water temperature control valves.

A hot water heating system should be designed so that air can be eliminated from all high points, as well as from radiation and coils. The system is often used with all-air air conditioning systems and particularly with variable volume systems. The control of the radiation can be integrated with the air conditioning control for each zone to avoid simultaneous heating and cooling of the space.

Hot water heating systems are particularly adaptable to resetting temperatures to a lower point during unoccupied hours of operation and for heat reclaim purposes. The heat can consist of condenser water from an air conditioning cycle or heat pump cycle, or it can be waste heat from a separate operation in the facility.

Steam Heating Building steam heating system pressures rarely exceed 125 lbs. Low-pressure steam heating systems offer the advantage of moderate steam temperatures when compared with higher pressure steam and low-pressure boilers; they require less stringent monitoring than do high-pressure boilers under most codes. Steam heating systems using radiators directly within the space are not common in modern buildings but are found quite frequently in older buildings. Objections to steam include excessively hot radiation temperatures, difficulty of pipe installation because of the need to drain condensate, excessively large supply pipe sizes, and the tendency of return lines to corrode. An additional disadvantage is that the entire system, including piping, is at excessively high temperatures most of the time.

Systems exist that can operate under subatmospheric and vacuum conditions and produce steam at lower temperatures. These systems require good maintenance to maintain the vacuum. Orifices installed in the inlet to the radiation restrict the flow of steam, creating subatmospheric conditions within the radiation unit itself because of the vacuum maintained at the radiator outlet. In effect, the steam temperature can be "reset" down by varying the pressure within the radiator. This can be effective in a properly maintained system but currently is not in common use. In many early heating systems, the main steam riser is extended from the basement or boiler room level up into the attic space. There the steam piping system is divided into branches, and the steam is down-fed to the radiation around the perimeter of the building. The main reason for one or more large risers is the economics of pipe sizing resulting from a single riser. When steam is supplied vertically, the condensate, by its very nature, flows down against the steam. This requires oversized steam pipes to prevent water hammer. A single oversized riser is much cheaper than multiple risers.

Figure 25-30 shows an up-feed system, common in current buildings. A horizontal main runs around the perimeter of the building, connecting

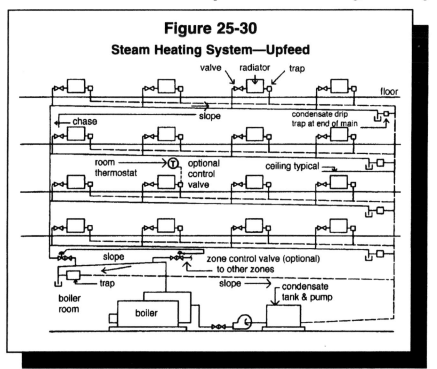

Figure 25-30
Steam Heating System—Upfeed

to the vertical risers. Steam can be zoned in much the same manner used in air systems. Typically, it would be zoned by weather exposure because its primary function occurs only during the heating season. Separate mains are provided for each exposure of the building, with a motorized valve controlling the flow of steam to that particular exposure. An outdoor thermostat with solar compensation or a thermostat located in a "typical" room provides control for the zone. Additional control for the spaces within the zone can be provided at the radiation with manual valves, self-contained thermostatic valves, or wall-mounted thermostats. Existing steam systems with perimeter radiation can provide the heat necessary for the installation of a variable volume system or air conditioning system in a remodeled building. Low-pressure steam can furnish the heat necessary for coils found in air handling units that provide heat to the zones by means of the air temperature supplied.

A major disadvantage of a steam heating system is the space required by horizontal runs. The pipe should be pitched uniformly, approximately 1 in. in 50 ft. Every time the horizontal steam main encounters an obstruction, the gravity flow of liquid condensate within the pipe presents complications that have to be accommodated.

Steam systems require careful maintenance of the steam traps. Typically, radiation found in spaces has a thermostatic trap. Large coils often have a float and a thermostatic trap, in the case of a low-pressure steam coil, and a bucket trap for high-pressure steam coils. One of the operating characteristics of low-pressure steam heating systems is that the air trapped within the radiation must be eliminated. Traps are typically designed to provide for the removal of air. Standing cast iron radiation can be provided with a separate air vent because of the large volume of the steam chamber.

When steam traps leak, the entire heating system supply and return operates at essentially the steam supply temperature, creating excessive fuel use. It is not unusual for the energy loss to amount to 20 to 25 percent of the energy consumption of a building. A facility without a steam trap maintenance program will probably have in excess of 50 percent of the traps not working. With modern electronic temperature-sensing devices, it is relatively easy to determine whether a steam trap is functioning properly.

Steam used in heating coils has a tendency to freeze up rapidly when used for heating fresh air. In a multizone unit, or any air handling unit with a long steam coil, the end of the coil located farthest from the steam inlet tends to be colder. In the case of a multizone air handling unit, if a small zone is attached to the hot deck on the far end of the coil, the coil will often have insufficient heating capacity because of a lower air temperature provided by the colder end of the coil. Double-feed coils are available for situations where long coils are used. This eliminates most of the problems of cold spots in the coil.

One of the major advantages of steam is that, unlike hot water heating systems, it can be distributed without pumps. As a result, steam heating systems have low energy transportation costs. Typically, the only energy required, other than that for the boiler burner, is the pump required to return the condensate back into the boiler. In the case of a gravity system, pumps are not required.

Other Systems

Humidification/Dehumidification Specific levels of humidity may have to be maintained for comfort or process environments. Because maintenance of humidity levels can have a major impact on energy consumption, the need for specific control must be carefully considered. Achieving acceptable levels with low initial costs and acceptable operating costs requires careful planning and design.

Humidification is relatively simple with most air conditioning systems. In all-air and many air-water systems, the humidity can be introduced directly into the airstream going to the spaces. It can be provided by the introduction of steam or by vaporizing or misting moisture into the airstream. Steam can come from a central steam plant, from an in-house steam boiler, or from package-type electrically heated steam generators. The package-type steam generators have the maintenance problem of scaling and mineral deposits.

In certain environments, chemicals used in water treatment in steam generating plants are unacceptable. This includes areas involving biological, pharmaceutical, and organic work; artwork; and historical document storage, where chemical deposits may have an undesirable effect. These conditions require a pure steam generator—either a package-type electric generator or a generator using steam from a central source to boil water containing no boiler compounds to form the pure steam.

Humidification lends itself easily to zoning by the introduction of the humidity into the airstream serving a specific zone. Humidifiers can also be located within the space, independent of any airstream. These are generally wall-mounted and can be provided in multiple units.

Dehumidification in most buildings is achieved by condensation of the moisture in the air on the cooling coil during the cooling cycle. In most climates, this moisture removal during air conditioning cycle operation provides the supply air with acceptable humidity levels.

Under light load conditions or when a cooling system is oversized, the cooling does not cycle on long enough for the appropriate amount of moisture to be removed. The space dry bulb temperature might be satisfied, but high moisture results. In a space where the moisture gain is relatively high compared with the typical heat gain, special measures must be taken to remove the moisture. This can be accomplished with a reheat system where the cooling coil is kept on long enough and is cold enough to remove the moisture. The air may then be too cold to introduce into the space without reheating. This method provides an accurate means of controlling the humidity within the space but requires higher operating costs.

Dehumidification can also be achieved with desiccants, both liquid and dry. Silica gel, the material often packed with valuable photographic and electronic equipment in shipping containers, is the most common type of dry desiccant. Large rotary wheel mechanical dry desiccant systems are available that circulate space air through the desiccant. As the desiccant absorbs the moisture, it is rotated through a hot airstream that dries and regenerates it. This type of operation requires heat for the drying process. The equipment is often used in areas where air conditioning is not required, as the warm desiccant introduced in the main airstream has a heating effect. There are package systems that provide heating, cooling, and desiccant dehumidification.

Liquid-type desiccants consist of a desiccant spray chamber with a piping system and spray nozzles, a basin containing liquid desiccant, and pumps to recirculate the desiccant throughout the system. A separate regenerator section warms the liquid desiccant to boil off the moisture picked up from the airstream passing through the spray chamber. Lithium bromide is a typical absorbent-type salt used for this application. Liquid-type dehumidifiers can be used in areas where air conditioning is not required or in conjunction with air conditioning systems. The application provides accurate control of space conditions and can remove large amounts of moisture. The airstream must be provided with good filtration to minimize dirt deposits in the spray system.

Water Spray Chamber   When moisture in the air comes in contact with a surface colder than its dew point, moisture will condense and be removed from the air. This principle applies not only to the metal surface of the coil, but also to any surface, including glass, masonry, plant life, and water. If a water spray is chilled to a point below the dew point of an airstream passing through the spray, the moisture in the airstream will condense on the water droplets, and the moisture will be removed from the airstream. This method of humidity control should be considered only in special applications and requires careful design. Airstreams contain dust and dirt, and sprays used for either dehumidification or humidification provide a filter action that introduces dirt and foreign matter into the liquid system. Care must be taken to prevent fouling of the spray nozzles and eventual deterioration of the system. This problem can be minimized by appropriate air filtration prior to introduction of the airstream into the spray chambers.

Multiple Package Unit/Unitary Systems   Package-type unitary equipment provides the advantages of low cost and a wide variety of types, styles, and applications. They require only the extension of basic utilities such as gas and electricity and are relatively simple to maintain. They also offer the advantage, in some installations, of no failure of the entire system when only one unit fails. A disadvantage involves maintenance of numerous pieces of dispersed equipment. The life expectancy of package equipment is less than that of high-quality standard air conditioning components such as centrifugal chillers, pumps, and central fans.

The major objection to package equipment is the effect of equipment location on maintenance. The equipment is generally compact and can be difficult to maintain above ceilings, in tight closets, or outside in inclement weather. Despite these disadvantages, appropriate situations exist in which this equipment can be used effectively. One application is in remote or rural areas where the maintenance staff and factory service required for more sophisticated equipment are not available.

The heating capacity of many package types of unitary equipment has little relationship to the cooling capacity. This should be carefully checked to adjust the capacity to the proper size.

Window Units   Window air conditioning units are the most common and easily recognized type of package unitary air conditioning equipment. They offer the advantage of low initial cost, portability, adaptability to existing wiring systems, individual space control, low maintenance because of hermetic design, and simple controls. Often a simple solution for cooling a limited area, this type of spot cooling can be used for human comfort or for another functional needs.

Window units have numerous disadvantages, including on-off control with no modulation, high noise level, localized drafts, unsightly appearance, and relatively high operating costs. Despite these disadvantages, they provide effective air conditioning for such spaces as isolated small security buildings, maintenance shops, and offices. Window units are available with electric heating coils.

Through-the-Wall Units   A through-the-wall unit is actually a variation of a window unit that was developed as the need for air conditioning expanded rapidly throughout the country. They are, in effect, fan coil units with individual compressors and heating elements. The units are available in attractive cabinets that can blend in with the architectural design of the space. They have many of the same disadvantages as a window unit; however, they are generally of higher quality, resulting in a lower noise level, improved operating economies, variable fan speeds, improved controls, and better access for maintenance.

Through-the-wall units are often used in offices that are occupied throughout the week when the remainder of the building is shut down (e.g., as in a church). They offer individual room control, and the loss of a unit does not affect other spaces. Through-the-wall units cannot provide cooling with outside air.

Through-the-wall units are similar to window units in that the condensing portion is located outside the exterior wall and uses the air as a heat rejection medium. Ventilation air in nominal amounts can be obtained through the exterior louver or by ducting the air to each space from a central unit.

Heat Pumps, Air Source   All air conditioning refrigeration cycles are, in effect, heat pumps. Heat available from room air, products, and objects is extracted by the refrigeration cycle and transported to a point of discharge through the condenser. A heat pump is able to reverse the cycle and discharge heat to the space in the heating season. To do this, the heat pump must have a source of heat. In the air source heat pump, the outside air is the heat source or "heat sink." During the heating cycle, the unit cools the outside air from the ambient to a lower temperature prior to discharging it back to the outside. The heat extracted from the outside air is then discharged into the space by means of the compression cycle with the cooling coil being used as the condenser. If a window unit was simply removed from the window and mounted in reverse, it could function as a heat pump. Air source heat pumps can be found in such forms as window units, through-the-wall units, or residential furnace-cooling combinations, and larger commercial package-type units including rooftop units.

Heat pumps have the advantage of requiring only one utility source. They can provide individual space control or serve larger areas depending on the type of equipment in which they are incorporated. The air source heat pump's ability to heat with outside air declines dramatically as the outside air temperature drops to a point where it is ineffective as a heating unit. This occurs at 0°F to approximately 10°F. To overcome this, the units are provided with supplementary electric heating coils. These coils can be multiple-staged or single elements. The heat pump's primary advantage, relative to energy costs, is the ability to provide heat from the electrically operated refrigeration cycle, thereby greatly reducing the heating costs when compared with straight resistance electric heating units. Depending on utility rates, air source heat pumps can be competitive in a properly designed system. Figure 25-31 shows the typical air source heat pump cycle.

All heat pumps operate at relatively low room supply air temperatures during heating, requiring careful design of air distribution to avoid drafts.

Heat Pumps, Water Source   Some of the disadvantages of air source heat pumps are overcome by the use of water as a heat sink. The water can come from a closed-circuit condenser water system connected to a heat source, such as a boiler in the winter, and a heat rejection point such as a closed-circuit cooling tower system in the summer. Other sources can include looping of pipes in the earth to provide a heat sink, the installation of

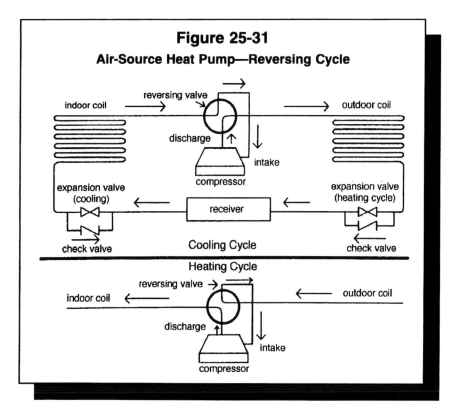

Figure 25-31

Air-Source Heat Pump—Reversing Cycle

pipes in small low-volume vertical wells, standard well water, lakes, and condenser water from the refrigeration cycle of a unit used to provide cooling year-round. The water source generally operates at a relatively constant water temperature. In the case of ground and well water, the temperature ranges from 55°F to 65°F. In the case of mechanical systems, the water temperature can vary from approximately 85°F to 105°F. Generally, water source heat pumps do not require supplemental resistance heating and offer a lower electric heating cost throughout the entire heating season. Water source heat pumps can also be a form of energy conservation and operate efficiently in a well-designed system in which heat removed from a process or an interior space is used to heat perimeter spaces or processes at the same time.

A unitary closed-circuit, water source heat pump can be similar in appearance to an air source through-the-wall system or a fan coil unit. The units can also be located in closets and above ceilings. One of the advantages of a closed-loop system is the ability to heat and cool simultaneously throughout the building. Spaces requiring cooling reject the heat into the closed-loop water circuit, which then becomes available for extraction in spaces that require heating, such as occurs in the spring and fall when the south zones of a building need cooling and the north zones need heating. Water source heat pump systems of the small individual type cannot provide cooling with outside air directly. Depending on the quality of the equipment, noise may be a problem. Ventilation air can be provided by a ducted system to each space or to the intake of each unit.

Outdoor Unitary Equipment   As the name implies, this type of equipment is generally located outside of the building, usually on the roof. The equipment is complete within itself and is connected to the ductwork within the building. This equipment can vary from small systems of approximately 5 tons for individual spaces, to large systems well over 100-ton capacity. Depending on the size and sophistication, outdoor unitary equipment has the ability to provide cooling with outside air, can have return fans for positive control of ventilation and building air pressure, and can be equipped with sophisticated controls and air filtration systems, making it suitable for providing environmental control for critical operations. Outdoor unitary equipment can be single-zone, multizone, or connected to double-duct and variable volume systems.

The location of exterior rooftop units and the questionable quality of construction often found in rooftop units are major objections to their use on buildings considered permanent with long life expectancy. To use rooftop units to avoid providing adequate mechanical space within the building is false economy. The cabinet of a rooftop unit is, in effect, the mechanical space building or envelope. It is often fragile and can present major maintenance problems and significantly reduce energy efficiency. The maintenance of rooftop equipment in inclement weather can be difficult and, under certain circumstances, dangerous. When rooftop units are used in applications requiring sophisticated control of environmental conditions, the location becomes a negative factor. It is not uncommon for first-time users to go through a cycle of pleasant surprise over the low initial cost; mild concern over higher operating costs; realization of maintenance difficulties; and, ultimately, total disillusionment over the condition of the equipment after years of operation. There are many places where the use of these units can be justified, but the consequences must be carefully considered before making the decision.

There are rooftop units that, in effect, are penthouses complete with corridors and large doors, allowing personnel access to all parts of the system. These units are of unusually high quality and do not exhibit the usual disadvantages of outdoor unitary equipment. They can be field assembled.

Outdoor unitary equipment can be completely self-contained, with compressors, direct expansion cooling coil and air-cooled condensers, gas furnaces, and electric heat, or they can be connected to a remote source of chilled water for cooling, steam, and hot water for heating. Heat can also consist of reclaimed heat available from another operation.

Indoor Unitary Equipment With indoor unitary equipment, the complete package is generally located within the building and is connected only to an exterior source of utilities and to a condensing media source for the refrigeration cycle. Utilities can be electric or gas, and the condensing media can be a remote air-cooled condenser or water from an appropriate source, such as a cooling tower or well. Indoor unitary equipment can consist of a typical residential furnace with air conditioning or larger sophisticated systems capable of handling double duct and variable volume systems. The size of such systems generally ranges from 1 to 2 tons to more than 50 tons. The typical installation has an air-cooled condenser. Small units are available for location above ceilings and can be connected to a ductwork system serving several rooms. The air-cooled condenser would be located outside on the ground or on the roof near the unit. These are generally referred to as *split systems*. Units can be arranged to provide cooling only or heating and cooling, or cooling with outside air in larger systems. Unitary equipment has the same advantages and disadvantages of other systems. They generally have a low initial cost compared with conventional systems utilizing air handling units and chilled water/hot water heating and cooling media.

Unit Heaters. Unit heaters are most commonly used in industrial areas such as warehouse facilities, storage facilities, shops, and other areas that are not air conditioned. The unit heater consists of a fan and a heating coil mounted in a cabinet, with discharge louvers or, in certain cases, duct collars for connection to a ductwork system. Unit heaters are controlled by a space thermostat that cycles the fan and the heat source control valve. The heat source for unit heaters can be steam, hot water, natural gas, or electricity. Because the fan can throw the airstream a relatively long distance, these units are generally used for heating large areas. Unit heaters have the advantage of requiring no floor space. Their use is limited to areas where aesthetics is not a major concern.

Cabinet Unit Heaters. Cabinet unit heaters, generally speaking, are floor- or ceiling-mounted, surface-mounted, or recessed. They are similar to fan coil units. They generally use steam or hot water for heating media, although electrical applications are common. Their main function is to provide heat in a space where their aesthetic appearance makes them far more acceptable than the typical unit heater. They are used in vestibules, stairways, and other areas that need a high-capacity heat source in a relatively compact unit. Generally, the controls cycle the fan and the coil heat source valve. If a valve is not provided on the heat source, the cabinet can become a convector and cause overheating of the space.

Furnaces. Furnaces often serve as a heat source for a building that does not require cooling or can be used in an installation where cooling will be provided at a later date. They can be mounted vertically on the floor or horizontally in a space above the ceiling. Furnaces have the advantages of low initial cost; relatively high air circulation rates; and, depending on the fuel, economies of operation. They can be gas fired, oil fired, or contain electric heating elements.

Makeup Air Units. Certain operations and functions within a building require replacement of air. This can be air exhausted from a cooking operation, maintenance operations, painting, chemical processes, or where special air cleanliness is required. Makeup air units can use steam, hot water, gas, oil, or electricity as a heat source. The most efficient type is a direct-fired gas unit in which the combustion process takes place in the airstream supplied to the space, providing 100 percent efficiency of energy use. Units of this type are designed and approved in accordance with appropriate code authorities with an acceptable ratio of combustion products and air. Makeup air units should be installed to operate only when required by the exhaust cycle. Although low initial costs can be a temptation to use the makeup unit to provide heat during the off-hour periods, this results in high operating costs because of the use of all outside air. A separate supplemental heat source should be provided to maintain minimum temperatures during off-hour periods.

Computer Room Air Conditioners   Sophisticated unitary package equipment has been developed for computer applications requiring close control of the space environment. These units are generally of high quality and provide complete and accurate control of temperature, humidity, and, where required, air filtration. They are designed to be totally independent of other portions of the building air conditioning system and are designed to operate year-round. Although they can be connected to a central plant, the most common installation is the individual package type. The condensing source can be a remote air-cooled condenser or a closed-circuit loop with an antifreeze solution and an outside radiator, where required.

Humidification is usually provided by evaporation of water by the use of electric resistance heating elements. The evaporation of water introduces maintenance problems because of the collection of scale and flushing of the system. Steam from a remote source can also be used for humidification.

Of necessity, a computer room unit incorporates a reheat cycle for maintenance of precise humidity levels during the cooling operation. Computer room units are often tied into emergency power sources that serve the computer system complex. In the sizing of computer room units, it is appropriate to include redundancy. Most units are of the multiple-compressor, refrigerant circuit style in each package unit. Selection of enough redundancy to permit one or more failures of a refrigeration circuit will protect ongoing operation of the computers.

Radiant Systems   Radiation generally can be considered as a heat source located directly within the space. The word *radiator* is a common term applied to the old cast iron radiation found in buildings built before World War II. Cast iron radiation has little use in current modern buildings. There are, however, various other types of radiation available.

Radiant heat from the sun is the earth's fundamental energy source and provides the necessary heat for human survival. Radiant heat operates by virtue of the temperature difference between the hot surface and the cold surface, and is independent of the intervening atmosphere. Radiant heat transfer varies as the difference of the fourth powers of the temperatures of the surfaces involved. This is more easily understood when one compares the effect of radiant heating under a direct and extremely hot source, such as a lamp or radiant coil, and the effect of such heating under a surface at a moderate temperature, such as a convector. Generally, radiant heat requires little or no space within the occupied area and is easily controlled. It can be used for spot heating purposes or throughout an entire area.

Ceiling Panels. Ceiling panels are one of the most common forms of radiant heat in modern buildings. They can be integrated into the ceiling system to match the textures and support system. Electric radiant heat panels are flexible and can be relocated easily. Radiant heating can also consist of metal ceiling pans connected to copper pipes circulating hot water. A space thermostat modulates the water valve or the electric circuit to maintain space temperature. Ceiling panels have the advantage of quick recovery of surface temperature to provide radiant heat to the space occupants in a relatively short period of time. The low mass of a ceiling panel also prevents overheating of the space because of excessive heat retention when cycled off.

Floor Panels. Floor panels generally consist of hot water pipe circuits buried in a concrete slab. They are expensive to install, are slow to respond to load changes, and can cause overheating because of the mass involved. They are inflexible with regard to space rearrangements and changes in zone control. Their use is limited to such areas as warehousing and maintenance facilities. They offer greater comfort to personnel working in a building where the air temperature is kept relatively low.

Gas-Fired Radiant Heat. There are two forms of gas-fired radiant heat: high-intensity and low-intensity. High-intensity heating units have a radiant element that is exposed to the space occupants and is made incandescent by the combustion process operating directly against the ceramic radiant element. These units are simple to install and are easily relocated and provide quick, high-intensity spot heating. The products of combustion are introduced into the space, and therefore, a ventilation system is required. Maintenance is required to keep the surfaces free of foreign matter that will reduce the radiant characteristics of the ceramic materials. The ceramic material is often fragile and easily damaged during maintenance operations.

Low-intensity gas-fired radiant heat generally consists of a tubular combustion chamber and a long, horizontal, flue gas pipe used to conduct the flue gases from the combustion chamber to the exterior of the building. The system operates in lower radiant temperatures, as opposed to the incandescent range. It is extremely efficient; products of combustion can be cooled to temperatures as low as 100°F or less prior to discharge to the atmosphere. Most radiant systems are proprietary with regard to the design and installation techniques. At present, ASHRAE has no specific standards for calculating capacity requirements for radiant heating equipment, other than the standard heating load calculations available for other heating systems. Materials of construction are extremely important in the installation, particularly where flue gas temperatures are low enough to permit moisture condensation, which can result in corrosion from the formation of sulfuric acid. The heaters can be mounted in many configurations and are adaptable for warehouses, maintenance shops, industrial arts classrooms, and certain multipurpose activity areas. In building areas where the atmosphere is contaminated, combustion air should be drawn directly to the burners from the outside. Typical installations range from 60,000 to 120,000 Btu per hour for each burner, with multiple burners being installed.

Electric Radiant Heating. Electric radiant heating has been available for many years. These units may consist of incandescent heating elements directly exposed to the space, with an appropriate protective wire guard, or glass or metal heating surfaces that offer a more attractive appearance, ease of cleaning, and greater safety. Controls either are integral or consist of a space thermostat. Electric heating panels are versatile as to location and relocation. They are useful for spot heating, allowing the temperature in the remainder of the building to be lowered. Radiant-type heaters with an easily cleanable glass surface are used in hospital operating rooms to provide heat when the main air conditioning system is off.

Electric radiation can take on the same form as convectors or fin tube units. It most commonly consists of baseboard heat or individual cabinet electric heaters mounted in walls (or in ceilings in certain applications). Electric baseboard heat provides individual space temperature control. Controls can be incorporated in the unit or consist of a wall-mounted thermostat.

Radiant Cooling. Radiant cooling is not widely used but can be effective in certain applications. It requires no space in occupied areas. Metal cooling panels mounted in the ceiling can be used in hospital rooms for burn patients who cannot be covered and should not be subjected to air motion. Radiant panels cannot be permitted to become sufficiently cold to cause condensation; the water must be kept warmer than the dew point of the room air. Dehumidification of the space is provided by introducing a relatively small amount of air that has been dehydrated in a cooling coil. This air also provides the required ventilation for the space. Radiant cooling panels are expensive to install compared with other systems and require careful monitoring of operation.

Radiant cooling coils that can be concealed behind valances provide relatively good comfort as a result of radiant cooling and the conductive effect of the air circulation over the valance. Radiant cooling systems generally are high in initial cost.

Electric Heating Electric heating is the most versatile heating system available, as it is flexible, requires little building space, and has a relatively moderate initial cost when compared with steam or hot water. Electric heating devices are available in all the configurations found for steam or hot water. Individual control is relatively simple. In the cases of high-quality equipment, temperature controls can provide multiple steps for varied supply temperatures, similar to the controls found in hot water heating systems. Where electric heat is provided within individual spaces, it is generally 110 or 120 V, and in some cases 230 V, all single phase. Where electric heating coils are installed within air handling units, they can be 230 to 440 V and connected to a three-phase system. In the typical coil installation, the heating elements are subdivided into steps to provide modulation of heating capacity based on the load.

Some of the major disadvantages of electric resistance heat are its higher operating cost and its impact on natural resources because of generation and transmission losses. When electricity is provided from a local hydropower system or from another operation (cogeneration), electric heat can be practical and appropriate. Electric heating systems have relatively low maintenance costs. Properly designed and selected equipment will have a relatively long life.

NOTES

1. American Conference of Governmental Industrial Hygienists. "Industrial Ventilation." In *Manual of Recommended Practice,* 16th edition. Lansing, Michigan: American Conference of Governmental Industrial Hygienists, 1980.
2. Committee on Industrial Ventilation. *Industrial Ventilation—A Manual of Recommended Practice,* 16th edition. Lansing, Michigan: American Conference of Governmental Industrial Hygienists, 1980.
3. Scientific Apparatus Makers Association. *Laboratory Fume Hood Standard: LF 10-1980.* McLean, Virginia: Scientific Apparatus Makers Association, 1980.
4. American Society of Heating, Refrigerating, and Air Conditioning Engineers. *1995 ASHRAE Handbook—HVAC Applications.* Atlanta: ASHRAE, 1995.
5. Kaplan, K.J., and G.W. Knutson. "Laboratory Fume Hoods, A Performance Test: RP70." *ASHRAE Transactions,* Vol. 84, No. 1, 1978, p. 511.

6. American Society of Heating, Refrigerating, and Air Conditioning Engineers. *1993 ASHRAE Handbook—Fundamentals.* Atlanta: ASHRAE, 1993.

ADDITIONAL RESOURCES

Codes and standards can be found in the following publications:

- American Society of Heating, Refrigerating, and Air Conditioning Engineers. *Method of Testing Laboratory Fume Hoods Standard: 110-1985.* Atlanta: ASHRAE, 1985.
- American Society of Heating, Refrigerating, and Air Conditioning Engineers. *ASHRAE Standard 62-1989.* Atlanta: ASHRAE, 1989.
- Committee on Industrial Ventilation. *Industrial Ventilation—A Manual of Recommended Practice,* 16th edition. Lansing, Michigan: American Conference of Governmental Industrial Hygienists, 1980.
- Fuller, F.H., and A.W. Etchells. "The Rating of Laboratory Hood Performance." *ASHRAE Journal,* October 1979, pp. 49–53.
- *Industrial Ventilation,* 17th edition. Lansing, Michigan: American Conference of Governmental Industrial Hygienists, 1982.
- Kaplan, K.J., and G.W. Knutson. "Laboratory Fume Hoods, A Performance Test: RP70." *ASHRAE Transactions,* Vol. 84, No. 1, 1978, p. 511.
- Moyer, R.C. "Fume Hood Diversity for Reduced Energy Consumption." *ASHRAE Transactions,* Vol. 89, Parts 2A and 2B, 1983, p. 552.
- Occupational Safety and Health Administration. "User safety." *Federal Register,* 29 CFR 1910.1003–0.1016, July 1, 1995.
- Research and Development Division, National Safety Council. *Survey Report on Lab Fume Hoods.* Chicago, Illinois: Research and Development Division, National Safety Council, 1984.
- Saunders, G.T. "Updating Older Fume Hoods." *Journal of Chemistry Education,* Vol. 62, 1978, p. A178.
- Saunders, G.T. "A No-Cost Method of Improving Fume Hood Performance." *American Laboratory,* June 1984, p. 102.
- Scientific Apparatus Makers Association. *Laboratory Fume Hoods Standard: LF 10-1980.* McLean, Virginia: Scientific Apparatus Makers Association, 1980.

CHAPTER 26

Building Electrical Systems

Mohammad H. Qayoumi
University of Missouri-Rolla

26.1 INTRODUCTION

The electrical distribution system in most college and university buildings is 13.2-kV, 4,160-V, or 2,400-V three-phase alternating current (AC). The voltage is transformed to 480/277 V or 208/120 V for distribution within the building. This chapter covers low-voltage (less than 600 V) building distribution systems.

For building electrical systems, usually one transformer is used that has 480-V three-phase secondary windings for motor control centers and 277-V single-phase windings for fluorescent lights. A second transformer is used that has 208-V three-phase windings for the three-phase 208-V loads and 120-V single-phase windings for general purpose power outlets.

From the transformer, the power goes to the main distribution panels, an assemblage of circuit breakers (Figure 26-1). Standard distribution panels have four, six, eight, twelve, sixteen, eighteen, or twenty-four circuits, each equipped with circuit breakers or fuses. A main circuit breaker is required for the main distribution panels. The power from the main panel is distributed to secondary panels and eventually to the loads.

A good design allows for at least 25 percent system growth in distribution panels and includes a directory of the circuits and the levels served. If panels are in public areas, they should be inaccessible to all but authorized personnel.

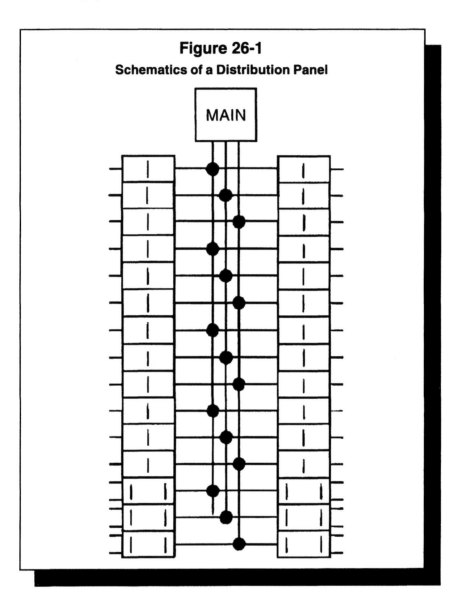

Figure 26-1

Schematics of a Distribution Panel

26.2   ELECTRICAL WIRE

A *conductor* is a material that easily conducts electricity, requiring little electrical stimulation to induce an electric current. Copper and aluminum are common conductors. The current-carrying capacity of a conductor is determined by its physical size and the ambient temperatures. Therefore, in the design of a circuit, it is important to ensure that the rate at which heat is dissipated in the conductor is equal to or greater than the rate at which heat is generated in the conductor.

All conductors have resistance, although this resistance is small. When current passes through a conductor, heat is generated. The amount of heat generated varies directly with the square of the current, and the amount of heat dissipated is a function of the circumference of the wire and the ambient conditions. To ensure a proper heat dissipation rate, wire size must be increased at a higher rate than wire capacity.

For example, to maintain a higher heat dissipation rate than heat generation rate, doubling the conductor size results in a current-carrying capacity that is less than doubled. Doubling the wire cross-section area increases the circumference by 41.4 percent. However, doubling the current increases the amount of heat generated by four times. Moreover, lower heat dissipation at higher temperatures requires derating the current capacity of the wire.

Wire Sizes

Wire size is measured in circular mils (CM). A mil is one thousandth of an inch, and a circular mil is the area of a 1-mil-diameter circle. The American Wire Gauge (AWG) is the most commonly used measure in the United States. However, AWG should not be confused with the gauge used to measure steel wire for nonelectrical applications. The AWG scale consists of wire numbers, which are even, starting with the number 40, representing a wire diameter of 0.003 in. The smaller the wire number, the larger the cross-sectional area.

Building wiring applications commonly use wire sizes of 14, 12, 10, 8, 6, 4, and 2. Wires larger than 2 are called 1/0 (one-naught), 2/0, 3/0, and 4/0. A wire larger than 4/0 is not designated by a numerical size, but rather by its cross-sectional area in mil circular mil (MCM). One MCM is equal to 1,000 CM; sample designations are 250 MCM, 500 MCM, or 750 MCM. These size designations are used for building wiring. Odd-number size designations are commonly used for magnetic wires, such as those used for motors and transformers. Number 8 or smaller wire can be solid or stranded, but number 6 and larger wire must be stranded to achieve the desired flexibility.

Wire Colors

Grounded wires are designated only by white or gray insulation. The grounding wire is identified by insulation that is green or green with yellow stripes; in some cases, the wire will be uninsulated. Hot wires have blue or yellow insulation, in that order, depending on the number of hot conductors.

Wire Insulation

Indoor building wire is suitable for voltages up to 600 V. The most common wire insulation, thermoplastic, comes in many different types. Moisture-resistant thermoplastic (type TW) and moisture- and heat-resistant thermoplastic (type THW) are the most common and can be used in wet or dry applications. When type TW is used, wire temperatures should not exceed 140.5°F; when type THW is used, temperatures should not exceed 167.5°F. Heat-resistant rubber (RHH) and heat-resistant thermoplastic (THHN) types are used only in dry locations, and the wire temperature should not exceed 194.5°F. THHN and heat- and moisture-resistant thermoplastic (THWH) types have an oil-resistant final insulation layer that adds strength and greater insulating capacity. Moisture- and heat-resistant cross-lined synthetic polymer (XHHW) type has a cross-linked synthetic polymer, which also adds strength and even higher quality insulation.

Wire Size Selection

In addition to ampacity, an important factor in wire size selection is voltage drop. The voltage drop on branch circuits should be kept below 2 percent, and that on the feeder and branch circuit, below 3 percent. It is important to keep the voltage drop to a minimum, because in addition to the energy loss, voltage drop has a negative effect on electrical appliances. For instance, a 5 percent voltage drop on an electric motor means a 10 percent drop in output power. Similarly, for incandescent lamps, a 5 percent voltage drop means a 16 percent drop in light output; for fluorescent lamps, a 10 percent voltage drop means a 3 percent drop in light output.

The following formula can be used to determine the appropriate wire size in circular mils (*Note:* 22 is the ohms per CM foot value of copper wire.):

$$CM = \frac{(\text{One way distance in feet}) \ (\text{Amperes}) \times (22)}{\text{Desired voltage drop}}$$

For three-phase applications, multiply the above formula by 0.866. Higher voltages are preferred for larger loads because they limit the voltage drop. For example, 277 V is preferred over 120 V for fluorescent lamps, and 480 V is preferred over 208 V for larger motors.

Wires in Parallel

In some instances, two or more wires can be used in parallel instead of using single wire. The National Electric Code (NEC) specifies that parallel wires must be larger than 1/0, be of the same material, have the same length and cross-sectional area, and terminate in the same manner.

Wire Splices and Terminations

The weak links in building wiring systems are usually wire splices and terminations. It is important to ensure that splices and terminations are electrically and mechanically correct. Screw-type terminals should have more than two-thirds wrap in the clockwise direction, with no overlap. Copper wires or copper-clad aluminum wires commonly require solderless connections, since aluminum rapidly forms aluminum oxide— a poor conductor—when exposed to air. The aluminum connectors should be able to penetrate the oxide layer. Currently soldering connections are seldom used for building wiring.

26.3   CONDUIT SYSTEMS

A conduit wiring system provides a high level of mechanical protection for the electrical circuits. This system reduces the probability of fire from overloaded or short-circuited conductors. With a conduit system, the circuit wires are easily replaced and easily removed, and new circuits are easily pulled if there is space. Conduits may be buried in walls or surface mounted. The ambient conditions determine the type of conduit, the type of coating, and the type of fitting. Dust-tight, vapor-tight, or water-tight conduits are available in 10-ft. lengths. The size is determined by the internal diameter in inches. The standard sizes are $^1/_2$ in., $^3/_4$ in., 1 in., 1 $^1/_2$ in., 2 in., 3 $^1/_2$ in., 4 in., 4 $^1/_2$ in., 5 in., and 6 in. The most common types of conduits are rigid galvanized conduit, intermediate metal conduit, electric conduit, rigid polyvinyl chloride (PVC) conduit, and flexible conduit.

Rigid Galvanized Conduit

Rigid galvanized conduit (RGC) provides the highest level of mechanical protection. RGC is made of heavy-wall steel that is either hot-dipped galvanized or electrogalvanized to reduce the damaging effects of corrosive chemicals found in insulations. RGC differs from wafer-type conduits in that the interior surfaces are prepared so that wires can be easily pulled. The wall is approximately 0.109 in. thick. The disadvantages of RGC are its high cost, its heavy weight, and its difficulty of installation (i.e., cutting and bending).

Intermediate Metal Conduit

Intermediate metal conduit (IMC) is similar to RGC, but the wall thickness is about 70 percent less, making it lighter, less expensive, and easier to install. However, managers should check local ordinances before installing IMC.

Electric Metal Tubing

Electric metal tubing (EMT) has a wall thickness that is about 40 percent less than that of RGC, making it lighter and less expensive. EMT is used mostly for branch circuits above suspended ceilings. Unlike RGC and IMC, EMT is not threaded into a fitting or box but mostly uses compression or set-screw fitting joints. EMT can be jacketed with PVC to make it resistant to corrosive chemicals. Proper care must be taken to prevent damage to the PVC jacket during cutting or bending.

Rigid Aluminum Conduit

Rigid aluminum conduit (RAC) is lightweight and rustproof and provides a better grounding system than RGC. Because aluminum is a nonsparking metal, it is safe when used near explosive gases. Because of RAC's relatively fragile nature, it should not be installed in concrete slabs. Rigid steel elbows usually are used with RAC.

Rigid PVC Conduit

PVC conduit is lightweight and works well even in highly corrosive areas or places where moisture and condensation are a problem. Two advantages of PVC conduit are that it has no voltage limitation, and it resists aging from ozone and sunlight exposure. Because PVC is not conductive, a grounding conductor may also be required.

Flexible Conduit

Flexible conduit should be used when a connection is needed with vibrating or moving parts, such as motors, or when rigid conduit cannot be formed to a required contour. Flexible conduit normally is used for short distances of no more than 60 ft.. PVC-jacketed, liquid-tight, flexible conduit is used for damp locations.

26.4 ELECTRIC MOTORS

Electric motors are rotating machines that convert electrical energy into mechanical energy. The two main elements in motors are the stationary elements (the starter, brushes, yoke, armature winding, and motor housing) and the rotating elements (the field winding rotor and the slip rings). Most motors can be classified as either synchronous-type or induction-type motors.

Synchronous Motors

Synchronous motors, rarely found in modern building systems, have been in use since 1890 and are almost identical to synchronous generators. The only difference between the two is the direction of flow of electrical and mechanical energy. A schematic appears in Figure 26-2. A synchronous motor has two types of rotor: the salient pole for slow speeds, and the round rotor for faster machines. Synchronous motors require AC power for the starter and DC power for the rotor.

Synchronous motor speed is directly related to the power line frequency and to the number of poles. The number of revolutions per minute equals 120 F/P (F = line frequency, and P = number of poles). A unique characteristic of a synchronous motor is that it has an average nonzero torque only at synchronous speed. At steady conditions, it runs only at one speed, making synchronous motors ideal when constant speed is required. Despite the favorable characteristics, synchronous motors are viewed as special-purpose motors because they are more expensive than induction motors and require both AC and DC power.

Three-Phase Induction Motors

In an induction motor, the stator uses alternating current. The rotor power is supplied by the stator through transformer action. Induction motors have two types of rotors: wound and squirrel-cage. A wound rotor has a polyphase winding with the same number of poles as the stator. A squirrel-cage rotor

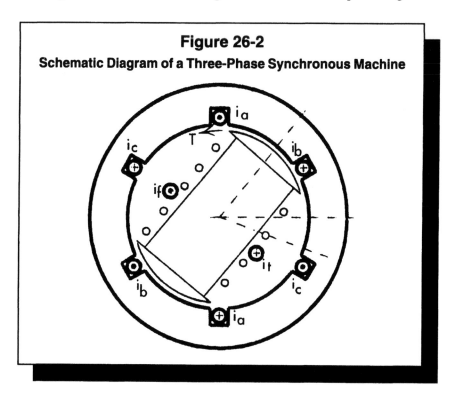

Figure 26-2

Schematic Diagram of a Three-Phase Synchronous Machine

has conducting bars in the rotor cove, which is shorted in both ends by conducting rings.

Unlike the synchronous motor, an induction motor runs slower than the synchronous speed. The per-unit difference between the actual speed and the synchronous speed is called the *slip,* which equals

$$S = \frac{n - n_1}{n}$$

where n = synchronous speed and n_1 = motor speed.

In a synchronous machine, the stator and rotor fields are stationary compared with each other, whereas in an induction machine, the stator field rotates at a slip frequency (SF). Because the slip is normally 3 percent to 10 percent, for a 60-Hz line frequency, the slip frequency is between 2 and 6 Hz. Because of the low slip frequency, the rotor impedance is mostly resistive, implying that the rotor current will be in phase with and proportional to rotor voltage. The torque characteristics of the motor are shown in Figure 26-3.

An important note: At synchronous speed, the machine torque equals zero, implying that an induction machine will never run at synchronous speed and will always run at a slower speed. Furthermore, if the mechanical torque is applied from an outside source, the motor will operate at a speed faster than the synchronous speed and operate as an asynchronous generator. This condition is called the *motoring region* of the torque curve.

If the motor is running in a particular direction and two of the phases are interchanged, the motor will produce a torque in the opposite direction of the motor's rotation. This action will slow down the motor, which, if it is still energized, will start rotating in the opposite direction. This is the *braking region* of the torque curve and is a technique for slowing down large three-phase induction motors (Figure 26-4).

The torque characteristic curve of induction motors is a function of the rotor resistance. The higher the rotor resistance, the higher the slip the motor will be running. Also, as shown in Figure 26-5, the maximum breakdown torque is achieved at a slower speed. Therefore, varying the resistance of the rotor changes the starting and running torque.

Motor designs are divided into four classifications: A, B, C, and D, as follows:

- *Class A.* Class A motors are designed for normal starting and torque applications. Class A rotor resistance is the lowest of the four classifications. These motors have low slip. The starting torque is about twice as high as the full-load torque, and the maximum torque is even higher than that. The full-voltage starting current can be as high as five to eight times the rated current, a disadvantage of class A motors. If a high starting torque is not necessary, a reduced-voltage start is achieved using an autotransformer. Another disadvantage of class A

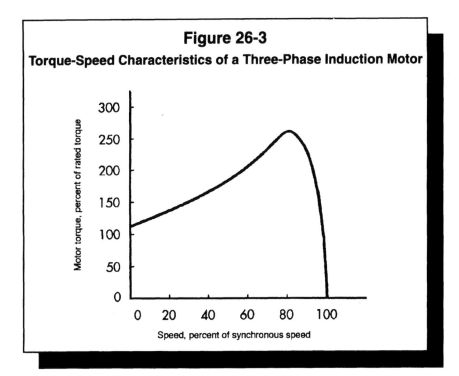

Figure 26-3
Torque-Speed Characteristics of a Three-Phase Induction Motor

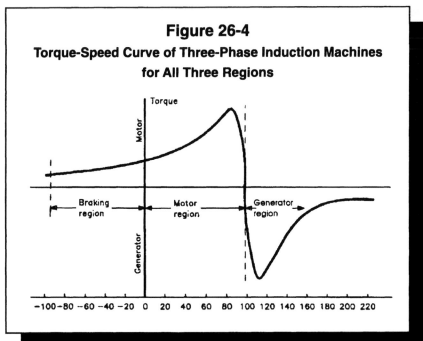

Figure 26-4
Torque-Speed Curve of Three-Phase Induction Machines
for All Three Regions

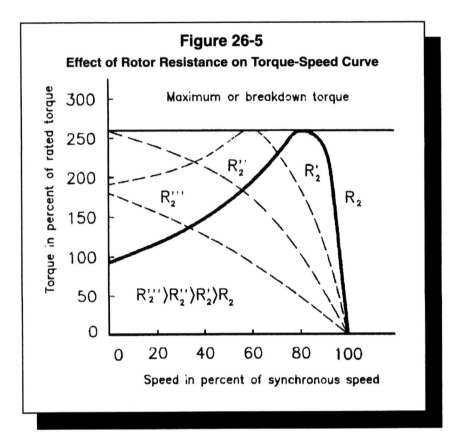

Figure 26-5

Effect of Rotor Resistance on Torque-Speed Curve

motors is that, after starting, the torque makes a small dip before beginning to increase. Under high load, the motor might not speed up beyond the dip in the torque curve. This phenomenon is called *cogging*. Cogging creates high losses in motor power, and sustained operation will heat up the motor and eventually will burn up the windings.

- *Class B.* Class B motors are designed for normal starting and running torque and low slip. Class B rotors have higher reactance than class A rotors, but the maximum torque is lower than class A. The starting torque is the same, but with 25 percent less starting current. Full-load efficiency and slip are good, but because of high reactance, the efficiency and power factor will be a little lower than class A.

- *Class C.* Class C motors are designed for high starting torque with low starting current. Class C rotor resistance is higher than that of class B, and the running efficiency is lower than that of class A and class B. These motors are usually used for high starting torque applications, such as compressors and conveyors.

- *Class D.* Class D motors are designed with the highest torque. Class D rotor resistance is higher than that of class C, and the full-load slip is about 10 percent. Class D motors are mainly used for intermittent loads with high starting torques, such as punch presses, shears, and centrifuges.

The most desirable motors have high starting torque and high running efficiency (Figure 26-6). At startup, rotor resistance is high, and as the motor approaches running condition, rotor resistance drops to a minimum value. To accomplish this, wound rotors add external resistance to the rotor circuit between the slip rings during startup and short the slip rings when the motor approaches full speed. Wound rotors also can be used for adjustable-speed drives. The main disadvantage of wound rotors is their cost.

Squirrel-cage rotors, as discussed thus far, have only one rotor resistance value, implying a compromise between high startup torque and high running efficiency. To overcome this problem, double squirrel-cage rotors with an inner and an outer cage are used. The inner cage is deep in the rotor core and has lower resistance and higher reactance. The outer cage is at the face of the rotor and has a higher resistance and lower reactance. During startup, because the rotor current frequency is close to line frequency and the outer cage has the lowest reactance, the outer cage's impedance is much lower than that of the inner cage, and most of the current flows through the outer cage. Therefore, the rotor exhibits high resistance. As the rotor approaches running speed, rotor current frequency is low, which means that reactances will have little effect on the impedance of the total circuit. Because the two cages are parallel, the effective rotor resistance is equal to the parallel resistance combination of the inner and outer cages. Therefore, the rotor exhibits low resistance at running condition.

Electric Motors and Energy Management

The total electrical energy consumed in the United States is about 20 trillion kilowatt-hours each year. Electric motors consume more than 13 trillion kilowatt-hours each year. About 60 percent of the electric motor energy consumption occurs in commercial and industrial applications. If the efficiency of all commercial and industrial motors was improved by an average of 3 percent, the energy savings would be 23.4 billion kilowatt-hours. An average of 6 cents per kilowatt-hour results in an annual savings of $1.4 billion. For example, a 30-horsepower motor with an efficiency improvement of 3 percent that works 10 hours a day for 250 days a year means a savings of $167 per year, or a total of $3,340 over the average 20-year life of the motor.

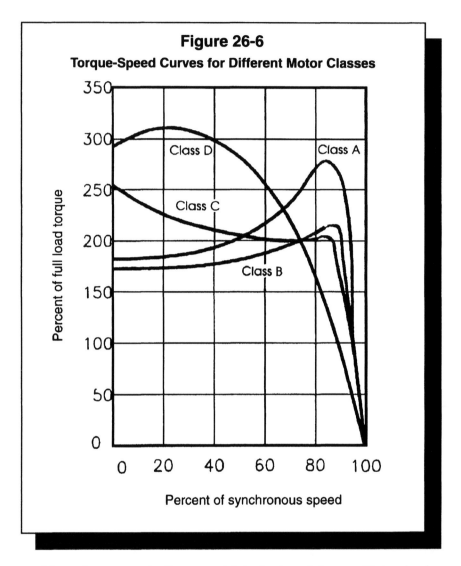

Figure 26-6

Torque-Speed Curves for Different Motor Classes

Motor losses eventually convert into heat, and lower efficiencies increase the winding temperatures, thus reducing the useful life of a motor. Motor energy losses come from three sources—namely, copper losses, iron losses, and mechanical losses.

In addition, the electrical system conditions can cause more motor energy losses. A motor operated outside its rated frequency and voltage has greater losses, and its power decreases. Voltage unbalance also increases motor energy losses. A 3.5 percent voltage unbalance means an additional 20 percent motor energy loss, as well as additional motor vibration that increases the wear and tear to the motor. Oversizing motors also increases the motor energy losses, because the motor's rated efficiency will not be achieved; it will always be lightly loaded. Oversizing motors also means higher operating costs and higher capital outlay to acquire the motor and its associated equipment, such as transformers, switchgear, wire, and conduits.

Motors should be properly sized for a given load. Oversizing motors means lower power factors for induction motors, and a power factor correction is necessary. If this correction were not made, the electric utility would charge a penalty for a low power factor, because the utility must supply a higher current to produce the same real power. To the utility this translates into larger generators, transformers, switchgear, and distribution systems and greater line losses.

Efficiency

Efficiency is a measure of a motor's effectiveness in converting electrical energy to mechanical energy. A motor's efficiency is expressed as the ratio of the motor's output to its input. Efficiency is also expressed as the ratio of a motor's output power to its output power plus losses. The lower the losses, the higher the motor's efficiency. Also, the higher the output load, the higher the motor's efficiency, which means the highest motor efficiency is achieved at a rated load, and the efficiency is lower for partial loads. Generally speaking, large motors have higher efficiency than small motors. Also, high-speed motors have greater efficiency than low-speed motors. Therefore, a higher speed motor with a reducing gear system might be more energy efficient than a lower speed motor, despite the mechanical energy losses created by the reducing gear.

In induction motors, efficiency is also a function of the motor slip. The higher the slip, the higher the motor energy losses, which causes heating of the windings and reduces the useful life of the motors. Multiple-speed motors have a lower efficiency than single-speed motors. Single-winding multiple-speed motors are more efficient than two-winding multiple-speed motors.

A small percentage increase in motor efficiency greatly reduces motor losses. For instance, an increase in motor efficiency from 85 percent to 88 percent translates into a 20 percent reduction in losses and a longer operating life for the motor.

Proper care increases the useful life of motors and ensures maintenance of good motor efficiency. A basic maintenance program should include periodic inspections and correction of unsatisfactory conditions. Inspections should incorporate checking lubrication, alignment of motor and load, belts, sheaves, couplings, tightness of the belts, ventilation, presence of dirt, input voltage, percentage of unbalance, and any changes in load conditions. Dust buildup on fans, misalignment of gears and belts, and insufficient lubrication increase motor friction, thus reducing the efficiency and life of the motor (Figures 26-7 through 26-10).

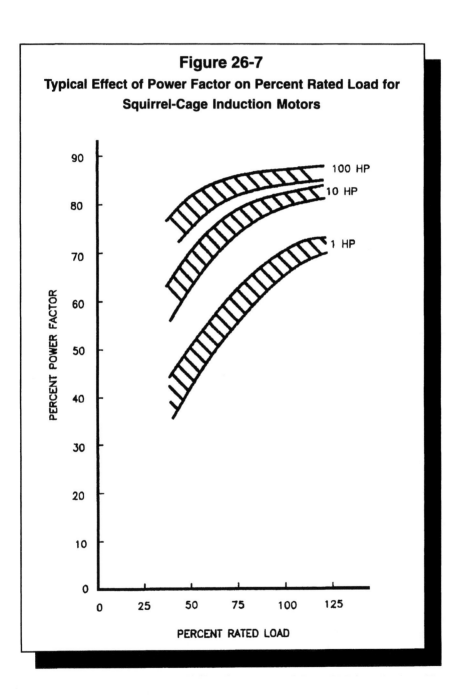

Figure 26-7

Typical Effect of Power Factor on Percent Rated Load for Squirrel-Cage Induction Motors

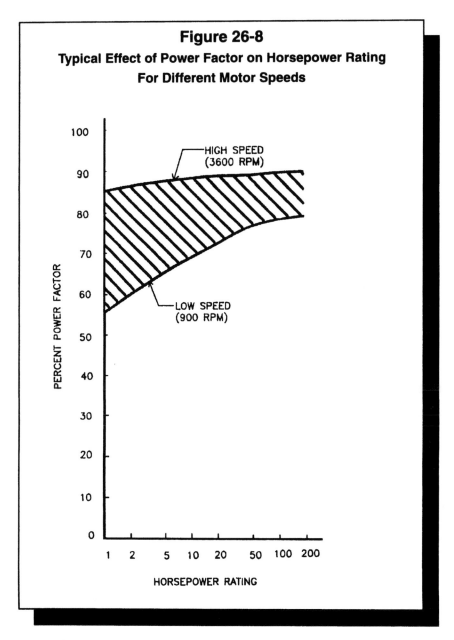

Figure 26-8

**Typical Effect of Power Factor on Horsepower Rating
For Different Motor Speeds**

Motor Control Centers

There are two types of motor control switches: manual and magnetic. Manual switches are used for fractional horsepower motors and typically consist of a toggle switch that turns the motor on or off. Such switches are simple and inexpensive.

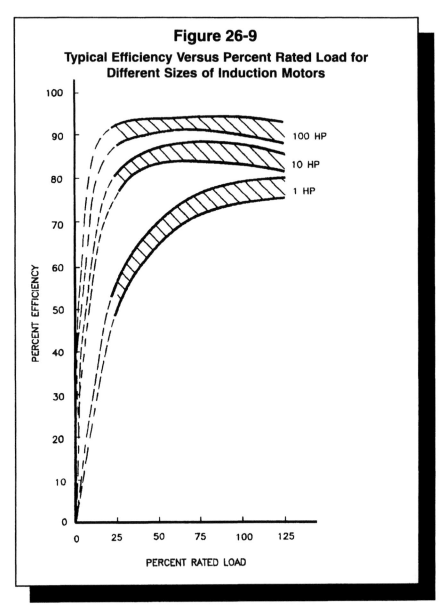

Figure 26-9

**Typical Efficiency Versus Percent Rated Load for
Different Sizes of Induction Motors**

The magnetic switch has three advantages. First, it has a low-voltage protection feature that disengages the electromagnetic unit if the line voltage drops below a certain level, thereby preventing the motor from overheating. The unit remains disengaged even if the voltage level goes up again. The second advantage of the magnetic switch is that it can be operated by a variety of methods. Finally, for larger motors and higher voltage applications, the magnetic switch allows the use of low voltage for the control voltage.

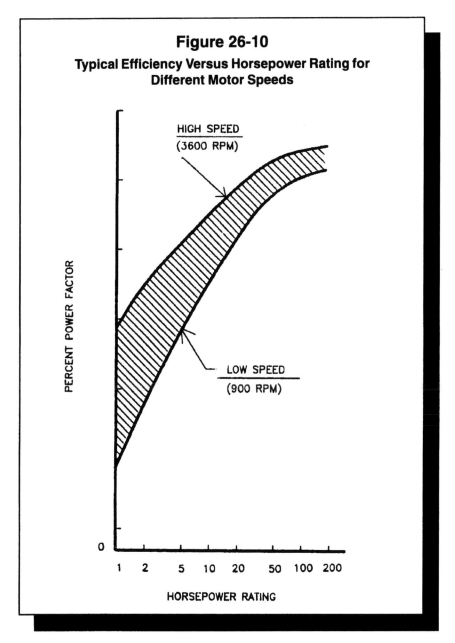

Figure 26-10

Typical Efficiency Versus Horsepower Rating for Different Motor Speeds

Motor startup is accomplished by full-line voltage or reduced voltage. For larger motors, if initial torque is not a problem, reduced startup voltage is used, especially if system capacity is limited. Currently, with the advent of power electronics, motor starters are mostly solid-state devices. These starters are more reliable and robust than the old electromagnetic contactors.

Variable-Frequency Drives

Most industrial processes require variable-speed motors. The speed of an AC motor is determined by the line frequency of 60 Hz. To effectively vary motor speed, a variable-frequency drive (VFD) must be used. A VFD consists of a DC rectifier, a filter circuit, and an invertor (Figure 26-11). Sixty-hertz line voltage is converted to DC voltage and then to AC voltage. The control circuit determines the output frequency and, thus, the speed of the motor.

The invertor circuit changes DC voltage into three-phase variable-voltage/variable-frequency AC output. The invertor consists of at least six thyristors, each conducting 180 degrees per cycle. The switching sequence produces a three-phase output voltage. The output frequency is determined by control circuitry.

The three common types of variable-frequency drives are adjustable voltage input (AVI), current source invertors (CSIs), and pulse width modulation (PWM). The voltage in AVI is controlled by the DC input to the invertor and requires the use of either a silicone-controlled rectifier (SCR) or diode rectifiers in conjunction with a chopper to carry input voltage. AVI is the simplest of the three invertor types in .erms of control circuitry and also has regeneration capability (Figure 26-12).

CSIs receive DC input voltage from an SCR bridge in series with a large inductor, creating the current source (Figure 26-13). In a PWM invertor, the output voltage wave form has a constant amplitude with periodically reversing polarity that provides the output frequency. The output voltage also is varied by changing the pulse width. PWM has the most complicated control logic and the lowest efficiency, but it also has fewer harmonic problems compared with the other two types. PWM is becoming popular for many applications (Figure 26-14).

26.5 LIGHTING

Introduction

Lighting deals with providing visibility so that humans can visually experience their surroundings. Natural lighting is provided by the sun, and artificial lighting is predominantly provided by converting electrical energy into lighting energy in the frequency range visible for humans. There are many reasons why lighting is needed for an area; these reasons can be divided into three general categories: functional lighting, safety and security lighting, and ornamental or architectural lighting.

Functional lighting is needed to complete a particular human task, such as working in an office, walking along a sidewalk, or driving a car in a parking lot. Lighting level and the quality of light necessary for people to perform the required task have been the subjects of many management

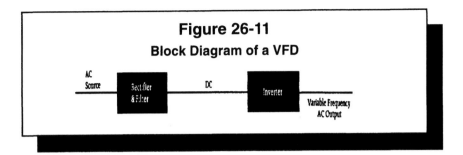

Figure 26-11
Block Diagram of a VFD

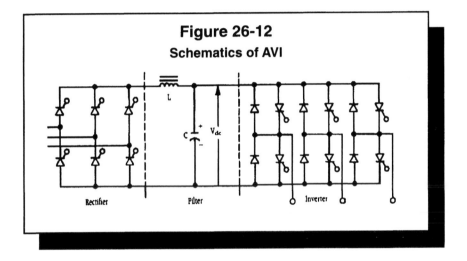

Figure 26-12
Schematics of AVI

studies since World War II. The reason for the high interest in this topic is that labor costs are a large percentage of most industrial production costs, so even a small percentage gain in productivity can have a significant impact on the net revenue of an enterprise. Because lighting interacts with the work environment, lighting designers must be sensitive to the needs of the people who are affected.

Safety and security is the next major category of lighting needs. Many individuals associate lighting level with safety. This perception has been the source of long debates. Both the quantity of light and lighting quality are important. In outdoor applications, especially in parking garages, improper lighting can result in a variety of problems such as shadow zones, reduced visibility, a loss of direction, a feeling of claustrophobia, and a sense of insecurity. Both horizontal and vertical illumination levels must be considered for these applications. Another important factor for safety and security lighting is the location of lights in relationship to landscaping.

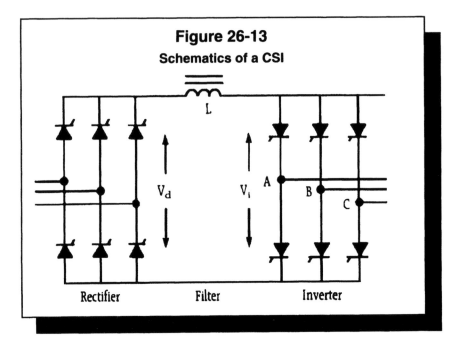

Figure 26-13

Schematics of a CSI

Rectifier   Filter   Inverter

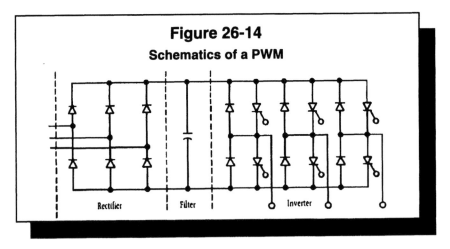

Figure 26-14

Schematics of a PWM

Rectifier | Filter | Inverter

Sometimes large quantities of light may be blocked by tree leaves, large bushes, or other vegetation, and therefore lighting locations should be properly coordinated with the site landscaping.

The third general category of lighting deals with architectural lighting. The main purpose of this category of lighting is aesthetic lighting, which primarily deals with a particular mood or visual impact the designer is trying to create. Because of the ornamental nature of this type of lighting, functionality is not a major concern.

Elements of Lighting

Electrical lighting is generated when electrical current flows through either a resistance filament or a gas. In the first case, the filament is heated to such high temperatures that it starts glowing; in the second case, the gas molecules' excited atoms emit radiant energy. There are basically three types of lighting: incandescent, fluorescent, and high-intensity discharge (HID) (e.g., mercury, metal halide, high-pressure sodium, and low-pressure sodium). All gaseous lamps, such as fluorescent and HID lamps, need a ballast to start the lamp and regulate the current. The entire lighting assembly—lamp, ballast, lens, and associate housing—is referred to as a *luminaire.*

The quantity of light in an area is related to *luminous flux,* the amount of light emitted by a lamp, and *illuminance,* the amount of light reaching the surface area. The initial lumen output of a new lamp will be high, but after the lamp operates for about 100 hours, the light output stabilizes for the useful life of the lamp. When the lamp gets old, the light output drops rapidly over time. Other factors that affect the light output level of a lamp are the amount of dirt and dust accumulated on the fixture, the reflectance of the room surfaces, room proportions, and surface materials.

In addition to the quantity of light, the color of light is important. *Color rendering* refers to how accurately colors and color shades are rendered and is measured using a color rendering index (CRI) ranging from 0 to 100. Typical fluorescent lamps have a CRI rating of about 65.

Lighting efficiency is measured in lumens per watt. The most common types of lamps—incandescent, fluorescent, metal halide, high-pressure sodium, and low-pressure sodium—are discussed in the sections that follow.

Incandescent Lamps Incandescent lamps are the oldest and simplest type of lamps that consist of a tungsten alloy filament. When the filament is raised to a high temperature, it glows, and light is emitted. Because of their ease and simplicity of construction, incandescent lamps are the simplest and least expensive lighting system. They have a high CRI rating, which makes them desirable for many indoor functions where color rendition is important, such as art galleries and painting studios. Incandescent lamps have an efficiency of 10 to 15 lumens per watt, which is the lowest of all commercially available lighting systems. Moreover, the average useful life of a lamp is about 1,000 hours. For these reasons the use of incandescent lamps has continually decreased for commercial and industrial applications. Many retrofit kits are available to convert existing incandescent lighting systems to more efficient systems.

Fluorescent Lamps The fluorescent lamp has been the most widely used type of lamp for commercial applications. The main reasons are its relatively high efficiency, its moderate cost, and the maturity of the technology, which has been around for many decades. Moreover, unlike other gas-filled lamps (e.g., HID), fluorescent lamps can be turned on and off frequently and will reach their full brightness rapidly. The typical efficiency of a fluorescent lamp ranges from 60 to 75 lumens per watt, and the average lamp life is between 8,000 and 10,000 hours. However, color rendition is not as good as that of incandescent lamps. The typical CRI of a fluorescent lamp is 65, which is acceptable for most applications, except where color rendition is particularly important. Fluorescent lighting needs a ballast and starter for the fluorescent lamp to operate.

Metal Halide Lamps The metal halide lamp is an HID lamp that is becoming popular for indoor applications. Such lamps are commonly sized from 32 to 1,000 W and produce excellent results for color rendering applications. Their high efficiency and long life make them a good choice for both indoor and outdoor applications. The efficiency of these lamps can range from 78 to 110 lumens per watt and the lamp life ranges from 10,000 to 20,000 hours. Generally, the larger wattage lamps have a higher efficiency.

There is a safety precaution that must be remembered with metal halide lamps to avoid possible shattering and early failure of the lamp. These lamps are constructed of an outer bulb with an internal arc tube made of quartz. The arc tube operates under high pressure and at approximately 1,100°C. The arc and outer bulb may unexpectedly rupture as a result of system failure or misapplication. To safeguard against any hazards, the following three precautions must be taken:

1. Lamps must be operated only in fixtures with lenses or diffusers that can contain fragments of hot quartz or glass up to the above mentioned temperature.
2. If a lamp is supposed to operate continuously, it should be turned off at least once a week for at least 15 minutes. Otherwise, the risk of rupture will increase greatly.
3. Certain types of metal halide lamps will automatically extinguish if the outer skin is punctured. This is a desirable feature, because if a lamp with a broken outer bulb is allowed to operate, the ultraviolet light emitted from the lamp may cause serious skin burns and eye inflammation.

High-Pressure Sodium Lamps The high pressure sodium (HPS) lamp is a high-intensity discharge lamp that can be used both for indoor and outdoor applications. Nominal lamp sizes range from 35 to 1,000 W. HPS has a high efficiency (typically 64 to 140 lumens per watt) and relatively long life (10,000 to 24,000 hours). The main objection to HPS lighting is its low CRI rating, which is about 22. With color correction this factor can be increased to 65, but as would be expected with better color rendition, there is an efficiency drop as well as a higher cost. Despite their poor color rendition, these lamps are desirable because of relatively higher efficiency and long life. HPS lamps can be used for parking garages, gymnasiums, swimming pools, transportation centers, malls, walkways, parks, and general security lighting.

Low-Pressure Sodium Lamps   The low-pressure sodium (LPS) lamp has the highest efficiency compared with other types of lamps. The typical lumens per watt ranges from 100 for 18-watt lamps to 163 for 135-watt lamps, and the average lamp life is 16,000 hours. The major shortcoming of LPS lighting is its poor color rendition (the CRI rating for LPS lighting is 0), and for this reason these lights are not desirable in areas used by people; such light can be used only in areas where color rendition is not an issue. LPS lighting is commonly used for extremely large areas such as warehouses and general security lighting.

Effect of the Energy Policy Act of 1992 on Lighting

Lighting is a major energy user in commercial and institutional buildings and is usually the first to be reviewed and reduced. Depending on the particular type of facility and the geographical location, the lighting load can account for 12 to 28 percent of a building's total energy use. The Energy Policy Act of 1992 created a number of requirements that fundamentally affected the lighting industry. In particular, lamps were affected under the provisions in two areas: efficiency labeling and efficiency standards. The motivation was to establish minimal efficiency standards for various types of lamps and enable consumers to select the most energy-efficient lamps to meet their application requirements.

The labeling standards became effective in 1995 along with some minimum efficiency standards established for certain types of fluorescent and reflector and parabolic reflector (R & PAR) incandescent lamps. Lamps that do not meet the efficiency standards can no longer be manufactured in the United States or imported. This will eliminate many widely used incandescent and 35-watt fluorescent tubes. For HID lamps, the Department of Energy (DOE) has been directed to investigate whether establishment of standards is technologically feasible, and whether such standards will result in significant energy savings. The second review of efficiency standards will be initiated in April 1997, and these standards will become effective three years later. Any revisions during this review process are to be completed by April 2002 and will become effective three years later.

Exit Lighting

Exit lighting is addressed separately because it may provide a unique savings opportunity compared with other types of lighting. The underlying reason is that exit signs operate 24 hours a day, 365 days a year, and therefore, any energy efficiency improvement will have significant financial impact. Traditionally, exit signs were lighted with a 40- to 60-W incandescent lamp. Aside from their high energy use, they had to be replaced several times a year. Later, after such fixtures were retrofitted with fluorescent lighting, the energy consumption for many facilities dropped to approximately 10 W per fixture. Another advantage of the retrofit was the higher average life for fluorescent technology, thus reducing lamp replacement frequency.

Since the 1980s, self-luminance exit signs have become popular for some installations. These fixtures contain a very low level radioactive material that radiates energy in the visible light spectrum without the use of any power source. The advantage of self-luminous exit signs is that no power connection to the fixture is required, and thus the installation cost is lower. The main disadvantage of these fixtures is the high disposal cost. Because of the radioactive material inside the fixture, the fixture must treated as hazardous radioactive waste. Because of this shortcoming, such lighting is not recommended.

More recently, a new type of exit sign has gained popularity. These units, which utilize light-emitting diodes as the source of light, consume about 5 W per fixture and have an average life of about 30,000 hours. In addition, they can have a battery pack in the unit so that the exit sign will continue to operate during short-term power outages. Although the cost of these units is relatively high, they have dropped in price since 1990.

Lighting Application

Placement of lighting deserves careful planning. The lighting levels provided should be uniform, with the maximum ratio between the highest and lowest level being less than 2:1 to prevent glare and undesirable shadows.

Indoor lighting can be direct or indirect. Indirect lighting is softer to the human eye and more desirable for many indoor commercial and residential applications. However, it requires surfaces with good reflectance and is more costly. For these reasons, even in most commercial applications and all industrial applications, direct lighting is used.

For outdoor lighting, an important phenomenon that must be considered is lighting spill to surrounding areas, which could become a source of nuisance for neighboring customers.

To maintain an acceptable lighting level, lamps should be replaced when they reach their average useful life, even if they are still functioning. When practical, group relamping is preferred. In the long run, group relamping is recommended over spot relamping. Another important practice is to clean the lighting fixtures periodically. As dirt and dust accumulate in a fixture, the quantity of light delivered depreciates significantly. This is a function of the ambient conditions. In a clean environment the light loss is about 10 percent, whereas in a dirty environment it can be as high as 50 percent. Cleaning light fixtures on a regular schedule is recommended.

In the past, lighting designers added large safety margins to compensate for lamp dirt depreciation and lamp aging factors. Currently most lighting systems are designed with relatively smaller safety margins. Therefore, if a few fixtures are burned out or the fixture becomes dirty, the drop in lighting level will be significant. Therefore, it is important for facilities managers to be sensitive to this fact and, where possible, to institute a light fixture cleaning program.

Lighting Design Calculations

Two definitions are helpful in understanding lighting design calculations:

- *Lumen:* The quantity of light striking an area of 1 sq. ft., all points of which are 1 ft. from a 1-candlepower lighting source.
- *Foot-candle:* A unit of measurement representing the intensity of illumination on a surface that is 1 ft. from a 1-candlepower light source and at right angles to the light rays from that light source. One foot-candle (FC) is equal to 1 lumen per square foot.

The lighting levels recommended by the Federal Energy Administration are 75 FC for prolonged office work that is visually difficult; 50–80 FC for classrooms, libraries, and offices; and 5–15 FC for hallways and corridors. For comparison, the illumination levels of some natural light sources are as follows:

- Starlight: 0.0002 FC
- Moonlight: 0.02 FC
- Daylight: 100–1,000 FC
- Direct sunlight: 500–1,000 FC

Several methods exist for calculating the number of fixtures required for a particular area. One of these methods—the lumen method—is summarized as follows:

1. Determine the required lighting level.
2. Calculate the room cavity ration (CR):

$$CR = \frac{5 \text{ x room height x (room length + room width)}}{\text{Room length x room width}}$$

Using the lighting tables, the cavity ratio, combined with wall reflectance and ceiling cavity reflectance, gives the coefficient of utilization (CU). The CU is a function of the ceiling and wall reflectance, as well as of room geometry; it is the measure of the ratio of the lumens reaching the working surface to the total lumens generated by the lamp. The higher and narrower a room, the larger the percentage of light absorbed by the walls, and the lower the CU value.

3. Determine the illumination loss factors. From the time a luminaire is installed and energized, a number of factors contribute to loss of illumination. The major factors contributing to loss are age, dirt depreciation, and voltage-to-luminaire variation. Variation in voltage to luminaire is a bigger problem for incandescent lamps than for fluorescent lamps. For an incandescent lamp, a 1 percent change in line voltage causes a 3 percent change in lumen output. For a fluorescent lamp, a 2.5 percent change in line voltage causes a 1 percent change in lumen output.
4. Calculate the number of fixtures required:

$$\text{Number of fixtures} = \frac{\text{FC x Area}}{\text{Lumens per lamp x CU x Light loss factor}}$$

5. Determine the location of lights based on the general architecture, task, and furniture arrangement.

The quality of light is just as important as the quantity of light. Quality of light involves proper light color, proper light distribution, and lack of glare. A relatively bright area within a relatively dark or poorly lighted area causes glare.

Lighting Control

A pole switch is the simplest form of light control. A dimmer switch supplies illumination control and can be surface mounted, flush mounted, or controlled by a pull chain. For safety, the control switch should be mounted on the hot wire rather than the neutral wire. If light control is needed from two locations, then three-way switches should be used, as shown in Figure 26-15. If additional controlling points are required, then two three-way and four-way switches should be used (Figure 26-16). The "on/off" designation is not found on three-way and four-way switches, because the lights can be on or off depending on the position of all switches.

A lighting contacting should be used for controlling large banks of light. The circuitry is similar to that used for a direct-start motor control center. Here the control voltage can be different from the lighting voltage, and the lights can be controlled remotely.

Lighting is an involved discipline, and a short discussion cannot cover the topic. However, it should be mentioned that lighting consumes nearly 35 percent of the nation's electricity, which is why improving lighting efficiency was greatly emphasized in the 1992 Energy Policy Act. Managers should refer to this legislation regarding specific recommendations concerning lighting and expected minimum efficiency standards. The references at the end of the chapter provide a more comprehensive treatment of this topic.

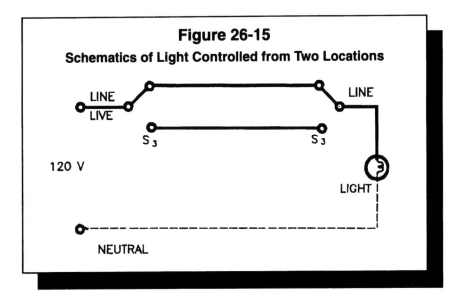

Figure 26-15

Schematics of Light Controlled from Two Locations

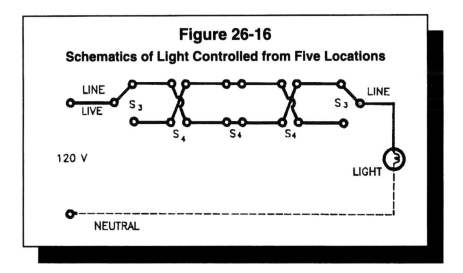

Figure 26-16

Schematics of Light Controlled from Five Locations

26.6   ELECTRIC POWER QUALITY

Currently the electrical distribution system is cluttered with a wide variety of nonlinear devices that generate power disturbances and interferences. These disturbances, commonly referred to as *electrical noise,* are the result of transients and harmonics. Both can have a damaging effect to sensitive electronic equipment. Power quality problems cost the industry about $2 million a day. Moreover, it is estimated that by the year 2000, 50 percent of power will be consumed by nonlinear load.

There are two types of noise: normal-mode and common-mode. Normal-mode noise exists when a noise voltage appears equally in each line-to-line and line-to-neutral connection. Common-mode noise occurs between line-to-ground or neutral-to-ground connections. Both can negatively impact sensitive electronic devices, resulting in system failure, component damage, database corruption, disk head crashes, and logic errors, among others.

Transients

A *transient* is a high-amplitude, short-duration electric pulse that is super-imposed on the normal sinusoidal AC voltage. The duration of a transient can range from 0.5 to 200 ms, and the voltage can rise thousands of volts per microsecond. In other words, a transient is a high-amplitude noise that can result in severe hardware fatigue and failure, in addition to other problems mentioned earlier. There are two principal sources of transients: lightning and switching surges. The impact of transients can be minimized with surge suppressors and active power line conditioners.

Harmonics

Ideally, AC electricity is a pure sinusoidal wave of one single frequency and can be referred to as *clean power.* More specifically, in the United States this frequency is 60 cycles per second, or 60 Hertz (Hz). In a power distribution system, as long as the circuit devices consist of linear elements, (resistors, unsaturated inductor, and capacitors), the AC power wave shape will remain the same. However, as soon as nonlinear equipment (switching devices, asymmetrical devices, saturated inductor) are introduced into the distribution system, the wave shape of the AC power will be distorted. The distortion wave can be mathematically analyzed as the summation of integer multiples of the fundamental AC power, known as *harmonics.* In other words, for a 60-Hz system, the frequencies for the second, third, and fourth harmonics are 120 Hz, 180 Hz, and 240 Hz, respectively.

One of the common measures of calculating the influence of harmonics is total harmonic distortion (THD), which is the summation of the root mean square of all harmonics as a relative percentage of the fundamental frequency; there are many devices available to analyze power harmonics. There are three major classes of nonlinear devices that are sources of harmonics in power distribution systems: switching devices (e.g., variable-frequency drives, synchronous generators, converters), ferromagnetic devices (e.g., transformers, motors, reactors), and arcing devices (e.g., fluorescent, mercury vapor, and sodium vapor lighting and arc furnaces). Approaches that can reduce the impact of harmonics are discussed in the sections that follow.

Derating Distribution Transformers

In the absence of any harmonics, transformers can be fully loaded to their rated value, under normal ambient conditions, without any problems. If there are harmonics in the system, then the transformer must be derated accordingly. For instance, if the third and fifth harmonics are 20 percent each, then the transformer should be derated by 8 percent. If the third and fifth harmonics are about 70 percent each, then the transformer will be overloaded even when connected to a small load.

Oversizing the Neutral Conductor

In a three-phase wye system, the neutral conductor carries only the unbalanced current. Therefore, if the system is balanced, the current flowing in the neutral wire is zero. However, this is only true in the case of the fundamental frequency current. If there is any third harmonic (or any other triplen) current in the three-phase system, the neutral current, instead of canceling out for the fundamental frequency, will add up. The neutral current can potentially be as high as 1.73 times the phase current. Because the neutral circuit is not normally protected by an overload device, such an overload will likely result in burned wires and electrical fires, especially at the connectors and splices. Therefore, facilities personnel should avoid the use of shared neutral wire when supplying a single-phase nonlinear load, or double the size of neutral wire when shared neutral must be used.

Loading Circuit Breakers

Harmonics can result in nuisance trips of circuit breakers operating near their design trip point because of the peak current heating of the contacts and the vibrations induced by the higher harmonic currents. It is recommended that when serving nonlinear loads, including computers, the panel circuit breakers should not be loaded above 80 percent of their continuous loading capacity.

Power Factor Considerations

The presence of harmonics reduces the system power factor to a lower value. Adding more capacitors could potentially cause resonant conditions, attracting high-frequency currents, and result in overheating or failure of the capacitors. To avoid such problems, a harmonic trap, which is a series LC circuit tuned to the lowest harmonic, can be installed.

Finally, to the degree possible, use low-impedance distribution transformers connected in a delta-wye configuration. This way the third harmonic will be trapped in the delta winding.

Power Quality Considerations

As facilities personnel experience further proliferation of sophisticated electronic equipment, the problem of power quality will become more prevalent. In most situations, determining the source of the problem is the challenging part. The first question is determining whether the problem is on the line side or the load side. A simultaneous increase in current and a decrease in voltage indicate that the problem is downstream. However, if there is no change in the current, the problem is upstream. If the voltage THD is more than 6 percent for branch circuits or above 4 percent at service entrance, then all THD is from one harmonic; if THD is concentrated at higher harmonics, there should be a cause for concern.

Lack of attention to power quality problems can result in equipment downtime, premature equipment failure, and high service cost. There is no standard solution that can work in every situation. This is yet another challenge that today's facilities manager must overcome.

26.7   POWER FACTOR

Power factor is the ratio of real power (watts) to apparent power (volt-amperes) and is determined vectorially as the cosine of the angle between the line voltage and the current (Figure 26-17). With alternating current and the presence of nonresistive loads (motors and transformers), the real power will be less than apparent power, so power factor will be less than unity.

In other words, line voltage multiplied by line current results in apparent power, or VA. Apparent power can be graphically broken into two components: real power (watts) and reactive power (or VARs) (Figure 26-18).

Notice that for the same apparent power, real power varies inversely with the size of the angle. Moreover if the angle is positive, it is called a *leading power factor;* and if the angle is negative, it is called a *lagging power factor.* In most power systems, the current lags relative to the voltage principally because of motor load, but also, to some extent, because of other loads. Therefore, more current is required to provide a given amount of real power if the power factor is less than unity.

A low power factor implies lower system efficiency. Most electrical utilities include a power factor penalty in their rate structure to discourage customers with low power factors. In addition, a close-to-unity power factor is important to reduce system loss, transformer size, cable size, and power cost and to help stabilize the system voltage.

Power factor improvement is accomplished in two ways: by operating equipment at unity power factor and by using auxiliary devices to supply the magnetizing power, or kiloVARs, needed by the load. Equipment that operates at unity power factor are incandescent lamps, resistance heaters and unity power factor synchronous motors.

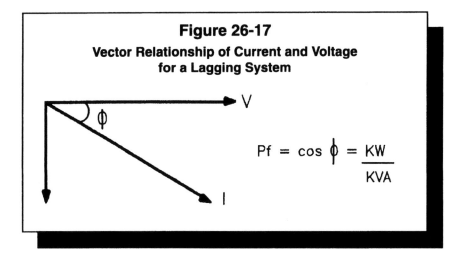

Figure 26-17

Vector Relationship of Current and Voltage for a Lagging System

$$Pf = \cos \phi = \frac{KW}{KVA}$$

Auxiliary devices for power factor improvement include the following:

• Overexcited synchronous motors
• Synchronous condensers
• Overexcited synchronous or rotary converters
• Capacitors

Capacitors are the best choice in most applications because they have no moving parts and the losses are less than 1 percent. Capacitors can be located at the loads, in the substations, or on the line. Power factor correction on the line is generally less costly, but it is better to position the power factor correction capacitor close to the device causing the low power factor. The capacity needed to improve power factor is calculated by the general formula shown in Figure 26-19.

However, it is usually not necessary to go through laborious calculations. Multiplication factors can be determined by using a power factor correction table.

To illustrate the system efficiency improvement produced by a power factor improvement, consider a 100-HP, three-phase, 208-V motor that has a power factor of 80 percent. If capacitors are added to improve the power factor to 85 percent, 90 percent, 95 percent, or 100 percent, line losses will drop by 12 percent, 21 percent, 30 percent, and 35 percent, respectively. If the motor is partially loaded, the power factor is even lower, and adding corrective capacitors will reduce line losses even more. Overcompensating for power factor should be avoided because in addition to wasting funds, it results in higher-than-system localized voltages, which can damage certain equipment.

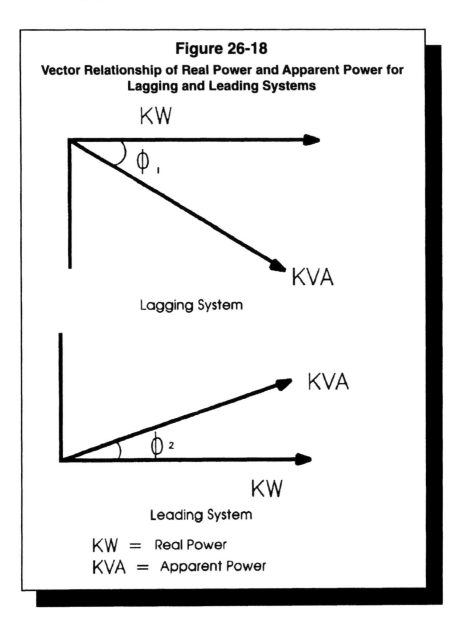

Figure 26-18

Vector Relationship of Real Power and Apparent Power for Lagging and Leading Systems

KW

ϕ_1

KVA

Lagging System

KVA

ϕ_2

KW

Leading System

KW = Real Power

KVA = Apparent Power

After capacitors are installed, proper care is needed. The capacitor's life is shortened by overheating, overvoltage, and physical damage. A capacitor inspection should include a check of fuses, ventilation, voltage, and ambient temperature. Prior to any capacitor maintenance, the capacitor must first be disconnected and discharged through a heavy-duty, 50-kΩ resistor between terminals and to ground. Discharging large capacitors by short-circuiting

Figure 26-19
Derivation of the Capacitor Table Formula

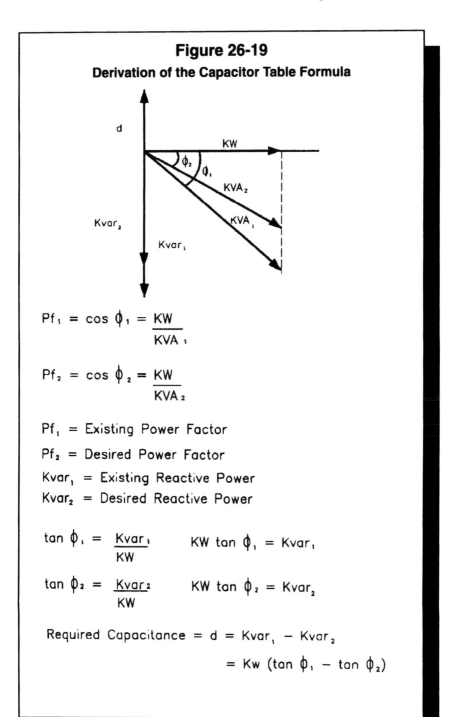

$$Pf_1 = \cos \phi_1 = \frac{KW}{KVA_1}$$

$$Pf_2 = \cos \phi_2 = \frac{KW}{KVA_2}$$

Pf_1 = Existing Power Factor

Pf_2 = Desired Power Factor

$Kvar_1$ = Existing Reactive Power

$Kvar_2$ = Desired Reactive Power

$$\tan \phi_1 = \frac{Kvar_1}{KW} \qquad KW \tan \phi_1 = Kvar_1$$

$$\tan \phi_2 = \frac{Kvar_2}{KW} \qquad KW \tan \phi_2 = Kvar_2$$

Required Capacitance = $d = Kvar_1 - Kvar_2$

$$= Kw (\tan \phi_1 - \tan \phi_2)$$

should be avoided, because the sudden release of a large discharge can injure the operator.

26.8 GROUNDING

Grounding eliminates the possibility of shocks and minimizes lightning damage; it is a required safety item for secondary distribution. There are two distinct grounding categories. First is system grounding, which means grounding one of the current-carrying wires to avoid a floating system. Second is equipment grounding, which consists of placing a non–current-carrying wire between the metallic frame of electrical equipment and the conduits, or armor of a ground rod. Therefore, a reference to "grounded wire" usually refers to a system ground, in which case, if the circuit is on, the wire is hot. However, "grounding wire" refers to the equipment wire that has no current under normal conditions.

Whenever the topic of electric shock is discussed, the question of danger normally arises. Shock danger is not related to the high voltage of the equipment, but to the current flowing through the human body. A current of one milliampere (mA) is not perceptible. A current of 1–8 mA causes mild to strong sensations. A current of 8–15 mA is unpleasant, but the shock victim can release the object producing the electrical current. A current above 15 mA causes a muscular freeze, preventing the victim from releasing the object producing the current, and a current of more than 75 mA is fatal.

Ground Fault Protection

The National Electric Code (NEC) requires ground fault equipment protection. Ground fault interrupters (GFIs) protect equipment from low-grade faults that are not large enough to be interrupted by conventional overcurrent protection devices, such as fuses or circuit breakers. The function of GFIs is often misunderstood. It is easier to state what a GFI is not. GFIs do not protect people or prevent shocks. They neither prevent ground faults nor protect equipment from a high-grade fault. GFIs only protect equipment from low-grade faults. The principle behind GFIs is simple: With the use of a current transformer surrounding the live wires, the current remains within the wires, and the net current is zero. However, if leakage occurs, the net current is not zero, and the GFI will trip the circuit breaker.

A GFI is required if the line-to-ground voltage in a grounded wye service is between 150 V and 600 V and service disconnect is rated at 1,000 A or more. For health care facilities, at least one additional layer of ground fault protection should be provided downstream toward the load. Like conventional current protection systems, GFIs should be coordinated to ensure selectivity. As mentioned earlier, a GFI does not prevent ground fault. To eliminate the possible causes of ground fault, required maintenance is recommended, which should include cleaning insulators, replacing cracked insulators, and tightening loose connections.

ADDITIONAL RESOURCES

Clidero, Robert K. *Applied Electrical Systems for Construction.* New York: Van Nostrand Reinhold Co., 1982.

Earl, John T. *Electrical Wiring Design and Application.* Englewood Cliffs, New Jersey: Prentice Hall, 1986.

Lighting Handbook. Bloomfield, New Jersey: North American Philips Lighting Corporation, 1984.

Richter, Herbert P., and Creighton W. Schwan. *Practical Electrical Wiring.* New York: McGraw-Hill, 1987.

Williams, Dan R. *Energy Policy of 1992.* Englewood Cliffs, New Jersey: Prentice Hall, 1994.

CHAPTER 27

Control Systems

John P. Sluis
Iowa State University

27.1   INTRODUCTION

Building heating and cooling control systems are independent systems that coordinate the activity of the building heating and cooling systems to create a pleasing environment for the activities in the facility. Control systems come in various types and levels of sophistication; however, the desired result is a facility where the environment is comfortable, and that comfort is achieved at the lowest possible energy cost.

As modern facilities and available technology have evolved, control systems have grown in sophistication. Current systems coordinate comfort, manage energy use, report problem areas, and coordinate information to assist facilities managers in facilities operation.

Control systems in university facilities differ significantly from those found in homes and small commercial buildings. University facilities are complex. Continuous and variable thermal and moisture loads occur. Loads change throughout the day, based not only on the weather, but on the influx of people and activities as well. In the commercial environment there is continuous exhaust of contaminated air to ventilate the occupants. There is a higher expectation for comfort than people expect in their homes. There is also a need to minimize the cost to operate the building systems

For these reasons, the heating and cooling systems in facilities have evolved to a high level of complexity. As systems have evolved, the need to coordinate the environmental systems has evolved. Building control systems are independent systems that coordinate the activity of the building systems to create a pleasing environment for the activities in the facility.

27.2 THE FUNDAMENTALS OF CONTROL SYSTEMS

Sensor to Controller to Controlled Device

Control systems in buildings all have a common concept. The concept is known as *sensor to controller to controlled device*. A control system is assigned to a building function, such as temperature, humidity, pressure, or cleanliness of the building environment. For example, the process of controlling a room's temperature is a typical and frequent situation in a building. To control a room, we must have something that can determine the temperature of the room, a *sensor*. The sensor, however, merely determines the room temperature. There is a need for a device to determine a set point (the desired temperature) and to determine how quickly and in what direction a change will occur if needed (how to get the desired temperature). This device is the *controller*. In the case of the control of a room's temperature, the device most often used is a thermostat. The thermostat is a sensor and a controller combined in one device. The sensor and controller cannot create a comfortable environment without a controlled device. A controlled device is the mechanism that actually causes a change in the environment. It may be an automatic valve or an electrical switch; it may be an airflow controller. In this example, it is the device that regulates the heating or cooling of the room.

Feedback

Imagine driving your car with a blindfold over your eyes. The odds of getting to your destination would certainly be poor. As we drive, we make adjustments to keep the car on course. This is *feedback*. Control systems must also have feedback to be effective. In the case of building control systems, we need to measure the result of a change that has been made. If the thermostat we used in the prior example were not in the room we were controlling but instead was in the garage, it would not be effective at providing comfort in our room. It would certainly sense the temperature and signal the controlled device, but without measuring the controlled area, without feedback, it would be ineffective.

Coordinated Operation

Another fundamental of control in buildings is coordinated operation. Again, we'll use the example of an office with a thermostat. In our example, we have a heating device at the exterior wall and a unit to cool the room. Imagine if we had the cooling system running when it was cold in the room and the heat at full output when the room was warm—not a comfortable thought. Alternatively, imagine if the heating and the cooling operated at the same time. Not only would there be questionable comfort, there would also be a tremendous waste of energy and dollars. A control system coordinates the operation of the system equipment with the need. If the room is cool, then heat is introduced. If the room is warm, then cooling is introduced. When the room is comfortable, the systems are shut down. At the level of a single room, the entire building, or the campus, the building control systems coordinate the operation of systems to maintain comfort, minimize equipment operation, and limit energy consumption.

27.3 A BRIEF HISTORY OF AUTOMATIC CONTROL SYSTEMS

The Original Systems

As buildings grew in sophistication, a need developed to control the environment. Early buildings were small, were not cooled in summer, and were heated in winter with stoves or fireplaces. Early control systems were not automatic and were designed to limit the rate at which coal or wood burned in the stove. As the Industrial Revolution progressed, buildings grew larger, and individual fireplaces and stoves became unworkable. To overcome these shortcomings, designers put in steam heat as a means to heat spaces, and individual windows for daylight, cool air, and ventilation. Now a fire could be stoked in one location, and the heat from the fire could be distributed. The distribution of heat created the need for the first control system.

A first system involved the use of a bell. When an occupant in a room was too cold, he or she would give the bell a loud ring. The fire stoker would increase the amount of heat flowing to the room by manually adjusting a valve. In 1885 Professor Warren Johnson of Milwaukee, Wisconsin discovered that certain metals bend in a peculiar way in the presence of heat. Professor Johnson saw that these metals could provide for the automatic sensing of temperature. When he combined a metal temperature sensor with an electrical switch, he created the first automatic temperature control: the *thermostat*. When he connected this new sensing device to a controlled device, an electric bell, he developed the first automatic control system. Thus, when a room was cold, the thermostat could automatically ring the bell!

As building size and sophistication increased, new systems were included in the designs. Early buildings allowed each person to be near a window. However, the widespread use of electricity in buildings soon brought numerous changes. Larger buildings were created with interior spaces without windows, and thus building designers needed to provide fans for ventilation. Electric motors were coupled to valves that had previously been manually operated. Thus, the electric thermostat could operate the valve directly and automatically, eliminating both the bell and the fire stoker workman. Control systems were no longer a luxury but were becoming a real need in buildings.

In these early systems, the controls operated controlled devices in an "on-and-off" manner. When a room was cold, the heat was on to the maximum amount. When the desired temperature was reached, the heat was entirely off. This resulted in widely fluctuating temperatures that would be unacceptable by current standards. However, it was a significant advance in the early years of the 20th century.

These widely fluctuating temperature conditions, particularly in mild weather, resulted in a need for better control. There was a need for systems that could put a small amount of heat in a room when there was a small need for heat. To respond to this need, pneumatic controls were introduced, and the concept of proportional control was developed for heating and cooling controls.

Pneumatic controls were a response to the need for proportional control, as well as the need for lower costs. Although electric actuation was possible, its cost was prohibitive. Air-operated pneumatic controls could be made inexpensively and could offer proportional control. Proportional control gradually opens or closes the controlled device in response to the gradual change in the sensed variable. Proportional control was truly a breakthrough in control systems. New types of controlled devices were needed to accomplish proportional control. It was found that certain metals could withstand repeated flexing without breakage. These metals were formed into designs similar to fireplace air bellows and were known as *sylphons*. Sylphons could utilize a low-pressure air supply, typically 15 lbs. per square inch gauge (psig.), and incrementally position a controlled medium (e.g., the flow of water in a pipe or the flow of air through a duct). With the advent of sylphons, the need for pneumatic thermostats became evident. Pneumatic thermostats were developed in approximately 1910. These devices also utilized metals that bent when heated, but instead of snapping an electrical switch, they allowed compressed air to gradually escape. The escape of compressed air was linearized to correspond to sensed changes in temperature. The result was an air pressure output from the thermostat that changed gradually in proportion to the sensed variable. These devices were the beginning of pneumatic proportional controls, which are still in use today.

The Pneumatic Controls Period

The relatively low cost of pneumatic controls allowed a much wider use of controls in buildings. Temperature control became a standard in buildings, and controls for other environmental needs were developed. For example, it was becoming apparent in northern climates that the introduction of humidity into a space would increase comfort. Sylphons were available that could gradually control the introduction of steam or water into the air. However, there was a need for a sensor and controller that could regulate moisture in the air. It was found that there were materials that could change their physical dimensions when subjected to moisture in the air. These materials—human hair, animal bone, animal horn, and wood—were the foundation materials in the development of the hygrostat, the product currently known as the *humidistat.*

An additional major development in pneumatic controls was the development of capillary sensing. As buildings grew in size, space heating, humidification, and ventilation were accomplished with centralized systems. Room-type thermostats were inadequate to sense air conditions in ducts and certainly could not be mounted in pipes to sense water temperatures. The capillary sensor was a means of using small diameter metal tubing, attaching a small-bellows sylphon to one end, and filling the tubing with a liquid or gas that would expand or contract in relation to the sensed temperature. This change in volume as a result of expansion caused the bellows sylphon to move. The motion could then be used to control the leakage rate of air from a controller to achieve proportional pneumatic signals. Similarly, controllers with external humidity sensors were developed to allow the sensing of duct humidity. Remote sensor controllers were developed for sensing the temperature of water in pipes. Additional sensor controllers were developed for pipe and duct pressure sensing. There was now a complete array of tools to sense and control building environmental conditions.

Armed with this wide array of control tools, building systems grew in complexity. Although early buildings may have had one thermostat for the entire building, modern buildings of the pneumatic period had individual thermostats for each room and thermostats and humidistats for central air and water systems. Major companies, such as Johnson Service Company, Minneapolis Honeywell Regulator Company, Powers Regulator Company, and Fulton Sylphon Company grew and became major suppliers to the building industry.

The quality of controls began to increase with the development of a controls industry. Improved accuracy of control was needed. Early sensor controllers, typically known as *regulators,* were first-order response devices. First-order devices do not have an internal mechanism to produce controller feedback. This type of controller creates a difference between the desired result (the set point) and the actual result; this difference is known as *offset*. Also, the speed with which controllers could respond typically was limited to achieve stability. When these devices were applied to systems that frequently experienced wide variations, response was slow and unstable. The market responded with the creation of second-order mechanisms.

Second-order feedback-type controllers were developed in the early 1920s. They went on to become the industry standard for pneumatic controllers into the 1990s. Second-order controllers use internal feedback of the output signal to adjust the output signal. If there is a great need for output, the output is allowed to go to maximum. As the sensed variable approaches the set point, the feedback loop slows the output. Typically, the rapid output changes cause the sensed variable to go beyond the controller set point; this is known as *overshoot*. The controller then responds by rapidly diminishing the output to achieve set point. Second-order controllers gave users more accuracy by diminishing offset and provided quicker response, which improved stability. In modern terminology, increased speed of response is known as *derivative control* and the elimination of offset is known as *integral control*. Thus, modern controllers can incorporate proportional-integral-derivative (PID) control.

In the early 1960s, building complexity continued to increase. Building size continued to grow, both vertically (the skyscraper) and horizontally (the sprawling campus environment). Centralized cooling of spaces became commonplace. Windows that had previously opened became sealed. Also, systems typically ran 24 hours a day. In the early 1960s building owners began to see value in turning systems off at opportune times. There was a need to centralize control of systems to allow oversight of operation, as well as to control system operation. The control systems that had served so well with capillary sensing could not meet the need of centralized information. In response, the control industry developed the concept known as *pneumatic transmission.*

Pneumatic transmission made the sensor a separate component of the sensor-to-controller-to-controlled device sequence. Pneumatic sensors still sensed the variable; however, instead of being connected directly to the controller, the two components were connected with a pneumatic air pipe. The air pipe could be as long as desired. For example, the sensor or pneumatic transmitter sensed the variable over a fixed range, such as 0°F to 100°F, and created a 3-psig. signal at 0°F and a 15-psig. signal at 100°F. The signal was linear between the range extremes. For each degree change in the example, the air pressure in the pipe would change by 0.12 psig. The control industry standardized pneumatic transmission and the 3- to 15-psig. pressure change for a sensed range. Now the same sensed variable could be used to control a process at a fan system on the roof and simultaneously to indicate the process condition on a master panelboard in the basement. Previously, capillary type sensors were unable to indicate the sensed variable. A separate thermometer or hygrometer was needed adjacent to the duct-mounted controller to observe the variable. With transmission, the same sensor could be connected to a separate receiver controller as well as to a specially calibrated gauge that read in units of the sensed variable rather than in pressure. The receiving controllers as well as the gauges could be centralized in mechanical equipment rooms in control panels. Pneumatic transmission was widely accepted in the building industry and currently is the most common type of control system installed in campus buildings.

The introduction of transmission and the flexibility of this concept introduced much greater complexity to buildings. A wide range of new control logic applications appeared as the sensed variable was made available for manipulation. Comparisons could be made of the lowest temperature, the highest pressure, and the humidity in various locations. Different receiver controllers, with different set points for daytime or nighttime, winter or summer, could be switched in and out of the control logic. The inclusion of time control clocks and interconnection to a wider range of electric devices, such as switches and lights, made the systems electropneumatic rather than pure pneumatic, as in earlier years.

As controller complexity increased, the ability to understand building system controls decreased. In years past, a typical building had six different types of devices that were used repeatedly to make up a control system. Now there were hundreds of choices of products. To address the change in technology, the building engineer who typically oversaw controls was replaced by control system specialists and technicians. As systems became more and more sophisticated, the shortcomings of pneumatic transmission became clear. Pneumatic controllers lacked long-term reliability, requiring frequent inspection and recalibration to achieve the needed reliability. Control rooms and centralized monitoring became important to oversee the proper operation of the controls.

A new need developed as control complexity grew and the need to centralize oversight became commonplace. The cost to bring hundreds of wires and air lines to one central location grew significantly. The industry began to offer a lower cost system that could accomplish centralization without the numerous individual components. This product was the *building monitoring system*. Building monitoring systems started as multiplexed wiring systems. Multiplexing allowed numerous signals to be sent over a cable made up of many small wires. The central multiplexer switcher would select a pair of wires to read a value and sense another pair of wires to connect the sensor. The multiplexer was then connected to one common gauge at the central control panel. In a similar manner, an operator could select a motor or function at a remote mechanical location and cause it to start or stop, all over the same sets of wires. It became apparent that initial multiplexers were dependent on the console operator as they became more and more complex. Again, the industry responded in the late 1960s with the marriage of the general purpose digital computer with the multiplexing systems.

The combination of the computer with the building monitoring function gave building owners new flexibility. Until that point, all knowledge was contained in the personal memory of the building maintenance staff. Now memory in the computer could maintain this knowledge. The computer could also reproduce the manual actions of operators to interrogate systems and to start and stop system functions. Systems could be interrogated, conditions compared to stored ideals, and exception reports prepared. Computerized systems could free building maintenance staff to perform other duties. The early systems were quite crude by current standards; however, the features these systems offered were needed, and further advancements created modern building monitoring and control systems.

Building monitoring and control systems that sprang from the early combinations of controls were a part of the continuing evolution of building control systems. Significant changes to systems occurred in the early 1970s. Computing power was increasing dramatically while computing costs were decreasing.. The integrated circuit helped lower the cost of developing sophisticated circuitry and improved software for systems became more user-friendly. Installation techniques for systems became less costly. Building monitoring and control systems were being recognized as a necessary part of managing large facilities.

Building Monitoring and Control Systems

The development of building monitoring and control systems was influenced most significantly by the development of system software. In the early 1980s, the use of a powerful central computer and software specially developed to manage facilities needs pushed facilities management into a wide array of new areas. Systems had previously managed environmental conditions but began to include the management of allied building systems, such as fire alarm, security reporting, access management, lighting control, and even facilities maintenance. Control had historically addressed comfort and the environment, but systems were now being applied more and more to facilities integration. The trend continued to the mid-1980s, when the development of the microprocessor, the availability of inexpensive computer memory, and the large-scale integration of electronic circuits ended the period of pneumatic control. The period of digital control had begun.

Until the mid-1980s, buildings typically had two control systems. The actual operation of the systems was accomplished by pneumatic control systems. The oversight of systems and the reporting of exceptional circumstances such as emergencies and off-normal conditions were typically accomplished by a second system. The two systems had points of interaction, but they could each operate independently of each other. Digital control changed this design totally. In early control systems, determination of the sensed variable, such as temperature and humidity, was done with pneumatic transmission. Pneumatic sensing still relied on the principle of temperature or humidity causing motion, which allowed a variation in a signal. Pneumatic controllers used levers and springs to create amplification of signals and comparison of conditions to set point to create control. The common theme was that mechanical motion was intrinsic to controls. Digital control used the microprocessor as the controller. Digital control relied on electronic sensors, whose electronic properties change in the presence of temperature, humidity, or pressure. There were no longer any moving parts. Digital control systems used one common digital controller for the numerous control loops present in an area. All sensors connected to one controller, rather than the numerous ones that were required for pneumatic control. Different manufacturers developed control panels to accommodate a few control loops or many loops. The significant difference was in the software. Where control logic was accomplished in pneumatic systems with the addition of components, now additional control logic could be created with software. In addition, digital control systems shared input data as needed. In the past the pneumatic system needed an input, and the building monitoring and control system needed a separate input. Now the digital controller could use the same input device, typically known as a *point,* to operate the local control loop and report to the central monitoring system. Digital control systems often still retained the pneumatic actuation used in pneumatic control, because it was inexpensive and reliable. Digital control systems typically sensed and controlled with digital electronic logic, and then transduced signals to provide pneumatic outputs to the controlled devices. The period where buildings had two systems for control ended with digital control.

Digital Control Systems

The first digital control systems typically were known as *stand-alone systems.* These systems were not dependent on a central master computer for their operation. As the systems progressed, networking capabilities began to enter into digital control. The first systems were dependent on the input points being connected to the controller. It became apparent that there were economies available if certain information could be shared. Input points such as the outdoor air conditions, pressures at various locations within the building, and fire safety information could be needed at all control locations. The first systems were unable to share this information. Networked systems made all information available to all systems.

27.4   MANAGING CONTROL SYSTEMS ON CAMPUS

Currently most campuses typically have three types of systems: pneumatic transmission systems, the overlay of a building monitoring and control system, and some digital control systems. The facilities manager must understand these systems, their functions, and resources used as aids in management. A facilities manager should have a strategy for control system management. Because these systems have the ability to affect comfort, environmental quality, and energy use, maintenance of these systems continues to be important. The maintenance strategy should include a strategy for understanding the systems, a strategy for maintaining the systems, and a strategy for replacing the systems at the conclusion of their useful life.

Understanding Control Systems

Campus systems have controls installed over a wide range of years. These control installations have typically been done by local control installation vendors representing major manufacturers. In many cases the manufacturers themselves have local installation branch offices and have installed the systems. The control system suppliers can be a facilities manager's largest resource in receiving value from the operation of control systems. The operating staff within facilities should have a well-developed relationship with the control system supplies. Suppliers offer a wide array of training opportunities, ranging from pamphlets and books to training at the vendor's headquarters.

In addition to training, control suppliers have traditionally documented the systems controls and the operation of the system in documents known in the industry as *control drawings*. These drawings are fundamental to facilities managers understanding system operation. In certain cases these drawings represent the only remaining documentation of original design intent and operation. Control drawings should be obtained for every campus system. They should be maintained in a library and made available to service and vendor staff. In addition, the drawings should be reproduced and placed at each system location so that technicians can more easily understand and more rapidly repair the control systems. In the library, staff should also create a summary of control components. Manufacturers prepare individual catalog description sheets for each control component they manufacture; these components change as years pass. The installed base of components should be maintained on site to allow the staff to learn the specifics of installed products.

Maintaining Your Control Systems

The control systems on campus do take maintenance. Older pneumatic systems, particularly those using pneumatic transmission, require the most frequent attention. At a minimum, pneumatic systems must be calibrated semiannually. Calibration is required for each sensor, each controller, and each controlled device. It is recommended that attention be given to each major system before the cooling season and before the heating season. Because of the intrinsic nature of the mechanical components in pneumatic devices, they drift away from their original settings and must be reset.

Building monitoring and control systems also need regular maintenance. Earlier systems had less reliable electronics and required maintenance of field components such as servo controllers, power supplies, and filters. Digital control systems and modern monitoring systems have been designed with greater reliability; however, they still must receive modest attention to maintain their value. To manage the systems, the facilities manager must decide whether to use campus staff to service the systems or to contract with suppliers for operation and maintenance of the systems. Each choice deserves thought. On small campuses staffing is often limited. The staff assumes the roles of generalists. They need to know about a wide range of topics and probably will not be experts on any one topic. Control systems require experts to be effective. When a campus cannot retain an expert, it is advisable to work with a local supplier to obtain maintenance services. Control suppliers will prepare a wide range of choices for selection. These choices range from total care of systems, including parts and emergency service, for one annual cost, to occasional system inspection and oversight of in-house staff.

On larger campuses there are typically enough systems to justify the establishment of in-house expertise. With the various vintages and suppliers found on campuses, continual training of in-house staff is needed to maintain competency. The facilities manager may wish to use a compromise strategy, employing a control system supplier for more complex technology and using in-house staff for simpler control applications.

A facilities manager must make a decision about the number of system vendors that the campus will permit. During the period of pneumatic controls, vendor parts were interchangeable between system controllers, pneumatic room thermostats, pneumatic transmission devices, and controllers. With the advent of building monitoring and control systems, this interchangeability ended. Manufacturers created systems that were proprietary, and the ability to interconnect one system to another was lost.

This has created quite a dilemma for facilities managers. Building monitoring and control systems and the more modern distributed digital control systems are often used for purposes other than simply controlling the environment. The systems relay fire alarm conditions, dispatch fire departments, alert safety staff to security breeches, alert service staff to alarms, and generate facilities condition reports. This process becomes more difficult with multiple suppliers of centralized systems. It can still be accomplished, but information and capability may be lost. In addition, the inclusion of each new and different system will create a need for greater training of service staff and a need to maintain an inventory of parts to support each system. The facilities manager must balance the need to maintain a competitive bidding environment with the greater internal costs for operation and maintenance multiple systems.

Replacement of Control Systems

An additional management strategy for campus control systems is the need to upgrade to higher capability systems. Older pneumatic transmission systems have a finite life. Controllers and sensors begin to fail as these systems age, and associated control components become less reliable and accurate. Eventually the system will need renewal. Many building monitoring and control systems were installed in the early 1980s, and these systems are being slowly made obsolete by their manufacturers in favor of higher capability distributed digital control systems. Some suppliers have created newer versions of old systems that enable older versions to be incorporated in overall system operation, allowing incremental upgradability. Other suppliers have not planned for upgradability, and systems must be totally replaced. A well-thought-out strategy can assist the facilities manager in maintaining facilities management continuity for the future.

27.5   APPLICATIONS OF CONTROL SYSTEMS

The application of controls to building systems is a career unto itself. This section cannot offer expertise on every strategy for the control of mechanical and electrical systems. For additional expertise, readers would be best advised to read the 1995 *ASHRAE Handbook of Heating, Ventilating, and Air Conditioning Applications.*[1]

Controls are best applied with a good knowledge of the intent of the design of the mechanical system. What is the expected result? What are the seasonal changes to the control logic? What are the occupancy implications that affect control? The facilities manager cannot expect to be an expert on every application. It is recommended that a competent engineer be used for specific application guidance. The facilities manager can take the more valuable role of representing the users and maintainers of the building.

In past years there was a temptation on the part of control suppliers and users to allow controls to dictate the design of a system. The facilities manager should avoid the temptation to overcontrol a space. One way to avoid odd control applications is by establishing facilities standards. How will controls be applied to similar situations? Nothing frustrates operating staff more than the random trial of solutions to similar needs. Find a solution that works well, and avoid the temptation to try the newest method. Experience has shown that building operation staff will reduce the complexity of a system to their level of understanding. If controls are allowed to become overly complex, the operating staff may negate some of the complexity in favor of increasing understanding. Know the knowledge level of the staff. The best strategy to impose on staff or consultants who are designing systems is simplicity.

A second role of the facilities manager is to create an understanding of the level of quality of the control systems. Like all purchases, there are choices in the level of quality available in controls. The quality of the controls can have significant impact. Accuracy is vital if the control system is being used to meter energy to a facility and this energy is a source of income to the institution. Similarly, the accuracy of the controls is vital to the success of the users of the building if the controls are preventing the migration of contaminants from one space to another in a research laboratory. The facilities manager should research the quality issues pertinent to a design in progress. What are the limitations of the proposed systems? Do the proposed controls provide the needed accuracy? Are proposed systems compatible with existing systems on campus? If there are wider reaching future plans for the campus or further building integration, can the proposed systems meet these future needs, or will they become "white elephants"? In addition to the selection of quality control system products, there is the analysis of the local service capability of the control supplier. The facilities manager should research service support available in the locality. Is there a local service commitment, or are there only a few service agents nationally? Are spare parts warehoused locally, or are parts available only with a wait of several days or weeks? Can the supplier provide the necessary training of your staff? Are there locally satisfied customers? Does the supplier have expertise in campus applications? The facilities manager must qualify the suppliers to be certain that in times of need, the supplier will be available for assistance.

As control systems are proposed, the facilities manager should understand the needs of the users to determine whether the systems are adequate. What are the needs of the users? Are there critical spaces that need to be monitored? Are there significant flows that must be maintained? What are the needs that are required in the event of an emergency? What systems must be integrated to accomplish the desired result in facilities operation? The facilities manager should also check to see that facilities standards have been followed in the proposed applications. Does every space where an employee works have some control? Does the design comply with appropriate regulatory issues, such as air exchange rates and fume removal?

The role of the facilities manager in the application of controls is best described as oversight.

27.6   ENERGY CONSERVATION STRATEGIES

Controls have been offered as a solution to higher energy costs on campus. The facilities manager should approach the use of controls to achieve energy use with caution. Controls can reduce energy use in facilities; however, the facilities manager should understand energy reduction can be achieved with a variety of strategies. In most cases, campus involvement and commitment will achieve the greatest energy reductions. Heating and cooling controls are only one strategy to be considered.

The best strategy for reducing energy consumption in building systems is through sound engineering practices. Controls can merely manage mechanical processes that are in place. Engineering evaluation can review the process. As an example, there is a control cycle known as *duty cycling*. Duty cycling turns air systems off for a fixed amount of time each hour, perhaps 10 minutes. The logic is that a 10-minute reduction will achieve a 16 percent reduction in energy. Duty cycling is applicable only to systems that serve spaces that can tolerate an outage. In this example, the first choice should be to re-evaluate the system to determine if there is excess capacity. For example, if the spaces served by the system have changed over the years, can the system be reduced? Re-evaluation often finds excess capacities that can be reduced to achieve savings every minute of the hour. Application of a duty cycle would have maintained the system inefficiencies and achieved some savings for only 10 minutes of each hour.

If a system is soundly engineered, there are legitimate control strategies that should be applied by facilities managers. The greatest savings available from controls is through coordination of energy use. Just as original control systems coordinated energy use in simple buildings, modern control systems coordinate energy use in complex buildings using the principle of feedback to create coordination. As various control strategies are applied, the facilities manager must always keep the phrase, "meet the needs of the occupant" in mind. The best energy conservation is to simply turn off energy-consuming equipment. This most likely will not meet the needs of the occupants. The facilities manager must determine the energy use philosophy for the campus. Will energy reduction have a high priority, or will the needs of the user have a higher priority? When it is appropriate to use control systems to reduce energy, there are a number of strategies a user can choose.

Occupied-Unoccupied

The best strategy for conservation is the implementation of an unoccupied period control. This strategy turns off equipment at times the building is not in use. Facilities are seldom used 24 hours per day, 365 days per year. When the staff is not using the facility, non-safety-impacting equipment should be turned off. This can be done through a simple time clock or preferably through the monitoring and control system. This strategy not only significantly reduces energy consumption, but also extends the life of the equipment. In areas where weather extremes may cause the building to become too cold or too hot during the unoccupied mode of operation, a feedback thermostat is needed to restart the equipment to keep building conditions within limits.

Unoccupied Setback

In a typical facility, each occupied room should have individual control. Typically there is a ducted system delivering air to the room and a locally controlled device such as a baseboard heater or cooler. Just as it is important to turn off central equipment, it is important to reduce or increase the set point of the individual room thermostats. These thermostats will reduce energy consumption during weather extremes.

Temperature Compensated System Startup

Equipment that has been turned off because the building has been unoccupied must be restarted at a time that will restore the building environmental condition before staff arrive. In a simple manner, a time clock can be set for a predetermined time. However, the time selected will always be the longest period that results in acceptable conditions. Compensated start-up observes the weather outdoors and the temperature indoors and determines the best start time. As weather conditions change, the start time follows to minimize system operation.

Demand Control

The electrical usage of a campus varies randomly. There are times when all users draw power at the same time. To meet the demand, the local power company must have power generation and power transmission capacity in place to meet this demand. The power bill a campus receives includes a charge for this reserve capacity. Control systems can minimize this billing by reducing demand during certain peak demand periods. This can be done only with a central monitoring and control system. The system measures campus demand and cycles off deferrable loads until the peak period has passed. The loads are automatically restored at the conclusion of the peak. Demand control strategy must be approached with caution, as it is only effective on campuses with a large number of deferrable loads and high peak demand charges. In general, the only campuses that have found this strategy to be effective are those that have large electric-driven chillers that can be duty cycled.

Duty Cycling (Load Cycling)

As previously described, duty cycling consists turning off loads during their normal operating times for a prescribed period and then restoring the loads. This strategy will result in savings if a review of the loads indicates that the loads are eligible for cycling. The facilities manager should exercise caution in the selection of eligible loads, as the frequent starting and stopping of loads in certain cases will cause premature motor and drive failures.

Zoning

Certain buildings cannot have the entire systems turned off. In these facilities it may be possible to create common-use zones. With proper modification of system piping or air distribution equipment, it may be possible to turn portions of systems off while keeping the remainder of the system in operation. Zoning should be reviewed by an engineer to be certain that airflows and water flows are properly operating if zoning is added to an existing building.

Individual Space Operation

With the advent of digital control, it is possible to add the individual room environment control to the building monitoring and control system. This offers managers a new option: individual space operation. These parameters can be adjusted with digital information available to the central system. Weather-related changes can be made to airflow minimums and maximums, and room set points can be altered for seasons. The space operation can be altered to match the operating schedule of the occupant. If a person lunches from noon until 1 pm, the system air can be turned off for this period without affecting the user. If a professor vacates an office during class and laboratory hours, the system can be adjusted. Individual space operation offers a new dimension in energy reduction possibility.

Automatic Set Point Adjustment

Systems in buildings often have operating set points that are determined by operation during extreme conditions. In milder weather and during nonoccupancy conditions, these settings can be automatically adjusted to minimize energy use. To accomplish this adjustment, a control system must read the need of select rooms within the building and signal the central controller to adjust the set points.

27.7   THE FUTURE OF CONTROL SYSTEMS

The control systems available give a good indication of the future direction of control systems. The facilities manager should review newer features to determine the value these features will offer.

Systems Integration

Control systems will continue to integrate systems within buildings. Facilities integration, with the end result of one universal campus system, will continue to incorporate features provided by suppliers. Currently system suppliers are creating pathways to allow communication between building systems that have not previously been integrated. Fire alarm system, boiler system, air compressor, and air distribution system manufacturers are configuring common control interfaces to connect to the common building monitoring system. As electronics continue to be applied to all aspects of society, the facilities manager should expect to see low-maintenance, high-reliability control systems. In the interest of further integration, the American Society of Heating, Refrigeration, and Air Conditioning Engineers (ASHRAE) has commissioned the establishment of the first common communication protocol: Building Automation and Control Net (BACnet). This protocol, although still under development, will give facilities managers the potential to have various suppliers of systems integrated onto one campus system.

Distribution of Access

Historically, control systems have been managed by facilities experts. With improvements in control products, the management of systems will become distributed to the building occupants. Currently there are personal environmental controllers made for desktops. These controllers place direct control of airflow, temperature, and humidity in the hands of the occupants. In addition, vendor software for systems is reaching out to users. Suppliers have developed direct interfaces for the personal computer networks common in offices. Software programs with pull-down menus give occupants the ability to interface with the system without interacting with facilities staff.

27.8 SUMMARY

In summary, heating and cooling controls play an integral part in the successful operation of campus facilities. It is important that the staff who operate and maintain these controls be properly trained and kept abreast of the latest developments. A preventive maintenance program must be in place to ensure continued accuracy and reliability. Control systems are only as good as the mechanical and electrical systems they control. Thus, standards should be created and applied early in a project's design to ensure compatibility with need. Finally, proper application of control systems depends on the needs of the occupants, which must be thoroughly understood. After all, systems exist for the needs of the occupants.

NOTE

1. American Society of Heating, Refrigerating, and Air-Conditioning Engineers, Inc. *1995 ASHRAE Handbook of Heating, Ventilating, and Air Conditioning Applications.* Atlanta: ASHRAE, 1995.

ADDITIONAL RESOURCE

Haines, Roger W. *Control Systems for Heating, Ventilating and Air Conditioning,* third edition. New York: Van Nostrand Reinhold Co., 1983.

CHAPTER 28

Fire Protection

Douglas Forsman and Matthew S. Manfredi
Oklahoma State University

28.1 INTRODUCTION

Most colleges, universities, and other institutions represent sizable investments by either private organizations or public entities. In either case, the sheer value of property at risk warrants a significant consideration of fire protection issues. Even more important is the concern for life safety in the work, study, and living environments operated by institutions.

This section will briefly discuss the process and tools involved with managing fire risk at an institution. Of major importance will be the many references to other documents such as National Fire Protection Association consensus standards. These documents provide in-depth and current information regarding a specific subject area.

28.2 MANAGING FIRE RISK

In the current institutional setting, the manager of a facility must be able to understand the concepts associated with fire risk. Fire risk and fire hazard are different concepts; they are not interchangeable terms. Fire risk is the expected loss from a fire of a given severity, whereas fire hazard is a condition or set of conditions that could lead to the start or spread of a fire. For example, the probability of a fire in a given environment may be high, but because of the presence of a sprinkler system, the risk of having a severe fire might be quite low. Describing fire risk, or any class of risk, therefore carries with it the need to describe both the likelihood and the severity of the event.

Risk management, to be completely effective, must maintain and identify potential losses so that their impact can be measured. This requires risk assessment of the probability of certain events happening at the facility. This is where managers decide what actions would be necessary to effectively lower the potential by cost-effective risk reduction methods.

Some of the benefits of implementing risk assessment include the following:

- It establishes an optimum balance among prevention, protection, and emergency response from within the plant; coordination with local emergency response personnel is also important.
- It prevents or reduces the probability that severe accidents will occur.
- It helps to facilitate and maintain compliance with applicable regulations and standards that apply to the appropriate risk.

28.3   FIRE PREVENTION

Fire prevention activities can be the most important aspect of the total fire protection effort. Well-planned fire prevention activities can save thousands of dollars by preventing destructive fires. In industry, the ability to cut fire losses is usually attributed to the amount of time spent on fire prevention.

For years the level of commitment to fire protection by corporate management was based on the loss suffered by the industry. As has often been stated, fires do not just happen; they are almost always caused by an unsafe act or condition. Thus, most fires can be prevented if the unsafe act or condition that contributes to the cause of fire and the possibility for injury or loss of life can be eliminated.

Justification for a fire prevention program (including budget, personnel, equipment, and time) can be proven by quantitative evidence. A proper record-keeping system of all activities, including inspections, hazards abated, fire protection systems installed, and educational programs, will prove valuable in this area. An analysis of the results in these areas can indicate the degree of success of fire prevention efforts.

Public Education

To maintain a year-round awareness of fire prevention, it is essential that numerous and varied public fire education programs be established. A variety of facilities and organizations should sponsor, conduct, or assist in these programs to ensure maximum coverage in all areas. The basic concept should be to increase the awareness of the public through its involvement in the fire prevention programs.

There are many fire prevention educational efforts that will benefit students and employees at home. Because employees are at home far more hours than they are at work, their exposure to fire is greater at home. For this reason, along with the economic considerations involved in replacing skilled employees that may be injured in home fires, it is cost-effective for an organization to devote some time to home fire safety in employees' fire safety subject matter.

Monthly or quarterly campaigns draw attention to specific fire hazards, and, each month or quarter, a different type of hazard is the focal point. Conditions vary with respect to the subject that is to be emphasized for that particular month or quarter. Certain subjects lend themselves to some months more readily than others.

During certain seasons, special emphasis should be placed on the hazards that are particular to that time of year. Special posters, spot announcements, films, and handouts for each of these periods are available from various sources. Some of the special subjects for seasonal campaigns include the following[1]:

- Spring and fall (clean-up campaigns)
- Cold weather (heating appliances, electric blankets)
- Hot weather (air conditioning, fans)
- Summer (camp fires, barbecues, fireworks)
- Holiday season (trees, decorations)

Codes

If fire prevention is to be effective, various types of codes must be enacted and enforced in the facility by the authority having jurisdiction. Major fires in early American cities inspired many of the building and fire codes. Fires such as the great Chicago fire of 1871 caused the insurance companies to charge outrageous rates for fire insurance. Then, even though they had charged high rates, the insurance industry still suffered great losses when fires spread out of control.

In 1866 the National Board of Fire Underwriters (now the American Insurance Association) was formed to combat unsafe structures that were being built. They began to emphasize safe building construction, control of fire hazards, and improvement in water supplies and fire departments. In 1905 the National Board of Underwriters published the first recommended building code, which was later known as the National Building Code. As the years progressed, this organization compiled and analyzed information collected from fire incidents. More and more information was compiled and codes were updated to reflect the information that had been collected in an effort to decrease the number of catastrophic fires America was experiencing.

Building Codes

A building code is a law that sets forth minimum requirements for design and construction of buildings and structures. Minimum requirements must be set to protect the health and safety of the general public. The minimum code requirements often establish how the builder and building owner will address structural design, fire protection, means of egress, light, sanitation, and interior finishes.

There are two types of building codes. *Specification codes* are used to tell the builders what materials can be used in the building. *Performance codes* establish objectives and criteria to be met by the designer and builder. They are fairly strict in the methods that are to be used but allow for suitable substitutions if the materials are proven to be adequate.

More recent building codes have focused on the minimum requirements for structural stability, fire resistance, means of egress, sanitation, lighting, ventilation, and built-in fire safety equipment. There are several building codes organizations that have published their versions of a building code. This section discusses several of the more well-known and national building code organizations.

International Conference of Building Officials   In 1927 the International Conference of Building Officials (ICBO) first published the Uniform Building Code, which is reprinted every three years. This code has been primarily used and adopted in the western United States.

Southern Building Code Congress International   The Southern Building Code Congress International (SBCCI) was first formed in 1940 and published its first code in 1945. This code is used throughout the southern United States and is updated every three years. The SBCCI develops, maintains, and promotes the standardization of model building codes and the enforcement of these codes.

Building Officials and Code Administration   The Building Officials and Code Administration (BOCA) published the first edition of the Basic Building Code in 1950. The name was later changed to the National Building Code and is used in the Midwest and Northwest. It is revised every three years, and changes to the code are printed in supplements.

American Insurance Association   Formerly the National Board of Fire Underwriters, the American Insurance Association (AIA) first published its National Building Code in 1905. The last National Building Code from the AIA was in 1976. BOCA has since taken over producing the codes.

National Fire Protection Association   Although the National Fire Protection Association (NFPA) does not publish building codes, the organization is worth mentioning here because it plays a major part in the development of all major building codes. The NFPA established the Life Safety Code (NFPA Standard 101), which was first developed in 1927. It establishes minimum requirements for providing a reasonable degree of safety in buildings and structures. The contents of the code address construction, protection, and occupancy features necessary to minimize danger to life from smoke, fumes, or panic. This document and other NFPA standards have been adopted by reference in several model building codes, as well as state and local government codes.

Fire Prevention Codes

The NFPA publishes more than 250 codes, standards, recommended practices, manuals, guides, and model laws. The complete collection of these documents, which addresses fire prevention and protection, is called the *NFPA National Fire Code.* There are several notable codes within the National Fire Code that have gained international acceptance, such as NFPA 70, the National Electrical Code; NFPA 30, Flammable Combustible Liquids Storage; NFPA 54, the National Fuel Gas Code; and of course, NFPA 1, Fire Prevention.

Life Safety Code

The Life Safety Code was formerly known as the Code for Safety to Life from Fire in Buildings and Structures. As stated before, the Life Safety Code addresses design and operation that affect safe egress from a building or structure. This code does not contain any measures for the protection of property or building safety requirements. The objectives of the Life Safety Code are based on the following twelve principles:

1. A sufficient number of unobstructed exits of adequate capacity and arrangement are provided.
2. The means of egress is protected against fire, heat, and smoke for the time that is required for the occupants to leave the building.
3. An alternate means of egress is provided in the event one of the exits is blocked by fire, heat, and/or smoke.
4. Compartmenting the building with adequate fire barriers creates areas of refuge where evacuation is not the primary means of providing occupant safety.
5. Protection of vertical openings limits the movement of fire and smoke to multiple floors.
6. Fire detection and/or alarm systems alert occupants and, in some cases, notify the fire department of a fire condition.
7. Adequate illumination is provided for the means of egress.
8. Signs indicate the path of travel to reach the exits.
9. Adequate protection is provided from unusual hazards that could impinge on the means of egress.
10. Adequate evacuation plans and exit drills are provided.
11. Adequate instructions to occupants facilitate their effective movement, particularly where crowding or severe fire hazards exist.

12. Interior finish materials are controlled to prevent fast-spreading fires and dense smoke production that could endanger occupants.

Building Classifications

The materials used to build a structure provide the basis for classification of the building. In the early years of codes, the only classifications were fireproof and non-fireproof. It was later realized that no material or building was totally fireproof. From those findings, a common idea was founded that the combination of materials used in the building and the contents and furnishings of the building all contribute to the severity of the fire. Optimum fire-resistive design, balanced against the anticipated fire severity, is the objective of structural fire protection requirements of modern fire codes. NFPA 220, Types of Building Construction,[2] is used by all three of the model building codes. The classifications are discussed in the following text.

Type I Construction Type I construction is construction in which the structural members of the building are noncombustible and have a specific fire resistance. There are two subtypes in this category, and they vary with respect to the amount of fire resistance to different types of structural members. This type of construction can provide reasonable fire protection for low and moderate combustibility contents in a building.

Type II Construction Type II construction is construction in which the structural members are made entirely of noncombustible or limited combustibility materials. There are three subtypes associated with this type of construction: type II, type II, and type II unprotected. This feature is valuable, because it prevents the spread of fire from entering concealed spaces or the structure itself.

Type III Construction Type III construction is the type in which all or part of the interior structural elements may be of combustible materials or any other material permitted by the particular building code being applied. The exterior walls must be made of noncombustible or limited combustibility materials. This type has two subgroups: protected and unprotected. Protected construction, type III (211), has a one-hour fire resistance for the floors and structural walls. Type III (200) construction has no resistive rating for the floors or structural elements.

Type IV Construction Type IV construction is a type of construction in which the structural members, columns, beams, arches, floors, and roofs, are basically all made of unprotected wood. No concealed spaces are permitted in the floors and walls in these buildings. During a fire, heavy timber resists failure longer than a conventional wood frame structure.

As these structures burn, a layer of char develops and creates a layer of insulation, thus resisting further burning and therefore continuing to carry the load for the structure.

Type V Construction Type V construction is construction in which the structural members are entirely of wood or any other material permitted by the code. Type V construction is probably more vulnerable to fire, both inside and out, than any other type of construction. There are two different subtypes: type V, which has a one-hour fire resistance, and type V, which has no fire resistance.

Mixed Types of Construction In mixed construction, two or more types of construction are mixed together in the same building. Area height and occupancy requirements would apply, at least for fire-resistive construction.

Occupancy Classifications

Occupancy classifications are intended to provide a distinction between uses of buildings and the requirements that are necessary for each. These classifications will drive many requirements in the Life Safety Code.

Light Hazard Occupancy Light hazard occupancies are occupancies or portions of other occupancies where the quantity or combustibility of contents is low, and fires with relatively low rates of heat release are expected.
   Examples of light hazard occupancies include:

- Museums
- Libraries
- Educational
- Office, including data processing
- Hospitals
- Residential
- Institutional
- Theaters and auditoriums (excluding stages and prosceniums)

Ordinary Hazard Occupancies Ordinary hazard occupancies are classified into three groups.
   Ordinary Hazard Group 1. Occupancies or portions of other occupancies where combustibility is low, quantity of combustibles is moderate, stockpiles of combustibles do not exceed 8 ft., and fires with moderate rates of heat release are expected. Examples of ordinary hazard occupancies group 1 include automobile parking garages, electronic plants, and laundries.
   Ordinary Hazard Group 2. Occupancies or portions of other occupancies where quantity and combustibility of contents is moderate, stockpiles

do not exceed 12 ft., and fires with a moderate rate of heat release are expected. Examples of ordinary hazard group 2 include machine shops, metal working shops, printing and publishing offices, libraries (large stack rooms), and wood product assembly business.

Ordinary Hazard Group 3. Occupancies or portions of the other occupancies where the quantity and/or combustibility of contents is high, and fires with a high rate of heat release are expected. Examples of ordinary hazard occupancies group 3 include repair garages, wood machining shops, exhibition halls, feed mills, and warehouses (those having moderate to high combustibility of content, such as paper, household furniture, paint, and general storage items).

Extra Hazard Occupancies Extra hazard occupancies are occupancies or portions of other occupancies where quantity and combustibility of contents are high, and flammable liquids, dust, lint or other materials are present, introducing the probability of rapidly developing fires with high rates of heat release. Examples of extra hazard occupancies include aircraft hangers and woodworking shops that use flammable finishes.

Under favorable conditions and subject to the approval of the authority having jurisdiction, a reduction of requirements to the next less restrictive occupancy classification may be applied to the following occupancies:

- Machine shops
- Metal working
- Feed mills
- Mercantiles

Judgment must be applied in using this guide. For instance, hotels are listed under light hazard occupancies, but while the lobby and guest rooms are in the light hazard class, the kitchen and laundry areas should be considered ordinary hazards.

Fire Suppression Requirements Based on Building/Occupancy Classification

The requirements for fire protection systems may be driven by the occupancy and building classifications. Some of the equipment and systems that may be required or desired are discussed in the following sections.

28.4 FIRE SUPPRESSION SYSTEMS

Fire Extinguishers

The portable fire extinguisher, one of the most common fire protection appliances currently in use, is found in fixed facilities and on fire apparatus. A portable fire extinguisher is excellent to use on incipient (early-stage) fires. In many cases, a portable extinguisher can extinguish a fire in much less time than it would take to deploy a hose line.

It is important that managers be knowledgeable in the different types of portable fire extinguishers and their correct usage. This section covers the various types of portable fire extinguishers in common use. Also covered is information on the rating, selection, and inspection of portable fire extinguishers. All portable fire extinguishers should meet the requirements set forth in NFPA 10, Standard for Portable Fire Extinguishers.

Extinguisher Rating System

Portable fire extinguishers are classified according to their intended use on the four classes of fire (A, B, C, and D). In addition to their letter classification, class A and class B extinguishers receive a numerical rating. The numerical rating system is based on tests conducted by Underwriters Laboratories, Inc. (UL) and Underwriters Laboratories of Canada (ULC). These tests are designed to determine the extinguishing potential for each size and type of extinguisher. Extinguishers for use on class C fires receive only the letter rating, because class C fires are essentially class A or class B fires involving energized electrical equipment. The class C designation merely confirms that the extinguishing agent is nonconductive. Class D extinguishers, likewise, do not contain a numerical rating. The effectiveness of the extinguisher on class D metals is detailed on the faceplate. Multiple letters or numeral-and-letter ratings are used on extinguishers that are effective on more than one class of fire.

Class A Ratings Class A portable fire extinguishers are rated from 1-A through 40-A, depending on their size. For a 1-A rating, 1¼ gallons (5 L) of water are required. A 2-A rating requires 2½ gallons (10 L), or twice the 1-A capacity. Therefore, a dry chemical extinguisher rated 10-A is equivalent to five 2½-gallon (10-L) water extinguishers.

Class B Ratings Extinguishers suitable for use on class B fires are classified with numerical ratings ranging from 1-B through 640-B. The test used by UL to determine the rating of class B extinguishers consists of burning the flammable liquid *n*-heptane in square steel pans. The rating is based on the approximate area (in square feet or square meters) of a flammable liquid fire that a nonexpert operator can extinguish.

Class C Ratings There are no fire tests specifically conducted for class C ratings. In assigning a class C designation, the extinguishing agent is tested only for electrical nonconductivity. If the agent meets the test requirements, the class C rating is provided in conjunction with a rating previously established for class A and/or class B fires.

Class D Ratings Test fires for establishing class D ratings vary with the type of combustible metal being tested. The following are several factors considered during each test:

- Reactions between the metal and the agent
- Toxicity of the agent
- Toxicity of the fumes produced and the products of combustion
- The possible burnout of the metal instead of extinguishment

When an extinguishing agent is determined to be safe and effective for use on a metal, the details of instruction are included on the faceplate of the extinguisher, although no numerical rating is applied. Class D agents cannot be given a multipurpose rating to be used on other classes of fire.

Multiple Markings Extinguishers suitable for more than one class of fire should be identified by multiples of the symbols A, B, and/or C. The three most common combinations are class A-B-C, class A-B, and class B-C. There is no extinguisher with a class A-C rating. A new portable fire extinguisher must be labeled with its appropriate marking. If a new extinguisher is not properly marked, the seller should be requested to supply the proper decals.

Selecting and Using Portable Fire Extinguishers

Selection of the proper portable fire extinguisher depends on numerous factors, as follows:

- Hazards to be protected
- Severity of the fire
- Atmospheric conditions
- Personnel available
- Ease of handling extinguisher
- Any life hazard or operational concerns

Extinguishers should be chosen that minimize the risk to life and property yet are effective in extinguishing the fire. Hopefully, these considerations will be addressed when choosing a fire extinguisher to be mounted in a particular area. For example, it would be unwise to place dry chemical extinguishers with a corrosive agent in areas where highly sensitive computer equipment is located. In these particular areas, halon or carbon dioxide extinguishers would be a better choice although halon systems become less available under the chlorofluorocarbon (CFC) regulations.

Portable extinguishers come in many shapes, sizes, and types. Although the operating procedures of each type of extinguisher are similar, operators should become familiar with the detailed instructions found on the label of the extinguisher. In an emergency, every second is of great importance; therefore, everyone should be acquainted with the following general instructions applicable to most portable fire extinguishers. These can be remembered with the acronym PASS.

- Pull the pin at the top of the extinguisher that keeps the handle from being pressed. Break the plastic or thin wire inspection band as the pin is pulled.
- Aim the nozzle or outlet toward the fire. Some hose assemblies are clipped to the extinguisher body. Release the hose and point.
- Squeeze the handle above the carrying handle to discharge the agent. The handle can be released to stop the discharge at any time. Before approaching the fire, try a short test burst to ensure proper operation.
- Sweep the nozzle back and forth at the base of the flames to disperse the extinguishing agent. After the fire is out, watch for remaining smoldering hot spots or possible reflash of flammable liquids. Make sure that the fire is out.

Types of Portable Fire Extinguishers

There are many different types of portable fire extinguishers. This section highlights some of the common extinguishers encountered by fire service personnel.

Stored-Pressure Water Extinguishers Stored-pressure water extinguishers, also known as air-pressurized water extinguishers, are used only on class A fires. These extinguishers are useful for all types of small class A fires and are often used for hitting confined hot spots during overhaul operations, as well as for extinguishing chimney flue fires. They are available in sizes of $1^1/_4$ to $2^1/_2$ gallons (5 to 10 L). Under normal conditions, their stream reach is 30 to 40 ft. (9 to 12 m). The discharge time is 30 to 60 seconds, depending on the size of the unit. NFPA 10 requires that these extinguishers be protected against freezing if they are going to be exposed to temperatures less than 40°F (4°C). Freeze protection is accomplished by the addition of an antifreeze solution to the water.

The water is discharged by compressed air that is stored in the tank with the water. A gauge located on the top of the tank shows when the extinguisher is properly pressurized. With this type of extinguisher, the pressure is ready to release the extinguishing agent at any time. When the shutoff device is released, a stream of water is expelled through the hose. Stored-pressure water extinguishers are designed to be carried to the fire in an upright position. The hose is held in one hand and the shutoff device in the other hand. When water is needed, the shutoff handle is squeezed, and the water is directed at the target.

Halon 1211 Extinguishers   Halon 1211 (dichlorodifluoromethane) extinguishers are primarily designed for class B and class C fires. However, halon
1211 extinguishers greater than 9 lbs. (4 kg) in capacity will also have a small
class A rating (1-A to 4-A, depending on the size). Hand-carried extinguishers
are available in sizes from $2\frac{1}{2}$ to 22 lbs. (1 kg to 10 kg). Larger halon 1211
extinguishers are available on wheeled units up to 150 lbs. (68 kg). These
extinguishers have a stream reach of about 8 to 18 ft. (2.5 m to 5.5 m). The
total discharge time is 8 to 18 seconds. The inherent features of halon 1211
extinguishers do not require special freeze protection.

Halon 1211 is stored in the extinguisher as a liquefied compressed
gas. Nitrogen is added to give the halon 1211 added pressure when discharged. When the shutoff handle is squeezed, the halon 1211 is released
in a clear liquid stream, giving it greater reach than a gaseous agent. How
halon 1211 or any of the other halon agents extinguish fires is not clearly
known. Research indicates that they interrupt the chain reaction of the
combustion process.

Halon 1211 extinguishers are designed to be carried by the top handle.
They have a limited range, and the discharge of the extinguishing agent
may be affected by draft and wind. The initial application should be made
close to the fire, and the discharge should be directed at the base of the
flames. The discharge should be applied to the burned surface even after
the flames are extinguished. Best results will be obtained on flammable
liquid fires if the discharge is directed to sweep the flame from the burning
surface. Apply the discharge first at the near edge of the fire, and gradually
progress forward while moving from side to side. It should be noted that
halon is damaging to the ozone layer of the atmosphere and is being phased
out in all respects, with production being eliminated.

Carbon Dioxide Extinguishers (Hand-Carried)   Carbon dioxide (CO_2) extinguishers are effective in extinguishing class B and class C fires. Hand-
carried units are available in sizes from 2 to 20 lbs. (1 to 9 kg). Because their
discharge is in the form of a gas, they have a limited reach of only 3 to 8 ft.
(1 to 2.5 m). Their total discharge time ranges from 8 to 30 seconds, depending on the size of the extinguisher. They do not require any special
freeze protection.

Carbon dioxide is stored as a liquefied compressed gas. When the
shutoff handle is squeezed, the carbon dioxide is discharged in a gaseous
form to smother the fire. The carbon dioxide is stored under its own pressure and ready for release at any time. The gaseous discharge is usually
accompanied by little dry ice crystals or "snowflakes." These flakes become gaseous shortly after discharge.

Carbon dioxide extinguishers are designed to be carried by the top
handle. The discharge expels through a horn on the end of a hose or short
metal fitting. *(Caution:* Often a "frost" residue will form on the nozzle horn.
Contact with the skin could result in frostbite.) The discharge horn should

be pointed at the base of the fire. Application should continue even after the flames are extinguished to prevent a possible reflash of the fire. On flammable liquid fires, best results are obtained when the discharge from the extinguisher is employed to sweep the flame off the burning surface. Apply the discharge first at the near edge of the fire and gradually progress forward, moving the discharge cone very slowly from side to side.

Carbon Dioxide Wheeled Units   Carbon dioxide wheeled units are similar to the hand-carried units, except they are considerably larger. These units are to be used only on class B and class C fires. Wheeled units range in size from 50 to 100 lbs. (23 to 45 kg). Because of their size, they have a longer stream reach, which under normal conditions is 8 to 20 ft. (2 to 6 m). The discharge time is usually no more than 30 seconds.

Wheeled units are most commonly found in industrial facilities. The principle of operation is the same for these larger units as for the smaller CO_2 extinguishers. They are designed to be wheeled to the fire and operated according to the instructions on the extinguisher. They have a short hose, typically less than 15 ft. (4.5 m), that must be unraveled before use.

Dry Chemical Extinguishers (Hand-Carried)   Dry chemical extinguishers are among the most common portable fire extinguishers currently in use. There are two basic types of dry chemical extinguishers: ordinary and multipurpose. Ordinary dry chemical extinguishers are rated for class B and class C fires, whereas multipurpose dry chemical extinguishers are rated for class A, class B, and class C fires. Unless specifically noted in this section, the characteristics and operation of both types are the same.

The following dry chemicals are commonly used in ordinary base fire extinguishers:

- Sodium bicarbonate
- Potassium bicarbonate
- Ammonium phosphate
- Potassium chloride

Monoammonium phosphate and barium sulfate are used in multipurpose agent extinguishers.

During manufacture, these chemicals are mixed with small amounts of additives that prevent the agent from caking, and this allows the agent to be discharged easily. Care should be taken to avoid mixing or contaminating ordinary agents with multipurpose agents, and vice versa.

Hand-carried dry chemical fire extinguishers are available in sizes from 2½ to 30 lbs. (1.5 to 11 kg). The stream reach under normal conditions is 5 to 20 ft. (2 to 6 m), although it is easily adversely affected by wind. The total discharge time is 10 to 25 seconds. Because these extinguishers use dry chemicals, no freeze protection is required.

Inspection of Fire Extinguishers

Fire extinguishers must be inspected regularly to ensure that they are accessible and operable. This is done by verifying that the extinguisher is in its designated location, that it has not been actuated or tampered with, and there is no obvious physical damage or condition present that will prevent its operation. Servicing of portable fire extinguishers (or any other privately owned fire suppression or detection equipment) is the responsibility of the property owner or building occupant.

During an inspection, the inspector should remember that there are three important factors that determine the value of a fire extinguisher: its serviceability, its accessibility, and the user's ability to operate it.

The following procedures should be part of every fire extinguisher inspection:

- Check to ensure that the extinguisher is in a proper location and that it is accessible.
- Inspect the discharge nozzle or horn for obstructions. Check for cracks and dirt or grease accumulation.
- Check to see whether the operating instructions on the extinguisher faceplate are legible.
- Check the lock pins and tamper seals to ensure that the extinguisher has not been tampered with.
- Determine whether the extinguisher is full of agent and/or fully pressurized by checking the pressure gauge, weighing the extinguisher, or inspecting the agent level. If an extinguisher is found to be deficient in weight by 10 percent, it should be removed from service and replaced.
- Check the inspection tag for the date of the previous inspection, maintenance, or recharging.
- Examine the condition of the hose and its associated fittings. If any of the items listed are deficient, the extinguisher should be removed from service and repaired as required. The extinguisher should be replaced with an extinguisher that has an equal or greater rating.

Extinguisher Maintenance Requirements and Procedures

All maintenance procedures should include a thorough examination of the three basic parts of an extinguisher: mechanical parts, extinguishing agents, and expelling means. Building owners should keep accurate and complete records of all maintenance and inspections, including the month, year, type of maintenance, and date of the last recharge. Maintaining this data in a computer database is an easy way to stay current and keep the records required by various codes. Software for cataloging fire and safety equipment and for charting its maintenance is commercially available.

Fire extinguishers should be thoroughly inspected at least once a year. Such an inspection is designed to provide maximum assurance that the extinguisher will operate effectively and safely. A thorough examination of the extinguisher determines whether any repairs are necessary or the extinguisher should be replaced.

Stored-pressure extinguishers containing a loaded stream agent should be disassembled for complete maintenance. Before disassembly, the extinguisher should be discharged to check the operation of the discharge valve and pressure gauge.

Stored-pressure extinguishers that require a 12-year hydrostatic test must be emptied every six years for complete maintenance. Extinguishers that have nonrefillable disposable containers are exempt.

All carbon dioxide hose assemblies should undergo a conductivity test. Hoses found to be nonconductive must be replaced. Hoses must be conductive because they act as bonding devices to prevent the generation of static electricity.

A final word is in order regarding fire extinguishers. It is important to realize that the use of a portable fire extinguisher may place the operator at some risk. Several institutions have made a conscious decision to eliminate fire extinguishers in all but the most controlled environments. Applicable laws may not permit this in all cases, or it may be considered unwise. However, the level of training available to a potential user and the level of risk present should be considered. Consultation with the local public fire service would be in order when considering portable fire extinguisher issues.

Automatic Sprinkler Systems

Automatic sprinklers have been providing fire protection to business and industry for more than 100 years. Originally automatic sprinklers were developed to protect industrial buildings. The early sprinkler systems were crude and unreliable. However, evolving sprinkler technology has improved systems so that today they are quite effective and reliable when properly installed and maintained. Sprinklers are installed in schools, health care facilities, high-rise buildings, and commercial and residential buildings. Automatic sprinklers are generally considered the most useful and reliable method of providing fire protection.

Despite a high degree of reliability, sprinkler systems are not perfect. Sprinklers can fail to control or extinguish a fire because of the following factors: closed valves, frozen water supplies, inadequate water supply, obstructed sprinkler discharge, and impaired sprinkler heads. For sprinkler systems to function properly, they must be designed, installed, and maintained properly.

Fires that involve large industrial properties pose a threat to the entire community and overtax fire-fighting resources. Building owners are motivated to invest in and install automatic sprinklers because of code requirements, for insurance purposes, and for general fire protection.

Building and fire codes frequently require installation of automatic sprinklers. Sprinklers have been mandated in buildings as a result of the need to protect the community as a whole or the occupants in individual buildings (e.g., schools, nursing homes, high-rise buildings, apartments, and residential dwellings). Local codes require the installation of automatic sprinklers in buildings based on their occupancy, construction type, and size. Typically, when a building exceeds a given area limitation (established in a building code), it is required to have sprinklers.

Standards Related to Automatic Sprinkler Systems

Throughout this chapter, various NFPA standards related to automatic sprinkler systems are referenced. This section lists each of those standards and gives a brief synopsis of the scope of the standard.

NFPA 13: Standard for the Installation of Sprinkler Systems   NFPA 13 provides the minimum requirements for the design and installation of all types of sprinkler systems found in most occupancies. The only occupancies not specifically covered in NFPA 13 are smaller residential occupancies. This standard covers all aspects of system design, including components, water supply, and fire pumps.

NFPA 13A: Recommended Practice for the Inspection, Testing and Maintenance of Sprinkler Systems   Although not a standard, NFPA 13A provides guidelines for those agencies that are responsible for the upkeep or inspection of automatic sprinkler systems.

NFPA 13D: Standard for the Installation of Sprinkler Systems in One- and Two-Family Dwellings and Mobile Homes   This standard lists the requirements for small, fast-response sprinkler systems that increase the life safety factor in private homes.

NFPA 13E: Recommendations for Fire Department Operations in Properties Protected by Sprinkler and Standpipe Systems   As with NFPA 13A, this is not a standard. However, it does provide excellent information on how fire departments should use sprinkler and standpipe systems in a given occupancy during fire fighting operations.

NFPA 13R: Standard for the Installation of Sprinkler Systems in Residential Occupancies up to Four Stories in Height   This standard provides requirements for residential-type sprinkler systems in low-rise, multifamily dwellings.

Sprinklers

The *sprinkler,* also known as the *sprinkler head,* is that portion of the sprinkler system that senses the fire, reacts to that sense, and then delivers the water to the fire area. There are three main components of a sprinkler that are of interest to managers: the heat-sensing device, the deflector, and the discharge orifice. The following sections highlight each of these components.

Components and Operation of Automatic Sprinkler Systems

Despite advances in other forms of fixed fire protection, automatic sprinkler systems remain the most reliable form of fixed fire protection for commercial, industrial, institutional, residential, and other occupancies. To date, the NFPA has never recorded a multiple-death fire in a completely sprinklered building in which the system was operating properly. It is proven that fires controlled by sprinklers result in less business interruption and water damage than those that have to be extinguished by traditional fire department methods. In fact, data compiled by Factory Mutual Research Corporation indicate that about 70 percent of all fires are controlled by the activation of five or fewer sprinklers.

Automatic sprinkler protection consists of a series of sprinklers arranged so that the system will automatically distribute sufficient quantities of water directly to the fire to either extinguish it or hold it in check until firefighters arrive. Water is supplied to the sprinkler through a system of piping. The sprinkler can either extend from exposed pipes or protrude through the ceiling or walls from hidden pipes.

There are two general types of sprinkler coverage: complete and partial. A complete sprinkler system protects the entire building, whereas a partial sprinkler system protects only certain areas such as high hazard areas, exit routes, or places designated by code or by the authority having jurisdiction.

Standards such as NFPA 13, Standard for the Installation of Sprinkler Systems, is the primary guide used for establishing sprinkler protection in occupancies. These standards have requirements regarding the spacing of sprinklers in a building, the size of pipe to be used, the proper method of hanging the pipe, and all other details concerning the installation of a sprinkler system. These standards specify the minimum design area that should be used to calculate the system. This design area defines the maximum number of sprinklers that might be expected to activate. This is done because most public water supply systems could not be expected to adequately supply 500 or 1,000 operating sprinklers. Thus, the design of the system is based on the assumption that only a portion of the sprinklers will be opened during a fire.

The automatic sprinkler and all component parts of the system should be listed by a nationally recognized testing laboratory such as UL or Factory Mutual. Automatic sprinkler systems are currently recognized as the most reliable of all fire protection devices, and it is essential for the firefighter to understand the system and operation of pipes and valves.

In general, reports reveal that only in rare instances do automatic sprinkler systems fail to operate. When failures are reported, the reason is rarely because of failure of the actual sprinklers. A sprinkler system may not perform properly owing to:

- Partially or completely closed main water control
- Interruption to the municipal water supply
- Damaged or painted-over sprinklers
- Frozen or broken pipes
- Excess debris or sediment in the pipes
- Failure of a secondary water supply

Sprinkler System Effects On Life Safety   The life safety of building occupants is enhanced by the presence of a sprinkler system because the system discharges water directly on the fire while it is relatively small. Because the fire is extinguished or controlled in the incipient stage, combustion products are limited. The few fire fatalities that have been recorded in sprinklered buildings have been attributed to asphyxiation from a small fire that did not generate sufficient heat to fuse a sprinkler, or fatal injuries suffered by the victim before the sprinkler operated. Sprinklers may also be unable to help sleeping, intoxicated, or handicapped persons whose clothing or bedding ignites early in the fire process. However, in these cases, the sprinkler system will protect the lives of people in other parts of the building.

Sprinkler System Fundamentals   The sprinkler system starts with a feeder main into the sprinkler valve. The *riser* is the vertical piping to which the sprinkler valve, one-way check valve, fire department connection, alarm valve, main drain, and other components are attached. The *feed main* is the pipe connecting the riser to the cross mains. The cross mains directly service a number of branch lines on which the sprinklers are installed. Cross mains extend past the last branch lines and are capped to facilitate flushing. System piping decreases in size from the riser outward. The entire system is supported by hangers and clamps. All pipes in dry systems are pitched to help drain the system back toward the main drain.

Sprinklers discharge water after the release of a cap or plug that is activated by some heat-responsive element. This sprinkler may be thought of as a fixed spray nozzle that is operated individually by a thermal detector. There are numerous types and designs of sprinklers.

The sprinkler used for a given application should be based on the maximum temperature expected at the level of the sprinkler under normal conditions and the anticipated rate of heat release produced by a fire in the particular area. The temperature rating is indicated by color coding on the frame arms of the head, except for coated sprinklers and decorative heads. Coated sprinklers have colored frame arms, coating material, or a colored dot on the top of the deflector. Decorative sprinklers, such as plated or ceiling sprinklers, are not required to be color coded; however, some manufacturers use a dot on the top of the deflector.

Sprinklers are kept in a closed position by various means. Four of the most commonly used release mechanisms are fusible links, glass bulbs, chemical pellets, and quick-response mechanisms that fuse or open in response to heat.

Sprinkler Design   There are three basic designs for sprinklers: pendant, upright, and sidewall.

Sprinkler Storage   A storage cabinet for housing extra sprinklers and a sprinkler wrench should be installed in the area protected by the sprinkler system. Normally these cabinets hold a minimum of six sprinklers for small systems. To prevent damage, use a sprinkler wrench when changing sprinklers. Typically this function should be performed only by representatives of the building's occupants.

Control Valves   Every sprinkler system is equipped with a main water control valve. Control valves are used to cut off the water supply to the system so that sprinklers can be replaced, maintenance performed, or operations interrupted. These valves are located between the source of water supply and the sprinkler system. The control valve is usually located immediately under the sprinkler alarm valve, the dry-pipe or deluge valve, or outside the building near the sprinkler system that it controls. The main control valve should always be returned to the open position after maintenance is complete. The valves should be secured in the open position or at least supervised to make sure that they are not inadvertently closed.

Main water control valves are the indicating type and are manually operated. An indicating control valve is one that shows at a glance whether it is open or closed. There are four common types of indicator control valves used in sprinkler systems: outside stem and yolk, post indicator, wall post indicator, and post indicator valve assembly.

Operating Valves   In addition to the main water control valves, sprinkler systems employ various operating valves such as globe valves, stop or cock valves, check valves, and automatic drain valves. The alarm test valve is located on a pipe that connects the supply side of the alarm check valve to the retard chamber. This valve is provided to stimulate actuation of the system by allowing water to flow into the retard chamber and operate the water flow alarm devices. An inspector's test valve is located in a remote part of the sprinkler system. The inspector's test valve is equipped with the same size orifice as one sprinkler and is used to simulate the activation of one sprinkler. The water from the inspector's test valve should drain to the outside of the building.

Water Flow Alarms Actuation of fire alarms by sprinkler systems is accomplished by operation of the alarm check valve, dry-pipe valve, or deluge valve. Sprinkler water flow alarms are normally operated either hydraulically or electrically to warn that water is moving within the system. The hydraulic alarm is a local alarm used to alert the personnel in a sprinklered building or a passerby that water is flowing in the system. This type of alarm uses the water movement in the system to branch off to a water motor, which drives a local alarm gong. The electric water flow alarm is also employed to alert building occupants and it can also be configured to notify the fire department. With this type of alarm, the water movement presses against a diaphragm, which in turn causes a switch to operate the alarm.

Water Supply Every sprinkler system should have a water supply of adequate volume, pressure, and reliability. In some instances, a second, independent water supply is required. A minimum water supply must be able to deliver the required volume of water to the highest sprinkler in a building at a residual pressure of 15 psi. (100 kPa). The minimum flow is established by the hazard to be protected and is dependent on the occupancy and fire loading conditions. A public water system that has adequate volume, pressure, and reliability is a good source of water for automatic sprinklers; this is often the only water supply. A gravity tank of the proper size also makes a reliable primary water supply. To give the minimum required pressure, the bottom of the tank should be at least 35 ft. (11 m) above the highest sprinkler in the building.

Pressure tanks, another source of water supply, are used in connection with a secondary supply. Pressure tanks are normally located on the top floor or on the roof of a building. This type of tank is filled two-thirds full with water, and it carries an air pressure of at least 75 psi. (525 kPa). An adequate fire pump that takes suction from a static source, such as a large reservoir or storage tank, is used as a secondary source of water supply.

Fire Department Connections. As stated earlier, the water supply for sprinkler systems is designed to supply only a fraction of the sprinklers actually installed on the system. If a large fire should occur or if a pipe breaks, the sprinkler system will need an outside source of water and pressure to do its job effectively. This additional water and pressure can be provided by a pumper that is connected to the fire department sprinkler connection.

Types of Sprinkler Systems The following section highlights the four major types of sprinkler systems in use, namely, wet-pipe, dry-pipe, pre-action, and deluge. Facilities managers should have a basic understanding of the operation of each type.

Wet-Pipe System. The wet-pipe system is used in locations that will not be subjected to freezing temperatures. A wet-pipe system is the simplest type of automatic fire sprinkler system and generally requires little maintenance. This system contains water under pressure at all times and is connected to the water supply so that a fused sprinkler will immediately discharge a water spray in that area and actuate an alarm. This type of system is usually equipped with an alarm check valve that is installed in the main riser adjacent to where the feed main enters the building. To shut down the system, the main water control valve should be turned off and the main drain opened. A pressure gauge on the riser should indicate system pressure. If it reads no pressure, chances are the system has been shut off at the main control valve.

Dry-Pipe Sprinkler System. A dry-pipe valve is a device that keeps water out of the sprinkler piping until a fire actuates a sprinkler. A dry-pipe sprinkler system should be used in locations where the piping may be subjected to freezing conditions. In a dry-pipe sprinkler system, air under pressure replaces water in the sprinkler piping above the dry-pipe valve. When a sprinkler fuses, the pressurized air escapes first, and then the dry-pipe valve automatically opens to permit water into the piping system. A dry-pipe valve is designed so that a small amount of air pressure will hold back a much greater water pressure on the water supply side of the dry-pipe valve. This is accomplished by having a larger surface area on the air side of the clapper valve than on the water side of the valve. The valve will be equipped with an air pressure gauge above the clapper and a water pressure gauge below the clapper. Under normal circumstances, the air pressure gauge will read a pressure that is substantially lower than the water pressure gauge. If the gauges read the same, then the system has been tripped, and water has been allowed to enter the pipes. Dry systems are equipped with either electric or hydraulic alarm-signaling equipment.

The required air pressure for dry systems usually ranges between 15 and 50 psi. (100 and 350 kPa). Air pressure that is needed to service a dry system may be derived from two different sources. These sources are either from a plant air service or from an air compressor and tank used exclusively for the sprinkler system.

Pre-Action System. A pre-action system is a dry system that employs a deluge-type valve, fire detection devices, and closed sprinklers. This type of system is used when it is especially important that water damage be prevented, even if pipes should be broken. The system will not discharge water into the sprinkler piping except in response to smoke or heat detection system actuation. A system that contains more than twenty sprinklers must be supervised so that if the detection system fails, the system will still operate automatically.

Fire detection and operation of the system introduce water into the distribution piping before the opening of a sprinkler. In this system, fire detection devices operate a release located in the system actuation unit. This release opens the deluge valve and permits water to enter the distribution system so that water is ready when the sprinklers fuse. When water enters the system, an alarm sounds to give a warning before the opening of the sprinkler. Inspecting and testing the system will be essentially the same as that for a deluge system.

Deluge Sprinkler System. This system is ordinarily equipped with open sprinklers and a deluge valve. The purpose of a deluge system is to wet down the area where a fire originates by discharging water from all open sprinklers in the system. This system is normally used to protect extra hazardous occupancies. Many modern aircraft hangars are equipped with an automatic deluge system, which may be combined with a wet- or dry-pipe sprinkler system. A system using partly open and partly closed sprinklers is a variation of the deluge system.

Activation of the deluge system may be controlled by fire and heat detecting devices or smoke detecting devices plus a manual device. Because the deluge system is designed to operate automatically and the sprinklers do not have heat-responsive elements, it is necessary to provide a separate detection system. This detection system is connected to a tripping device that is responsible for activating the system. Just as there are several different modes of detection, there are also many different methods of operating the deluge valve. Deluge valves may be operated electrically, pneumatically, or hydraulically.

28.5　FIRE DETECTION SYSTEMS

The early detection of a fire and the signaling of an appropriate alarm remain the most significant factors in preventing large losses from occurring. History has proven that delay in fire detection and alarm transmission will result in increased injuries, deaths, and property losses. Modern fire detection and signaling systems are a reliable method of reducing the risk of a large-loss incident.

Basic System Components

Fire detection and signaling systems that are used for private fire protection purposes are highly technical. The system includes many forms of equipment that are usually installed and maintained by a specialist in such systems. The components of the system should be listed by a nationally recognized testing laboratory, such as UL or Factory Mutual, to ensure operational reliability. Testing reports may address either an entire system or individual components that may be used in interchangeable applications. The installation of the system shall conform to the applicable provisions of NFPA 70, National Electrical Code, and the respective standard for that particular type of system. Those standards are addressed later in this chapter within the discussions of the various types of systems.

The following sections highlight each of the basic components that can be found on all types of fire detection and signaling systems.

System Control Unit   The system control unit is essentially the "brain" of the system. This unit is responsible for processing alarm signals from actuating devices and transmitting them to the local or other signaling system. In actual installations, the system control panel is often referred to as the *alarm* or *annunciator panel.* All the controls for the system are located in the system control unit.

Primary Power Supply   The primary electrical power supply will usually consist of the building's main connection to the local public electric utility. An alternative power supply is an engine-driven generator that will provide electrical power. However, if a generator is used, either a trained operator must be on duty 24 hours a day or the system must contain multiple engine-driven generators. One of these generators must always be set for automatic starting. Both power supplies must be supervised and should signal an alarm if the power supply is interrupted.

Secondary Power Supply   A secondary power supply must be provided for the detection and signaling system. This ensures that the system will be operational even if the main power supply fails. The secondary system must be able to make the detection and signaling system fully operational within 30 seconds of failure of the main power supply. The secondary power source must consist of one of the following:

- Storage battery and charger
- Engine-driven generator and a four-hour-capacity storage battery
- Multiple engine-driven generators, one of which must always be set for automatic starting

Trouble Signal Power Supply   There must be a source of power available for the trouble signal indicator. This source of power does not have to be the primary power supply; it can be the secondary power supply or a totally independent power supply, as long as it does not entail the use of dry cell batteries.

Initiating Devices Initiating devices are manual and automatic devices that are either activated or sense the presence of fire and then send an appropriate signal to the system control unit. The initiating device may be connected to the system control unit by a hardwired system, or it may be radio-controlled over a special frequency. Initiating devices include manual pull stations, heat detectors, smoke detectors, flame detectors, and combination detectors. These devices are covered in more detail later in this chapter.

Alarm-Indicating Devices Once an initiating device activates and sends a signal to the system control unit, that signal is processed by the control unit, and appropriate action is taken. This may include the sounding and lighting of local alarms and the transmission of an emergency signal to a central station service or the fire department dispatch center. Local alarm devices include bells, buzzers, horns, recorded voice messages, strobe lights, and other warning lights. Depending on the design of the system, the local alarm may sound only in the area of the tripped detection device, or it may sound in the entire facility.

Auxiliary Functions Some occupancies have special requirements in the event of a possible fire condition. In these cases, the fire detection and alarm system can be designed to perform the following special functions:

- Shut down or reverse the heating, ventilation, and air conditioning system for smoke control.
- Close smoke and/or fire doors and dampers.
- Pressurize stairwells for evacuation purposes.
- Override control of elevators and prevent them from opening on the fire floor.
- Automatically return elevators to the ground floor.
- Operate heat and smoke vents.
- Activate special fire extinguishing systems such as halon systems or pre-action and deluge sprinkler systems.

Types of Signaling Systems

The purpose of a protective signaling system is to limit fire losses involving life and property. Signaling systems vary from quite simple to very complex. A simple system may sound only a local evacuation alarm. A complex system may sound a local alarm, control building services, and notify outside agencies to respond. The type of system installed in any given occupancy will depend on many of the following factors:

- Level of life safety hazard
- Structural features of the building
- Level of hazard presented by contents in the building

- Local and state code requirements
- Risk management needs

The following sections examine each of the common types of signaling systems that firefighters are likely to encounter. It is important that personnel be able to recognize each type of system and understand how that system operates. This is particularly important when performing preincident planning. The major types of systems include the following:

- Local
- Selective coded system
- Auxiliary
- Remote station
- Central station
- Proprietary
- Emergency voice/alarm communications

Previously, each of these types of systems had its own NFPA standard (NFPA 72A, 72B, 72C, etc.). However, as of May 1993, the requirements for all fire alarm and protective signaling systems are contained in NFPA 72, The National Fire Alarm Code.

Manual Alarm-Initiating Devices Manual alarm-initiating devices, commonly called *pull boxes,* are placed in structures to allow occupants to manually initiate the fire signaling system. Pull boxes may be connected to systems that sound local alarms, supervisory notification alarms, or both.

Automatic Alarm-Initiating Devices An automatic alarm initiation device, sometimes simply called a *detector,* is a device that continuously monitors the atmosphere of a given area for the products of combustion. Then, when such products are detected, the device sends a signal to the system control unit. This device is typically quite accurate at sensing the presence of the combustion products that it was designed to detect. However, in many cases these products can be found in a given area, even when an emergency condition does not exist. For example, flame detectors may trip if a welder strikes an arc in a monitored area. These possibilities force fire protection system designers to take into account the normal activities that take place in a given area. System designers must then design a detection system that will minimize the risk of an accidental activation.

Fixed-Temperature Heat Detectors Systems using heat detection devices are among the oldest types of fire detection systems in use. They are relatively inexpensive in comparison to other types of systems and are the least prone of any system to false activation. Heat detectors are limited, however, by the fact that they are typically the slowest type of system to activate under fire conditions.

Heat detectors must be placed in high portions of a room, usually on the ceiling, to be effective. Detectors must also be selected at a rating that will give at least some small margin of cushion above the normal ceiling temperatures that can be expected in that area.

Heat is an abundant product of combustion. It is detectable by certain devices using three primary principles of physics:

- Heat causes an expansion action of material.
- Heat causes a melting action of material.
- Heated materials have thermoelectric properties that are detectable.

All heat detection devices are based on one or more of these principles. The various types of fixed-temperature devices used in fire detection systems are covered in the following sections.

Bimetallic Detector. A bimetallic detector uses two types of metal that have different heat expansion ratios. These metals are each formed into thin strips that are then bonded together. The fact that one metal expands faster than the other will cause the combined strip to arch when subjected to heat. The amount of arch depends on the characteristics of the metals, the amount of heat the strip is subjected to, and the degree of arch when the strip is in the normal position. All of these factors are taken into account in the design of the detector.

The bimetallic strip may be positioned with one or both ends secured in the device. When positioned with both ends secured, a slight bow is placed in the strip. When the strip is heated, expansion causes the bow to snap in the opposite direction. Depending on the design of the device, this either opens or closes a set of electrical contacts that in turn send an alarm signal to the system control unit.

Most bimetallic detectors are the automatic resetting type. They do need to be checked, however, to ensure that they have not been damaged.

Rate-of-Rise Heat Detector. A rate-of-rise heat detector operates on the principle that fires rapidly increase the temperature in a given area. The rate-of-rise detector will detect these quick increases in temperature. These detectors will respond at substantially lower temperatures than do fixed-temperature detectors. Typically, rate-of-rise heat detectors are designed to send a signal when the rise in temperature exceeds 12°F to 15°F (7°C to 8°C) per minute. This is because temperature changes of this magnitude are not expected under normal, nonfire circumstances.

Most rate-of-rise heat detectors are reliable and not subject to false activations. However, they can occasionally be activated under nonfire conditions. An example of this would be when a rate-of-rise detector is placed near a garage door in an air-conditioned building. If the garage door is opened on a hot day, the influx of heated air will rapidly increase the temperature around the detector and cause it to activate. These situations can be eliminating during the design process by avoiding such placement. There are several different types of rate-of-rise heat detectors in use; all rate-of-rise detectors are automatically reset.

Smoke Detectors   A smoke detector senses the presence of a fire much more quickly than does a heat detection device. The smoke detector is the preferred detector in many types of occupancies and is used extensively in residential settings. There are two basic types of smoke detectors in use: ionization and photoelectric. The following sections describe each.

Photoelectric Smoke Detector. A photoelectric detector, sometimes referred to as a *visible products-of-combustion smoke detector,* uses a photoelectric cell coupled with a specific light source. The photoelectric cell functions in two ways to detect smoke: beam application and refractory application.

In the beam application, a beam of light is focused across the area being monitored and onto a photoelectric cell. The cell constantly converts the beam into current, which keeps a switch open. When smoke blocks the path of the light beam, the current is no longer produced, the switch closes, and an alarm signal is initiated.

The refractory photocell uses a light beam that passes through a small chamber at a point away from the light source. Normally the light does not strike the photocell, and no current is produced. When a current does not flow, a switch in the current remains open. When smoke enters the chamber, it causes the light beam to be refracted (scattered) in random directions. A portion of the scattered light strikes the photocell, causing current to flow. This current closes the switch and activates the alarm signal.

A photoelectric detector works satisfactorily on all types of fires; however, it generally responds quicker to smoldering fires than does an ionization detector. This detector is automatically reset.

Ionization Smoke Detector. During a fire, molecules ionize as they undergo combustion. The ionized molecules have an electron imbalance and tend to steal electrons from other molecules. The ionization smoke detector, sometimes referred to as the *invisible products-of-combustion detector,* incorporates this phenomenon.

The detector has a sensing chamber that samples the air in a room. A small amount of radioactive material (usually americium) adjacent to the opening of the chamber ionizes the air particles as they enter. Inside the chamber are two electrical plates, one positively charged and one negatively charged. The ionized particles free electrons from the negative plate, and the electrons travel to the positive plate. Thus, a minute current normally flows between the two plates. When ionized products of combustion enter the chamber, they pick up the electrons freed by the radioactive ionization. The current between the plates ceases, and an alarm signal is initiated.

An ionization detector works satisfactorily on all types of fires; however, it generally responds quicker to flaming fires than does a photoelectric detector. This detector also automatically resets.

Flame Detectors A flame detector is sometimes called a light detector. There are three basic types: 1) those that detect light in the ultraviolet wave spectrum (UV detectors), 2) those that detect light in the infrared wave spectrum (IR detectors), and 3) those that detect both types of light.

Although these types of detectors are among the fastest to respond to fires, they are also easily tripped by nonfire conditions such as welding, sunlight, and other bright light sources. They must be placed only in areas where these possibilities can be avoided. They must also be positioned so that they have an unobstructed view of the protected area. If they are blocked, they cannot activate.

An ultraviolet detector can give false alarms when it is in contact with sunlight and arc welding. Therefore, its use is limited to areas where these and other sources of ultraviolet light can be eliminated. An infrared detector is effective in monitoring large areas. To prevent activation from infrared light sources other than fires, an infrared detector requires the flickering action of a flame before it activates to send an alarm.

Fire Gas Detectors When fire breaks out in any confined area, it drastically changes the chemical gas content of the atmosphere in that area. Some of the gases released by a fire include the following:

- Water vapor
- Carbon dioxide
- Carbon monoxide
- Hydrogen chloride
- Hydrogen cyanide
- Hydrogen fluoride
- Hydrogen sulfide

Only water, carbon dioxide, and carbon monoxide are released from all materials that burn. The other gases released are dependent on the specific chemical makeup of the fuel. Therefore, for fire detection purposes, it is only practical to monitor the levels of carbon dioxide and carbon monoxide. This type of detector responds somewhat faster than a heat detector, but not as fast as a smoke detector. It uses either semiconductors or catalytic elements to sense the gas and trigger the alarm. A fire gas detector is not in common use, compared to other types of detectors.

Combination Detectors Depending on the design of the system, various combinations of the previously described detection devices may be used in a single device. Sample combinations include fixed-rate and rate-of-rise detectors, heat and smoke detectors, and smoke and fire gas detectors. These combinations give the detector the benefit of both services and increase their responsiveness to fire conditions.

28.6   FIRE PROTECTION DURING CONSTRUCTION, ALTERATION, AND DEMOLITION

The most likely time for a structure to be destroyed or damaged is during construction, alteration, and demolition. The possibility of fire is greatly increased because of the impairment of fire detection and protection devices. The absence of fire-resistant walls, floors, and enclosures often enhances the possible occurrence of a fire. The use of safety-conscious construction and alteration crews will improve the chances of not having a fire. The contractor must provide alternate methods for fire protection. The safety of the construction crew must be taken into account because of increased risk. Fire protection and safety professionals should be consulted to ensure that all fire protection systems are not disabled at once. Hand-held fire extinguishers may be employed in case of localized or spot fires during construction. During construction, a security company should be employed to guard the site. A constant watch must be kept for possible problems that have developed. Additional personnel should be used in the development of emergency procedures. NFPA 271, Safeguarding Construction, Alteration, and Demolition Operations, can provide additional guidelines regarding the safeguards needed on a constructions site.

28.7   COMPUTER FACILITIES

The use of computers has become overwhelming. Even the youngest of children have access to advanced computers. The use of computers and related electronic equipment in every aspect of business, government, and industry has required that new and innovative methods of risk management be developed.

Because of the Montreal Protocol, an international agreement that bans the use of halogenated agents that could deplete the ozone layer, the main fire protectant for computer systems is no longer halon. This compound protected computers from fires and did not create any additional damage once the fire was extinguished. The use of halons 1211, 1301, and 2402 has been on the decrease for several years because of the protocol. The search for substitute fire protection agents has been extensive. The trend has led the fire protection industry back to the use of large carbon dioxide systems and selected sprinkler systems, which do not affect the ozone layer.

Although many companies have voluntarily taken halon fire protection systems out of service and replaced them with alternate means of fire protection, many companies cannot afford to replace these systems. Once the systems have discharged, a replacement must be sought to comply with the protocol.

Some of things to consider when designing a computer area with risk management in mind are:

- *Equipment and construction features:* Heating, ventilation, and air conditioning (HVAC) systems, noncombustible construction, and uninterruptible power sources.
- *Fire Protection:* Portable fire extinguishers, sprinkler systems, special extinguishing systems, smoke detectors, and emergency plans and procedures.

28.8   LABORATORIES USING CHEMICALS

Facilities managers must contend with the ever-increasing complexity of laboratories and the chemicals contained and stored in them. NFPA 45, Fire Protection for Laboratories Using Chemicals, is the standard for protecting facilities of this type. Buildings that have been specifically designed for laboratories should contain the latest in fire protection, such as special extinguishing and protection systems, standpipe and hose systems, and portable fire extinguishers.

All flammable and combustible liquids present hazards in laboratories. The severity of the hazards associated with these liquids depends on several factors:

- Quantity of flammable and combustible liquid
- Volatility of the liquid
- Flammable range of the liquid
- System open or closed to air
- Methods of containment
- Location of storage
- Outside ignition sources
- Building construction
- Fire protection

One or several of theses factors could combine to cause a serious incident involving flammable or combustible liquids. The use of correct safety procedures involving these liquids could prevent accidents in the laboratory. The correct handling procedures for these liquids can be found in NFPA 30, Standard for Flammable Liquids.

28.9 LIBRARIES AND MUSEUMS

From a fire protection point of view, libraries and museums present unique hazards that require special attention. Some of the problems include high fuel loads, high values, and public access with awkward hours of business.

Because of the unique risk to life and property, special attention must be paid to fire prevention and protection efforts. NFPA 910, Standard for Libraries, and NFPA 911, Standard for Museums, set out the recognized good practice for protecting these reliable assets and the people that use them.

28.10 ASSEMBLIES

An assembly occupancy is a building or portion of a building that accommodates gatherings of people, usually in groups of fifty or more. Churches, concert halls, bus stations, airports, and convention centers are all examples of assembly occupancies. Assembly occupancies have several important factors that require attention by facilities managers. Things such as means of egress and occupancy load are crucial because of the potential for loss of life. The Life Safety Code or another relevant code should be applied to prevent such events from becoming a potential hazards to life.

28.11 FIRE AND EMERGENCY SERVICES

Although a few colleges, universities, and other institutional occupancies maintain their own fire departments, this is a decreasing trend. In nearly all cases, fire and emergency services for a given facility are provided by the unit of local government that serves the immediate area. Maintaining viable and open communications between the fire and emergency service providers and the institution is an important role for the facilities manager. Regular and positive communications should be maintained so that the fire department and emergency medical service provider are fully aware of the idiosyncrasies of the facility and the management is clearly aware of the capabilities of the emergency agencies.

In those cases where a facility is either remotely located or of sufficient complexity that it must maintain its own emergency response capability, it is likely that the emergency responders will fall under the standards prescribed by the Occupational Safety and Health Administration (OSHA). These proprietary fire services or fire brigades generally represent a costly alternative to quality communications and relationships with local fire services. In most cases it will be far more cost-effective to spend time and effort in familiarizing local emergency responders with the unique requirements of a given facility than it will be to organize, maintain, and operate a proprietary fire service or a fire brigade.

Proprietary Fire Services

This section is an orientation to the general duties of the fire brigade member, the employer, and the requirements for each under OSHA regulations and NFPA standards. These duties range from simple evacuations to fire fighting and record keeping requirements. Included is an orientation to the organizational structure of a fire brigade and descriptions of associated job titles.

Employee Emergency Action Plan

OSHA regulations mandated by the U.S. Department of Labor under 29 CFR 1910.38, Employee Emergency Plans and Fire Prevention Plans, require that an emergency action plan be maintained in writing to cover those actions required of employers and employees during a fire or other emergency. The emergency action plan is independent of the fire brigade organizational statement. The plan is required to address all potential emergencies that could occur in the workplace, and it should be reviewed with each employee. It should include, as a minimum, the following elements:

- Emergency escape route assignments from individual work areas
- Procedures to be followed by employees who remain to perform or shut down critical plant operations
- Procedures to account for all employees after emergency evacuation has been completed
- Rescue and medical duties for those employees who are to perform them
- Preferred means for reporting fires and other emergencies
- Names or regular job titles of persons or departments to be contacted for further information or explanation of duties under the plan
- Housekeeping standards, guidelines for storage of fuels, and procedures for maintenance of fire equipment

Reporting Emergencies   Every employee in the plant or facility must know how to report an emergency so that others in the facility are aware of the emergency and appropriate personnel can respond to it. This reporting can be accomplished in several different ways, depending on the policies adopted by management. Reporting an emergency can be done by shouting for help, by sounding an alarm, or by other means as specified in the emergency action plan. Each facility may have a different way of reporting an emergency. It is important for employees to know how to report an emergency in their respective facilities and to practice the procedure.

Evacuation Planning   The employer must establish in the emergency action plan the type and method of evacuation for each work area. Detailed floor plans or workplace maps should be used to show the emergency exit routes and any alternative escape routes (Figure 28-1).

The evacuation plan should also contain designated safe areas where evacuees assemble for an accounting of personnel. To eliminate confusion during actual or simulated emergency situations, evacuation routes should be clearly marked on the emergency evacuation plan, and the plan should be posted in highly visible locations in each building.

Some method of accounting for personnel must be a part of any evacuation plan. Designated employees will report to the incident commander

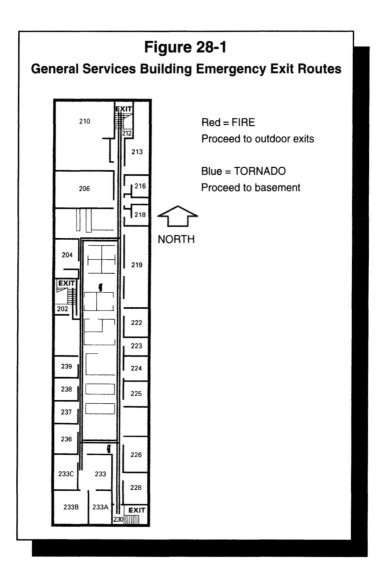

Figure 28-1
General Services Building Emergency Exit Routes

on the status of the evacuation and identify any individuals who are missing. This information is important so that a search for missing persons can be quickly initiated. Personnel should not leave the designated assembly area until instructed to do so by someone in authority.

Levels of Evacuation. Evacuation in an emergency situation can involve a specific work area, a single floor, or a total evacuation of the premises. It is important to establish policies for each level of evacuation needed, according to the anticipated complexity of the emergency. Management must consider and specify levels of evacuation and then incorporate them into the employee emergency action plan.

Alternative Assembly Areas. Management must specify assembly areas and should consider alternate assembly areas for employees whenever a building or area is evacuated during an emergency. All assembly areas should account for variable conditions such as weather and expected emergency conditions.

The following types of areas are typically designated:

- Open areas adjacent to the building
- Other buildings
- Parking lots
- Areas of safety that may be within a structure

Accounting for Evacuated Employees. Management is responsible for ensuring that all employees, visitors, and contractors have evacuated a building and for accounting for each person who was in the building. This process can easily be done by having a designated person check employee, visitor, and contractor names from established lists.

In university settings the occupancy is constantly changing, with no formal lines of authority over many of the occupants. In this case the duty may have to be shared, with classroom leaders accounting for the students in their classes (as is done in public schools) and other faculty and staff accounted for through a building or office supervisor.

Employees Who Remain Behind. During some emergencies it may be necessary for certain employees to remain behind and not immediately evacuate the area. These employees should remain in their work area only if the emergency does not put them in imminent danger.

It may also be necessary for other employees to remain in the area of the emergency to ensure that personnel have evacuated the area and are proceeding to the safe area. Designated employees may be trained to deal with specific emergencies and may be required to respond to the emergency area to perform critical process or emergency procedures. Other training may include special rescue and medical duties.

Explanation of Duties Under the Plan   To maintain employees' personal safety, it is very important that each employee know what his or her duties are during any emergency. If an employee is not clear on what the appropriate action should be, then he or she should get clarification from his or her supervisor or from the person in charge of fire and safety in the facility. It is management's responsibility to make this information readily available to employees and employees' responsibility to know their duties during an emergency. This information should be communicated in required orientation and training sessions and reinforced periodically thereafter.

Exit Identification   In industrial facilities, all exits should be identified, and all exits and exit corridors should remain unobstructed and accessible. Exits should remain unlocked during normal working hours or when otherwise occupied. Doors that must remain locked for security reasons must be able to be opened with one motion from the inside for emergency exiting.

Emergency Response Plan

Management must establish a written policy documenting employee responsibilities and actions in the event of a fire in the workplace. The policy must include an organizational statement, a definition of levels of emergency response, and a statement mandating a performance-based training program.

In defining levels of response, management must choose one or more of the following options:

- *Option 1.* All employees will evacuate from the workplace. No action will be taken by any employee to extinguish any fire, only to report it. Training on evacuation and fire reporting procedures is all that is required for this option.
- *Option 2.* Only designated employees are trained in the use of portable extinguishers and/or small hoselines to extinguish an incipient fire in their immediate work area. Initial training and annual refresher training are required.
- *Option 3.* All employees are trained in the use of portable extinguishers and small hoselines, and they are authorized to attempt extinguishment of an incipient (beginning) fire in their immediate work area, if it appears safe and prudent to do so. Initial training and annual refresher training are required.
- *Option 4.* Designated employees are assigned to an emergency response team that responds to fires and other emergencies in all areas of the facility. Training must be commensurate with the required duties and functions of the brigade, and the quality and frequency of training must ensure that the brigade member can perform required duties and functions in a manner that does not present a hazard to the member or others.
- *Option 5.* A full-time dedicated emergency response organization is established and carries the primary responsibility and training for response to fire and other emergencies in all areas of the facility. This option is functionally equivalent to operating a fire department and

is beyond the scope of this document. For further training information at this level, refer to *Essentials of Fire Fighting*.[3]

Fire Brigade

Organizational Statement OSHA requires under 29 CFR 1910.156, Fire Brigades, that "employers shall prepare and maintain a statement or written policy which establishes the existence of a fire brigade."

Basic Organizational Structure The NFPA's Technical Committee on Loss Prevention Procedures and Practices recognized the need for a standard for industrial fire brigades. NFPA 600, Standard on Industrial Fire Brigades, was developed to follow current OSHA regulations while expanding on the guidelines set forth in 29 CFR 1910.156 and 1910.157. NFPA 600 was developed to safeguard the occupational safety of employees who engage in mitigation of emergencies in the workplace.

The organizational statement for the emergency response organization must assign responsibilities to positions, establish operational limitations, and show how the fire brigade fits into the company emergency plan. In addition, the organizational statement should describe the relationship of the fire brigade to other response groups including municipal and contractual organizations.

For the fire brigade to function properly, certain individuals have to be assigned titles and leadership positions to manage the fire brigade operations. These positions are discussed in the sections that follow.

Senior Facilities Manager. The senior facilities manager is ultimately responsible for the actions of the fire brigade and for the definition of the level of response for the brigade. This person also bears the ultimate responsibility for ensuring that the organization is trained and equipped to follow the guidelines that have been established.

Fire Brigade Management Official. This individual is designated by the plant management, the facilities manager, or the corporate management as the emergency response brigade leader responsible for the organization, management, and operation of the industrial fire brigade. This position is analogous to a municipal fire chief, and he or she may be referred to as "Chief."

Brigade Leader. The brigade leader is responsible for directing the actions of the brigade, coordinating the training of members assigned to his or her shift, and ensuring that coverage is maintained at all times.

Brigade Member. The brigade member is responsible for carrying out actions as directed by the brigade chief or brigade leader, following standard operating procedures, and maintaining the knowledge and competency level required of the position.

Other Personnel. Other positions that provide support functions for the emergency response brigade include personnel from electrical, maintenance, security, environmental, and medical areas. Specific assignments and training programs should be developed for each function that supports the brigade.

Training If emergency response personnel are required to perform a particular task or evolution, those individuals must be provided with the training and education to effectively and safely perform that task (29 CFR 1910.156 [c][2]).

Training and drills are two distinct programs. Training should include both lecture and hands-on practice, with established and documented performance objectives and led by qualified persons. All emergency response brigade members shall be able to competently perform their functions to the established performance level in a safe and timely manner. Training should include the lessons, both positive and negative, learned from the experiences of previous emergency responses and drills. Postincident critiques are considered a valuable training tool.

For the fire brigade to maintain proficiency and effectiveness, regular drills should be held for each shift or team. A *drill* is a task or evolution to reinforce training that has already been received. Any deficiencies discovered in a drill can then be identified, resolved, and documented. After the critique, training should be scheduled and conducted to improve those areas found to be of concern.

All training and drills must be documented in each individual member's training file. Copies of the course outline and handouts should be maintained.

Size of the Emergency Organization The number of brigade members will depend on the type of fire brigade, the number of employees working in the facility, the size of the facility, the number and type of potential hazards in the facility, and the availability of emergency equipment in the facility. A brigade may be composed of employees who volunteer or employees who are assigned to the fire brigade. They should be physically capable, as determined by a medical examination, of performing expected assigned brigade duties or functions. Fire brigade members should come from each shift and department in the facility so that there will be an adequate number of members during any given period of the day or night.

Record Keeping Record keeping is an important aspect of ensuring compliance with any regulation or requirement regardless of whether the requirement is dictated by OSHA, the insurance carrier, or company policy. Records should be kept on at least the following items, plus any others that may be required for the individual organization.

- Training of emergency personnel
- Emergency response equipment maintenance
- Condition of fire and safety features of buildings

Documentation is critically important, because it is the only proof that the fire brigade member has been trained and is qualified to perform as indicated by the organizational statement. Drill documentation must show that the fire brigade member has practiced the skills taught in training sessions and should indicate individual levels of proficiency.

Interaction with Outside Agencies Industrial fire brigades may be required to interact with outside agencies at various times. These agencies may include the following:

- Local fire departments
- Formal mutual aid organizations
- ˥ocal law enforcement agencies
- Other industrial fire brigades
- Emergency medical services
- Outside contractors
- Government agencies
- News media

Effective communication with outside agencies must be maintained to ensure an adequate response to the facility when needed. The most important aspect of this relationship can be handled with an on-site visit by the outside agencies. Local fire officials should develop a preincident plan for the facility, whereas law enforcement agencies will have to develop a contingency plan to handle wide area evacuations. Effective communication usually begins with an invitation to view the facility and provide a full review of potential hazards. This type of evaluation should be part of any preincident planning. Without this type of cooperation, the ability for agencies to work together in an emergency could be impaired or could even collapse into chaos.

Once the decision has been made to ask local agencies to respond for assistance during an emergency, the person responsible for making this request must provide the responding agencies with the necessary information agreed on during the preincident planning process. This information is crucial for various outside agencies so they can properly determine the type of response required.

Environmental Awareness Fire brigade members should be aware that some incidents, as well as training sessions, may have an environmental impact. Training to provide the appropriate level of environmental awareness is beyond the scope of this document. Incidents with environmental impact must be reported or communicated to the proper persons within the organization. A good working relationship with the local fire department and hazardous materials team can assist management with valuable information and training in this regard.

Local Fire Department Liaison

Institutions should strongly consider the designation of a liaison to local government emergency services. In addition to fostering proper communications and facilitating awareness tours, the liaison officer should be prepared to represent the institution's interests in code issues, fire inspections, and emergency planning.

Many institutions have launched quality efforts with local fire services in providing high-quality public fire safety education programs. Adult audiences are often very difficult to deal with in attempting to modify behaviors that create hazardous situations. It will generally take a full commitment on the part of both the local fire department and the institution to provide a high-quality, fast-moving, focused public fire safety campaign that will be measurably effective. Organizations such as the Campus Safety Association section of the National Safety Council and the education section of the NFPA can offer assistance in this important area.

Incident Command System

The determination of who is in charge and how an emergency incident will be managed must be done prior to an emergency. Unfortunate consequences may result from power struggles when public fire services and institutional management both perceive themselves to be in charge at a serious emergency. Establishing an understanding of the roles of each agency is important for proper management of an emergency incident.

Without exception, public fire departments are mandated by a number of regulatory acts to operate in a strictly defined system that is generally referred to as the *incident management system.* Institutional officers must understand that the safety and effectiveness of public emergency responders are very much functions of the quality of leadership and the system in which that leadership operates during an emergency.

The incident commander of a fire or emergency scene is the officer in charge of the operation. This officer is at the top of a chain of command in a strictly defined incident command system and is ultimately responsible for everything that takes place at the emergency scene. This officer must coordinate and direct all efforts on the fireground or emergency scene to achieve specific goals or priorities. These priorities, in order, are 1) to ensure the safety and survival of emergency service workers and citizens, 2) to rescue and care for endangered occupants, 3) to control the situation, and 4) to conduct salvage operations to minimize loss.

Although there is only one incident commander, he or she works with an assistant who utilizes many resources. The institutional leadership should clearly plan ahead with the fire department or emergency service agency to provide a role within the incident command system for those who know the most about the institution.

Water Supply for Fire Services

Technology keeps advancing, with new methods and materials for extinguishing fires. However, water still remains the primary extinguishing agent because of its universal abundance and its ability to absorb heat. Two primary advantages of water are that it can be conveyed over long distances and it can be easily stored. These are the fundamental principles of a water supply system. Because water remains the primary extinguishing agent used by firefighters, it is important that firefighters have a good working knowledge of water supply systems.

Principles of Municipal Water Supply Systems

Public and/or private water systems provide the methods for supplying water to more populated areas. As rural areas increase in population, rural communities seek to improve efficient water distribution systems from a reliable source.

The water department may be a separate city-operated utility or a regional or private water authority. Its principal function is to provide water that is safe for human use. Water department officials should be considered the experts in water supply problems. The fire department must work with the water department in planning fire protection. Water department officials should realize that fire departments are vitally concerned with water supply and should work with them on supply needs, location, and types of fire hydrants.

The intricate working parts of a water system are many and varied. Basically, the system can be described by the following fundamental components:

- The source of supply
- The mechanical or other means of moving water
- The processing or treatment facilities
- The distribution system, including storage

Location of Fire Hydrants

Although the installation of fire hydrants is usually performed by water department personnel, the location, spacing, and distribution of fire hydrants should be the responsibility of the fire chief or fire marshal. In general, fire hydrants should not be spaced more than 300 ft. (90 m) apart in high-value districts. A basic rule to follow is to place one hydrant near each street intersection, with intermediate hydrants placed where distances between intersections exceed 350 to 400 ft. (105 to 120 m). This basic rule represents a minimum requirement and should be regarded only as a guide for spacing hydrants. Other factors more pertinent to the particular locale include types of construction, types of occupancy, congestion, the sizes of water mains, fire flows, and pumping capacities.

Fire Hydrant Maintenance

In most institutions repair and maintenance of fire hydrants are the responsibility of the facilities department, because this department is in a better position to do this work than any other agency. However, in many cases, the fire department performs water supply testing and hydrant inspections. Personnel should look for the following potential problems when checking fire hydrants:

- Are there obstructions, such as signposts, utility poles, or fences, that have been erected too near the hydrant?
- Do the outlets face the proper direction, and is there sufficient clearance from them to the ground?
- Has the hydrant been damaged by traffic?
- Is the hydrant rusting or corroded?
- Is the operating stem easily turned?
- Are the hydrant caps stuck in place with paint?

28.12 SUMMARY

Fire protection is more than just the design, purchase, and installation of a fire alarm or fire protection system. It is an integrated program of facilities design, facilities components, operations and maintenance, and support processes that provide the desired level of protection for the occupants, the building, and its contents. The fire protection program at a college or university should be examined to ensure that it encompasses all the necessary elements.

NOTES

1. Industrial Fire Protection Validation Committee. Taken with permission from the *IFSTA Industrial Fire Protection Manual,* first edition. Stillwater, Oklahoma: International Fire Service Training Association, 1982.
2. Arthur Cote (Ed.). *NFPA Handbook, 6-3-13.* Quincy, Massachusetts: National Fire Protection Association, 1991.
3. Essentials of Fire Fighting Validation Committee. *Essentials of Fire Fighting.* Stillwater, Oklahoma: International Fire Service Training Association, 1993.

CHAPTER 29

Elevator Systems

Bryan R. Hines
Lerch Bates & Associates, Inc.

29.1 INTRODUCTION

In any multistory building, the internal vertical transportation system typically comprises elevators, escalators, and/or moving walks. Because these units represent a significant expense, in terms of both initial construction and building operation, effective management of these vertical transportation systems becomes an important responsibility for all facilities managers. To assist in this responsibility, this section will discuss the different types of elevators, proper elevator design, code enforcement and safety inspection requirements, modernization and upgrade of existing elevators, and maintenance and operation of elevators.

29.2 TYPES OF ELEVATORS

Hydraulic Elevators

Hydraulic elevators are used in low-rise buildings of two to five stories and typically operate at a maximum speed of 150 ft. per minute (fpm). Hydraulic elevators receive their vertical motion from a hydraulic plunger fastened to the bottom of the elevator car. The plunger moves in a hydraulic cylinder that extends as deep into the ground as the vertical travel of the elevator in the building. A pump, driven by an AC electric motor, forces oil into the cylinder to lift the car. Gravity and the weight of the car force oil from the cylinder back into the oil reservoir to lower the car. Startup, slow-down, and leveling of the car are controlled by electrically operated valves.

Geared Traction Elevators

Geared traction elevators are used in midrise buildings of five to fifteen stories and typically operate at a maximum speed of 450 fpm. Geared traction elevators receive their vertical motion from an AC or DC motor that is directly coupled to a worm gear. The worm gear in turn rotates a ring gear. A drive sheave or pulley is attached to the ring gear. Steel cables run from the top of the car, over the drive sheave, to the top of the counterweight. The downward pull caused by the weight of the car and counterweight holds the steel cables tight in the drive sheave. When the drive sheave turns, the car is raised or lowered.

Gearless Traction Elevators

Gearless traction elevators are used in high-rise buildings of fifteen to one hundred stories and operate at speeds of 500 to 1800 fpm. Gearless elevators operate under the same principle as geared elevators, vertical motion derived by steel cables connecting the car and counterweight. However, the drive sheave of a gearless elevator is connected directly to the motor shaft, eliminating the worm and ring gears. Significantly faster speeds are possible with gearless elevators, because vibration is nearly eliminated in the absence of the worm and ring gears.

Figure 29-1 summarizes the three types of elevators, their applications, and their advantages/disadvantages.

29.3   PROPER ELEVATOR DESIGN

Proper elevator design begins with selecting the correct type of elevator and evaluating building conditions so that the proper number of elevators are installed. Typical office and academic buildings will utilize elevators in the 3,000- to 4,000-pound capacity range. Proper elevator speeds vary with the height of the building. Regardless of speed and capacity, elevators must provide smooth acceleration and deceleration and accurate leveling within ±1/4 in. of floor level. Elevators should be located within sight of the main building entry and should never open into pedestrian corridors. Elevators should be equipped with visual and audible signals in the car and in the lobby that indicate car position and travel direction. Waiting times for elevator service after registration of a corridor call should not exceed 30 seconds, and the elevator system should have the ability to move 10 to 15 percent of the total building population in a 5-minute period.

Figure 29-1
Types of Elevators

Elevator Type	Application	Maximum Speed	Life Cycle of Driving Machine	Advantages	Disadvantages
Hydraulic	Low-rise 2–5 floors, light use	150 fpm	20 years	Low cost Rapid installation No loads on building structure	Requires drilling, with possible oil contamination of ground water Slow speed Slow acceleration and deceleration rates
Geared traction	Mid-rise 5–15 floors, moderate use	450 fpm	25–35 years	Moderate cost Longer life cycle Higher speeds	Higher cost to maintain Imposes all structural loads on building structure Vibration may be noticeable at highest speeds
Gearless traction	High-rise 10–100+ floors, heavy use	1,800+ fpm	75+ years	Longest life cycle Rapid-floor-to-floor travel times Smooth acceleration and deceleration	Highest installation cost Highest maintenance cost Imposes all structural loads on building structure

29.4 CODE ENFORCEMENT AND SAFETY INSPECTIONS

In the United States, most states and localities have elevator codes based on some version of the American Society of Mechanical Engineers (ASME)/ American National Standards Institute (ANSI) A17.1 Safety Code for Elevators and Escalators. Copies are available from the ASME (Order Department, 22 Law Drive, Box 2300, Fairfield, New Jersey 07007, or call 201-882-1167). Because many local jurisdictions modify this national standard code and because the code is updated annually by an addendum with a complete new printing every three years, facilities managers are cautioned to determine the code in effect for their building.

In Canada, the CSA/B44.1 Code defines elevator safety rules and requirements; copies are available from the ASME Order Department (see address above). In Japan and other parts of Asia, elevator code requirements are defined in the Japanese Industrial Code (JIC).

In the United States, although the A17.1 Code strictly deals with elevators, escalators, and other forms of vertical transportation, other model building codes affecting elevators are 1) the Uniform Building Code (UBC), 2) the Building Officials and Code Administrators (BOCA) Code, and 3) the Southern Building Code (SBC). In addition to the model building codes, elevator code requirements are found in the National Electric Code (NEC), the National Fire Protection Association Code (NFPA-72E), and state and local ordinances. Accessibility requirements are generally outlined in the Americans with Disabilities Act (ADA), even though ADA is civil rights legislation and not a building code. In areas outside the United States, there are similar codes and standards that affect vertical transportation design, installation, and operation.

In the United States, facilities managers should be familiar with Part X of the A17.1 Code. Part X covers routine, periodic, and acceptance inspections and tests. Part X, Rule 1001.1 requires a semiannual routine inspection of all electric traction passenger and freight elevators. Rule 1002.1 requires an annual inspection of all buffers, safeties, governors, and all other safety devices on all electric traction passenger and freight elevators.

Rule 1004.1 requires a semiannual routine inspection and test of all hydraulic passenger and freight elevators. Rule 1005.1 requires a periodic inspection and test of all safety devices on all hydraulic passenger and freight elevators.

Part XII of the A17.1 Elevator Safety Code lists code requirements for alterations, repairs, replacement, and maintenance of elevators and escalators.

Because construction costs, attorney costs, and liability assessments in accidents involving vertical transportation equipment can be costly and time consuming, all facilities managers should be familiar with safety tests and inspections required by local and national code authorities.

29.5 MAINTENANCE

Most owners of elevators and other vertical transportation systems will contract with the elevator installer or another qualified elevator contractor to provide preventive maintenance and callback service. As elevators require routine maintenance to operate properly and safely, schools and universities with a large number of elevators may elect to provide in-house maintenance. This can substantially reduce costs, but the owner must be certain that the elevator technicians employed are qualified. Although elevators and escalators are the safest form of public transportation, accidents can and do happen to poorly trained personnel working on or around these units. If managers elect to maintain elevators in house, they may want to take advantage of training courses on elevator safety and maintenance practices offered by the ASME. In the United States, the National Association of Elevator Safety Authorities (NAESA) is the organization that certifies elevator inspectors and offers safety and code tests. The NAESA may be contacted at the address below:

NAESA International
4541 North 12th Street
Phoenix, Arizona 85014-4203
602-266-9701
602-265-0093 (fax)

Elevator contractors offer three types of elevator maintenance contracts: full or complete maintenance (CM); a parts, oil, and grease (POG) contract; and an oil and grease (OG) contract, as follows:

- *Full or complete maintenance.* For a stated monthly fee, the elevator contractor will perform all routine maintenance and repair and provide callback service when the elevator is out of service. There are no additional monthly charges for parts or labor, no matter how many times the elevator is out of service, unless the trouble is caused by vandalism or abuse.
- *Parts, oil, and grease.* Under the POG contract, the elevator contractor, for a stated monthly fee, provides a monthly (or other agreed-on frequency) routine examination. During this examination, the elevator is cleaned, oiled, and adjusted. The cost of small parts—usually relay contacts, motor brushes, light bulbs, etc.—is included in the monthly fee. All callbacks to place an out-of-service elevator back into service and all large parts are billed in addition to the monthly maintenance fee.
- *Oil and grease.* Under the OG contract, the elevator contractor, for a stated monthly fee, performs a routine maintenance examination. During this examination, the elevator is cleaned, oiled, and adjusted. Any parts or callback service are billed in addition to the monthly fee.

Although the cost of a CM contract is obviously higher than that of a POG or OG contract, monthly costs are known in advance, making it a much simpler process to budget for elevator maintenance.

Regardless of the type of maintenance contract, elevator maintenance can be broken into four general areas: housekeeping, lubrication, replacement or repair of worn components, and adjustments. These areas sometimes overlap but are sufficiently independent to allow separate evaluation.

Housekeeping

Housekeeping requires approximately 60 percent of the total time spent maintaining equipment. Although at first glance this would appear to be an excessive amount of time spent simply cleaning, it is time well spent. If the environment is kept clean, the fire hazard (especially in hoistways) is lessened. Potential troubles and worn components are often detected during routine cleaning operations. Dirt is a major cause of elevator malfunction; a speck of dust between relay contacts can shut an elevator down. Finally, a clean job makes routine inspection and maintenance easier.

Lubrication

Lubrication requires approximately 15 percent of the total time spent maintaining equipment. As with any mechanical equipment, proper lubrication minimizes wear, ensures proper operation, and lengthens the trouble-free life of components.

Replacement or Repair

Replacement or repair of worn components represents approximately 15 percent of elevator maintenance. By detecting and replacing worn components, it is often possible to prevent elevator malfunctions and unscheduled shutdowns. Systematic repair and replacement of components ensures optimum useful life of the elevator.

Adjustments

Adjustments require approximately 10 percent of elevator maintenance time. Proper, timely adjustment keeps the equipment working smoothly and quietly while optimizing performance.

The facilities manager should routinely tour the elevator machine rooms and ride the elevator cars. Does the elevator operate smoothly and level properly at the floor? Do all the call buttons illuminate when pressed? Is the elevator machine room clean, with all parts and lubricants stored in metal lockers? Are there adequate spare parts stocked on the job site? If your answer to any of these questions is no, some improvement in elevator operation and maintenance is possible and desirable.

29.6 MODERNIZATION

Often owners or facilities managers will want or need to upgrade older elevators. There are typically three major reasons owners want to modernize: to increase reliability, to increase traffic handling ability, and to incorporate current code and accessibility requirements.

- *Increasing reliability.* Old elevators often break down, and parts become increasingly difficult or impossible to find. Modernizing substantially improves reliability.
- *Increasing traffic handling ability.* The group control systems on older elevators are often inefficient compared with today's microprocessor controls. Modernizing older relay control systems with micro-processor control systems often results in a 25 to 40 percent reduction in waiting times as well as increased reliability.
- *Incorporating current code and accessibility requirements.* New accessibility guidelines, such as the Americans with Disabilities Act (ADA) dictate the height of call buttons and requirements for braille plates and visual and audible signals in the car. Most elevators installed before 1980 are not equipped with any accessibility features. Incorporation of these features is probably the most common reason for modernizing elevators.

Complete elevator modernization can be quite expensive, often costing as much or more than the initial installation. For this reason, it is important to have a clear understanding of what is to be accomplished in a modernization program. Part XII of the ASME A17.1 Elevator Safety Code outlines the requirements for repair and modernization.

29.7 CONSULTANTS

When impartial assistance regarding elevator issues is desired, that assistance is available through elevator consultants. These specialists have varying backgrounds and abilities. Active consulting firms have contacts within the elevator industry that may be valuable in solving a problem. Owners and property managers may utilize elevator consultants to gain a broader view of the purchase, modernization, and performance evaluation of elevators. However, always remember that it is important to check the past performance of consultants before contracting for their services.

29.8   SUMMARY

Elevators and other vertical transportation equipment are an integral part of all multistory buildings. Guaranteeing that these devices are safe and reliable is, in large part, the responsibility of the facilities manager.

ADDITIONAL RESOURCES

Elevator World Educational Package and Reference Library, Volumes 1, 2, and 3. Mobile, Alabama: Elevator World, Inc., 1990.

Elevator World Monthly Magazine. Mobile, Alabama: Elevator World, Inc. Telephone: 202-479-4514.

Strakosch, George R. *Vertical Transportation,* second edition. New York: John Wiley & Sons, 1983. (Available through Elevator World Education Services at 202-479-4514.)

SECTION II-C

PLANT SERVICES

Editor:
Norman H. Bedell (deceased)
Pennsylvania State University

INTRODUCTION

Plant Services

his section includes information on service delivery for typical services provided by a facilities organization. Chapter 30 provides information on organization concepts for service delivery and improved flexibility, effectiveness, and efficiency. Although other chapters will provide greater detail concerning the various components and techniques of a facilities organization, Chapter 30 discusses the interrelationships among these processes as well as the effects of these interrelationships. The remaining chapters in this section provide information for grounds maintenance and operations, custodial services, and solid waste handling and recycling. These chapters cover typical organizational structures, staffing, and process designs with highlights of important areas of consideration.

—Gary L. Reynolds

CHAPTER 30

Facilities Maintenance and Operations

J. Richard Swistock
Virginia Commonwealth University

30.1 INTRODUCTION

Facilities maintenance and operation are on the threshold of radical change at most colleges and universities. This change is being driven by many of the same forces that are reshaping all of higher education: changing demographics; the demand for accountability; and competition for resources, talent, and students, to name a few. Information technology, the need for sophisticated building environments to support research activities, dealing with antiquated facilities, and meeting ever-increasing environmental and safety demands are some of the changing new technical challenges facing facilities maintenance and operations personnel.

Meeting these changes while providing better customer services and becoming more cost-effective in this changing environment will require a new look at traditional facilities maintenance and operations organizations and the way they function. This chapter discusses this changing role and suggests organizational changes that may be appropriate.

30.2 RELATIONSHIP TO INSTITUTIONAL MISSION

The mission of the facilities management organization is to provide a physical environment that enables an institution to accomplish its broader mission. This environment consists of buildings, grounds, and other campus facilities. The maintenance and operation of these facilities can be justified only if they enable the university community to effectively pursue teaching, research, public service, and the supporting institution activities. The facilities are maintained not for their own sake, but to provide, within available resources, the best possible environment in which the university community can function.

The physical assets of a college or university represent a large investment, and this investment must be managed as other assets are managed. Because institutions of higher education are generally thought of as existing in perpetuity, there is a tendency to think that campus facilities should exist in perpetuity. Some buildings and campus facilities have historical significance and should indeed be maintained and preserved as such. Most facilities, however, exist to enable an institution to accomplish its mission and should be maintained, preserved, renovated, or disposed of based on the best business interests of the institution. To responsibly maintain and preserve the institution's facilities, management must meet the needs of the occupants and users of these facilities, not simply maintain and operate facilities.

Preventive maintenance is the cornerstone of preserving a college or university's real property assets and ensuring that physical facilities support the institution's mission in the most cost-effective way possible. Preventive maintenance should be the guiding philosophy of an effective maintenance organization. Because preventive maintenance is accomplished on a scheduled, predictable basis, it is the maintenance activity that is least visible to the university community, and its benefits are not always obvious in the short term. For this reason there is sometimes a tendency to minimize investment in the preventive maintenance program or to postpone elements of the program when budget or workforce resources are limited. Responsible facilities managers cannot allow this to happen. The details of preventive maintenance systems are discussed in Chapter 21.

In addition to maintaining buildings and grounds and structural, electrical and mechanical systems, an effective maintenance program provides housekeeping, grounds care, and maintenance activities such as painting and window washing that ensure the aesthetics and general attractiveness of facilities. Because college and university facilities are generally open to large segments of the public, and because much of the ultimate success of the institution depends on the perception that students, parents, faculty, staff, and visitors gain while on campus, the appearance of facilities is essential to the success of the institution's mission.

At most colleges and universities, renovations and construction activities are constantly in progress. The maintenance of these new facilities will be the responsibility of the facilities maintenance organization. To effectively manage these new or renovated facilities and ensure the maximum contribution to the university's mission, the maintenance management staff should be involved in the planning and construction of these facilities. This will help ensure compatibility with existing systems, allow maintenance forces to receive any special maintenance training, and expedite putting the facilities into service.

As institutions strive to become more cost-effective, cost concerns are properly addressed to the facilities management organization. This requires facilities managers to make cost-effective business decisions.

Again, plans and actions must support the larger university mission. Many institutions have traditionally provided basic maintenance services with in-house forces, even when specialty maintenance work is contracted out. In most cases the private sector can provide the same quality and levels of facilities management services as in-house forces, sometimes at a lower cost. An objective comparison between private sector costs and in-house costs for the same maintenance services can provide sufficient basis and motivation for improving in-house efficiencies. Such objective comparisons should periodically be made. When computing in-house costs, care must be taken to ensure that all overhead and administrative costs are taken into account. Similarly, when computing the cost of contractor-provided services, the cost of contract management must be taken into account. APPA: The Association of Higher Education Facilities Officers has several publications outlining the issues that should be considered.

Maintenance managers are usually responsible for institutional compliance with building costs, life safety codes, environmental regulations, Occupational Safety and Health Administration (OSHA) and Americans with Disabilities Act (ADA) regulations, and perhaps other external compliance and regulatory matters. Many of these regulations are constantly evolving and have a significant effect on the way maintenance activities are conducted and on the cost of these activities.

Regulations regarding control of chlorofluorocarbons (CFCs) have significantly increased the cost of maintaining air conditioning and refrigeration systems, and these cost increases will accelerate in the near future. Maintenance managers must be aware of these implications for current and future institutional budgets. In many cases large investments in capital improvements will be necessary to ensure compliance.

OSHA worker safety regulations, such as confined space entry rules, are requiring training and certification for some maintenance employees.

Asbestos abatement requirements, underground fuel tank management, above-ground storage tank management, ADA requirements, and Environmental Protection Agency (EPA) air and water emissions requirements all affect maintenance work. Noncompliance can cause significant problems and costs for the institution. Hazardous material management, solid waste disposal regulations, and recycling requirements are of concern to the maintenance manager, and failure to comply can have adverse consequences for the institution, ranging from bad publicity to significant fines and regulation of activities. Maintenance managers must be aware of the requirements, and regulations that affect maintenance operations. Managers must work closely with institutional safety material management and legal and other departments to ensure that the institution's interests are protected. A more complete discussion of these issues can be found in Chapter 36.

Facilities maintenance and operations reach far beyond the actual maintenance and operation of physical facilities.

30.3 CUSTOMER SERVICE AND COMMUNICATIONS

The reason for maintaining facilities and providing building services is to provide the environment in which the university community functions. The university community, students, faculty, and administrators are the direct customers of the facilities organization. To fully satisfy these customers, facilities managers must meet the requirements of those considered customers of the university community, sometimes known as the *stakeholders*. This usually includes campus visitors, parents of students, civic and community groups, government officials and benefactors. Although Chapter 5 provides a more complete discussion on communications with customers, there are a few key points that should be highlighted within the context of facilities management concepts and processes.

This diverse group of customers determines how adequately facilities management organizations accomplish their mission. For instance, if customers believe that building housekeeping is less than adequate, then housekeeping services are indeed inadequate. If management standards are being met in such a situation, then the standards are inadequate. The bottom line is that customers determine what level of maintenance and service is adequate. Facilities management must understand what customer expectations are and meet them. If customer expectations are unreasonable, uneconomical to achieve, or unattainable, then facilities managers must communicate with customers, educate and inform them, and ultimately convince them to accept the proper standard. However, facilities management cannot unilaterally set these standards, which include not only housekeeping standards, but also environmental standards such as lighting levels and temperatures, as well as maintenance standards for painting, carpet replacement criteria, and grounds and horticulture investment and maintenance standards. It even affects preventive maintenance by helping determine acceptable levels of risk of failure in building systems or components.

Communication with and education of customers regarding maintenance standards and level of service are critical to the success of the facilities management organization. This communication with and education of customers takes many forms from informal conversations between first-line workers and customers to formal institutional maintenance and service policies and directives. To ensure proper external communication with customers, good internal communication within the facilities organization is critical. Facilities managers must manage this communication at all levels.

One model for communication between the facilities management organization and customers can be viewed as a pyramid, as illustrated in Figure 30-1. At the top of the pyramid, formal directives and guidelines are the highest, most formal level of communication, and as communications with customers proceed downward through the pyramid in the facilities organization, they become less formal. However, the amount, intensity, and influence on the effectiveness of the organization increases as this communication proceeds downward through the pyramid. The most important communication occurs at the base of the pyramid. Communication between facilities management staff who directly provide services (e.g., housekeepers, air conditioning technicians, and work reception

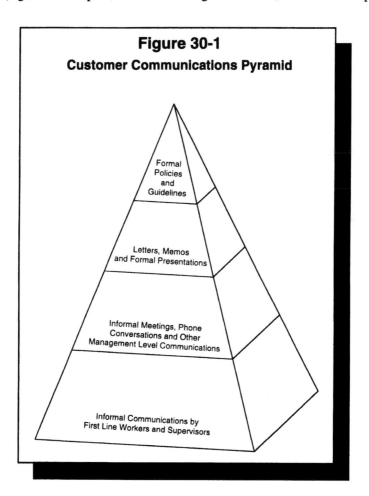

Figure 30-1
Customer Communications Pyramid

Formal
Policies
and
Guidelines

Letters, Memos
and Formal Presentations

Informal Meetings, Phone
Conversations and Other
Management Level Communications

Informal Communications by
First Line Workers and Supervisors

clerks) and customers who personally and directly receive the service (e.g., faculty, staff, and students) have the greatest effect on daily operations and the effectiveness of the facilities operation. It is obvious, then, that facilities management must communicate well with first-line supervisors and service providers to ensure that these individuals represent the facilities department well.

Current organizational structures usually include several layers of management/supervision between top facilities management staff and the service provider. Each level serves as a filter and interpreter of senior management communications. Even formal directives and guidelines are subject to such interpretation. Therefore, special efforts should be made to ensure that accurate and complete communications reach down through the organization. Written communications must reach everyone. For communication to be effective down through the organization, it must be effective coming up through the organization as well. If top management believes that full and correct information is not coming from the lowest levels of the organization, then it is a good assumption that full and correct information is not getting down through the organization. Although most communication must follow the chain of command, top management should periodically communicate directly with all levels of the organization. Visits to work sites and shop walk-throughs are opportunities for upper management to communicate informally with direct service providers. Periodic meetings with first-line supervisors and workers are a good opportunity to ensure that these important people receive unfiltered communications.

30.4 ORGANIZATIONAL CONCEPTS AND NEEDS

The hallmarks of an effective facilities organization are as follows:

- Effective leadership at all levels
- Full and open communications with a sense of trust throughout the organization
- Strong customer service orientation
- Authority and responsibility residing as close to the first-line worker as possible
- Adequate resources being employed by technically competent individuals

The importance of the first-line worker and his or her immediate supervisor cannot be overstated. Facilities maintenance and operational services are provided by first-line employees and are most directly influenced and guided by first-line supervisors. Supervisors and managers at higher levels establish policies and guidelines and provide resources, support, and guidance to these first-line supervisors and their employees.

Most facilities maintenance and operation activities take place in many different campus buildings, frequently without direct supervision. In this situation, employees are expected to be self-motivated, productive and competent in customer service.

Because facilities maintenance and operations work is frequently done with minimal direct supervision, supervisors must communicate well with and empower employees and follow up wherever necessary to ensure that customer needs are truly and effectively met.

Providing training and development of employees is the responsibility of management, and in a service organization where the first-line employees work with a high degree of independence and have frequent contact with customers, a comprehensive training program is essential. Training of the facilities workforce must include more than technical and job-related elements. Customer service communications and an understanding of the institution's mission and goals must also be included in the comprehensive training program.

30.5 ORGANIZATIONAL FORMS

The facilities maintenance and operations organization must be structured to provide administration and management for widely diverse work groups. Because facilities maintenance and operations are much different from teaching and research operations, the administrative and support functions in the facilities organization will usually be much different from those in other institutional departments. The facilities maintenance organization must provide for receipt of customer-requested work, proactive maintenance, facilities preservation, and renovations of facilities, as well as building-related services and utility systems and energy management. Chapter 3 provides a thorough review of facilities management organizational structures. Administration and management of the facilities organization usually involves cost accounting, construction and service contract management, and procurement of materials and services, as well as information/data systems management, facilities engineering, and architectural services.

Trade Shop Organizations

The traditional facilities maintenance organization comprises craft or trade shops. The electrical shop is traditionally responsible for maintenance of all electrical systems, and the carpenter shop, for all cabinet work and carpentry functions. These shops usually receive work input from a work control center that manages the preventive maintenance program and receives customer generated work. The control center establishes work orders and issues them to various shops, and makes appropriate input to the accounting system. The trade shops may be centrally located and managed, or on larger campuses, satellite shops may be established and geographically located to maintain portions of the larger campus.

In these trade jurisdictional models, each shop is responsible for all campus maintenance and repair work associated with its trade. In a variation of this model, a separate preventive maintenance shop is established to perform routine preventive maintenance such as filter changes, lubrication, and equipment checks. In this model, repair work identified during preventive maintenance is usually passed to the appropriate shop for accomplishment.

When shops are decentralized, management and work control may be centralized or decentralized with the shops. This model is frequently seen when an institution has two or more separate campuses and in effect operates two separate facilities operations. All of these models have separate trade shops as their central core. This trade jurisdiction is an outgrowth of the traditional trade jurisdictions of the construction industry.

Multicraft Organizations

A concept gaining more favor is that of multicraft maintenance teams. In one version of this model, maintenance teams consisting of trades from traditional crafts (e.g., carpentry; plumbing; electrical; heating, ventilation, and air conditioning [HVAC]; and occasionally others, such as sheet metal work) make periodic visits or sweeps in campus buildings and address preventive maintenance activities and other maintenance calls that have been identified since the last visit.

A more recent variation of this is the multiskilled craftsperson who addresses all routine maintenance requirements in a facility. An experienced multiskilled craftsperson can usually accomplish more than 90 percent of scheduled preventive maintenances and complete more than 80 percent of routine maintenance calls. The remaining preventive maintenance work and maintenance calls must be referred to specialty shops for accomplishment.

Zone Maintenance Organizations

A more sophisticated model of the multiskilled maintenance worker concept is zone maintenance. In zone maintenance, the campus is divided into conveniently sized zones, usually 300,000 to 1,000,000 sq. ft., with a team of multiskilled maintenance craftsmen assigned to each zone. A team leader usually reports to the director of maintenance or the assistant director of the facilities organization. Team size varies with the complexity, intensity of use, and size of the buildings in the zone. There should be six to fifteen workers to ensure a reasonable span of control for the team leader. The team leader should be the only supervisor on the team.

The team is responsible for all maintenance in their zones. All preventive maintenance and maintenance work are assigned to the team. Most of the assigned work is accomplished by the team, but team leaders may draw on support shops or contractual services to complete work beyond the capabilities of the team.

The major strength of zone maintenance is that the team is responsible for all maintenance work in its zone. Building managers and other customers who work with the facilities department soon learn that their zone team is the source of maintenance work in their spaces. The zones are usually headquartered in a building within the zone. The zone team members are field workers, not shop fabricators. Therefore, zone headquarters is not a traditional shop, but a location where team workers organize and from which they are dispatched. True traditional shop work such as carpentry shop functions and cabinet making, welding and sheet metal fabrication, and bench repairs to motors and other components are all done by the support shops outside the zones. Maintenance and repair work done by the support shops or by outside contractors are controlled by the zone team. Such work is funded from the zone maintenance budgets. The zone team leader, the first-line supervisor, is responsible for and has the authority to manage maintenance in his or her zone buildings. The traditional work control or customer service center still performs the same tasks (e.g., issuing and monitoring work orders) except that instead of having multiple shops perform work in a building and then designating lead shops for multitrade maintenance and repair jobs, the zone team is always responsible for completing required maintenance and repair work in zone buildings.

The zone should be structured so that the team consists of a leader and six to fifteen multiskilled workers at various proficiency levels. A typical ten-person zone team would have two to three highly skilled journey mechanics who could perform initial diagnosis and troubleshooting of building HVAC, electrical, and plumbing systems, and three to four less experienced mechanics who would be capable of addressing traditional carpentry maintenance problems such as window and door malfunctions, routine plumbing, and electrical HVAC problems. The third tier of multiskilled staffing would be capable of completing routine preventive maintenance work, such as filter changes, equipment lubrication, routine lighting problems, and painting, and would assist other mechanics as necessary. For team members, this organization provides a career job ladder from entrance level maintenance worker to highly skilled multidisciplined technician without requiring formal supervisory responsibilities.

The zone maintenance organization eliminates coordination between the shops; clearly fixes responsibility for work accomplishment; and encourages development of a better trained, more versatile workforce. Zone maintenance organizations usually require fewer supervisors and are more efficient than conventional trade shop maintenance organizations.

Although the concept of multicraft maintenance mechanics is not new to many private sector property management companies and small institutions, it represents a major conceptual change for many large, public institutions. The zone maintenance concept is contrary to the traditional trade shop maintenance organizations that developed from the construction industry. However, today's modern building technology with integrated building systems involving door locking and security systems, lighting and HVAC and energy systems, elevators, and fire alarm and control systems requires an integrated building maintenance concept such as zone maintenance.

30.6   CUSTOMER COMMUNICATIONS

Effective customer communication is essential for effective facilities maintenance and operations. Chapter 5 contains a more thorough discussion on quality communication, but some key points are made here. Most institutions have developed some form of building manager or facilities representative system. Facilities representatives are user representatives, frequently departmental business managers, staff, assistant deans or someone associated with the support of the educational, research, or patient care mission of a school or department. Facilities representatives, whether formally designated or not, exist in most major campus buildings. They are individuals who most frequently initiate trouble or service calls to the facilities organization, and they are one of the principal communication links between customers and facilities building managers. To be most effective and helpful to the facilities organization, these individuals should be formally designated and viewed by their departments as being influential in the facilities organization. Facilities management staff and supervisors should pay special attention to these facilities representatives. They should have advance information regarding work in their buildings and scheduled utility shutdowns. They are an invaluable source of information concerning customer expectation and perceived problems with facilities maintenance and services. For facilities representatives to be effective, they should receive copies of maintenance and repair work schedules and notices of completion of work. Semiannual or quarterly formal meetings should be held with them to advise them of upcoming facilities management seasonal efforts (e.g., inclement weather planning) and other issues (e.g., energy conservation initiatives and service enhancements) that may be forthcoming. These meetings are also an opportunity for facilities representatives to provide feedback directly to top management of the facilities organization. By keeping building managers informed in advance, they can become advocates of the facilities organization in customer departments. Regardless of the formality of the building manager system, it is a crucial element in facilities management–customer communications.

Customer service representatives are individuals on the facilities staff who have primary responsibility for ongoing facilities management–building manager and other customer coordination. This function may be performed by senior staff at a smaller institution, or by a specifically assigned individual at a larger institution. The important thing is that someone should have primary responsibility for coordination with customers. The person in this position acts as an "ombudsman" and should be viewed by customers as the one who can get potential problems solved before they become major customer issues. These individuals can be viewed as the customer service representatives of any large, effective service organization. Whether formally assigned as a full-time responsibility or as a collateral duty, this function must exist in any effective facilities organization.

All facilities organizations should have some form of work reception and control organization. This is where customer service requests for maintenance are first received. Because it is often the first contact a customer has with the facilities organization, it is important that this function be handled professionally and with a good customer service attitude. This is frequently the first place customers can routinely follow up on work requests. Good telephone skills and a good work order status and follow-up system will minimize customer frustration and improve communication with customers at all levels. Good communication between maintenance shops or zones is essential to the effectiveness of an organization.

30.7 MATERIEL MANAGEMENT

Having the correct material available to maintenance personnel when needed is essential for effective operations. Regardless of the system used, materials should be made available to field personnel with minimum effort. Preventive maintenance consumables, such as filters and drive belts, should always be readily available before work is started. When unpredicted requirements develop during the course of a job, a system that delivers the materials to the site is better than one that requires the mechanic to leave the job site to obtain materials.

Maintenance workers, when in the field, should rarely have a need to leave the work site to get materials. This requires prior planning to have materials on site, or the maintenance worker should take the materials to the job site. For most preventive maintenance work, materials such as lubricants, filters, drive belts, and fluorescent tubes can be pre-positioned in mechanical rooms or some secure space in buildings where they will likely be used, and maintenance vehicles can be outfitted with commonly required materials. A good measure of the effectiveness of the materials support system is how often a field worker has to return to the shop or warehouse for needed materials before a task is completed.

Field mechanics should rarely have to go to a vendor or distributor to get needed materials. If a review of material availability indicates that frequent trips are required from work sites to warehouses or storerooms, a system for delivering materials to the work sites should be explored.

Storeroom inventories are expensive to maintain. Modern materials management concepts embrace the "just-in-time" concept of material availability. This concept calls for reducing on-hand inventories and having arrangements or contracts with materials vendors to make materials available on short notice or on a scheduled basis when needed. In effect, vendors' inventories can take the place of much of the storeroom inventory. Chapter 16 has a more detailed discussion of storeroom operations and inventory management.

Because maintenance materials requirements are hard to accurately predict and many times small quantities of an item are needed, it is common practice for shops to have pre-expended materials on hand. Such inventories are difficult to control and manage and should be kept to a minimum.

There is a requirement in every maintenance organization for emergency spare parts and equipment. These items must be stored in secure areas but readily available to after-hours maintenance and repair crews. They should be carefully inventoried, and usage data should be maintained and periodically reviewed to ensure that only items used for emergencies are carried in this inventory.

All materials used in maintenance should be charged out and costed against the building and specific building systems and equipment, when appropriate. These costs, along with labor costs, should be accumulated in a hierarchy file to maintain accurate maintenance and repair costs for specific buildings and building components.

30.8   INDUSTRIAL ENGINEERING AND WORK CONTROL

Since the 1960s, most facilities management organizations have been applying engineering principles to the management of the facilities workforce. It is more difficult to quantify and predict maintenance and service activities than to schedule production activities. However, by quantifying a work unit, many of the principles of industrial engineering can be used to quantify maintenance and service work.

Preventive maintenance work is the easiest facilities work to quantify, measure, and analyze. Modern preventive maintenance programs define work units and quantify them. From this comes detailed work scheduling, quantifying the effort involved, comparing the work output to standards, and analyzing efficiency, which allows management to optimize productive efforts and improve the efficiency of the workforce involved.

The four techniques most often used in applying industrial engineering principles to facilities operations are work sampling, work measurement, value engineering, and work simplification. These technologies are useful tools for facilities managers, but their applications must be tempered with the importance of customer service and recognition of the empowerment of first-line workers and their supervisors.

Work sampling is a statistical approach to quantify work elements. With adequate random sampling of work efforts, it is possible to predict and identify areas where process changes, training, and management/supervisory intervention are required. With work sampling, facilities organization activities can be divided into productive work, indirect productive work, and nonproductive effort. These work efforts can be further refined into their respective elements.

Productive work is the direct effort used to produce the final desired results. Productive work includes items such as measuring, cutting, and fastening items in place. Indirect productive work includes such activities as planning, materials procurement and handling, and travel time. Nonproductive effort is time that adds no value and does not directly contribute to the final product. This includes such items as idle time waiting for transportation and materials, or anything that delays the production effort. It also includes clean-up time, breaks, training, and other time that does not directly support a productive job.

The principles of work sampling and categorizing should be familiar to all supervisors. Formal work sampling should be part of any major organizational or process changes that are contemplated in the facilities organization.

Work measurement involves the application of time standards to any element of work. Time standards may be published, universally accepted standards, or they may be established by first-line supervisors and informally applied to work elements. They should be quantified and agreed on by first-line workers, supervisors, and managers if they are to be meaningful in measuring and approving work performance. Local standards can be developed based on historical performance, current work practices, and judgment.

Value engineering is usually thought of as a technique to ensure cost-effective designs of buildings or other facilities and usually involves life cycle costing of a facility. The principles can be applied to component replacement in buildings, building energy use decisions, and maintenance decisions. For instance, value engineering principles may lead to a decision to perform preventive maintenance on window air conditioning units once a year rather than twice a year. Such an analysis may indicate the effectiveness of converting some lighting systems from fluorescent to high intensity discharges (HID) fixtures, or installing motion or ambient light sensors to turn lights on or off. Empowered first-line workers and supervisors can initiate valuable value engineering ideas.

Work simplification techniques review each element of a job task with the intent of doing it in a more effective manner. Again, empowered employees and first-line supervisors should be encouraged and rewarded for work simplification efforts.

30.9   USE OF CONTRACTUAL SERVICES

The mission of the facilities department, discussed earlier, involves providing the environment in which the institution functions in the most effective and efficient manner possible. In many cases this requires facilities management to provide maintenance and services completely with institutional employees. In other cases it is more effective to have outside vendors provide services. Chapter 17 provides a more detailed discussion on contracting for services; however, some key points are made here.

One of the most difficult tasks facing facilities management is deciding on the correct mix of in-house and contractual services. Facilities management is always responsible to the institutional customers for these services, whether provided in house or by a contractor. Contract services should be considered when:

- Technical or legal requirements cannot readily be met by existing in-house staff.
- Workload is seasonal or unpredictably variable.
- Costs of providing the same level of service are less than with in-house staff.

Maintenance and operations of modern buildings have become technically demanding, with integrated building systems, sophisticated building controls, and security and energy management systems. In many cases the technical capability to effectively manage and operate these systems does not exist in the facilities department. When such buildings and systems are put into service, facilities management must develop the needed capability in house or contract with capable vendors for the needed maintenance and operations.

Legal requirements for asbestos abatement, lead paint removal, and hazardous waste handling and disposal may be reasons for contractual services.

Seasonal functions such as grounds maintenance may be more effectively contracted, as contractors usually have more flexibility to quickly and easily vary staffing. One of the hidden costs in providing seasonal or variable services with in-house staff is finding truly productive employment for these people during the off-season or at slack times.

Services that are relatively low skilled and labor intensive can often be more economically provided by contractors. Painting and housekeeping are examples of cases where local labor markets and conditions may allow contractors to pay wages that are considerably less than those paid to institutional employees for the same work. Frequently the cost of employee benefits is less to these contractors. In evaluating the cost of in-house versus contractual services, the cost of contract administration and quality control inspections of contractor efforts must be considered.

As building systems become more sophisticated and environmental and life safety requirements become more demanding, specialized contractors should be considered to maintain the affected systems, rather than developing full in-house capabilities for these ever-changing technologies. Elevators with sophisticated control systems, automated building systems, fire alarm and suppression systems and other proprietary equipment, and life safety systems frequently are best maintained by vendors with highly trained technicians. Facilities management staff must develop adequate technical expertise and knowledge of these systems to properly manage the contracts.

Competitive-bid, single procurement is the traditional method of procuring construction and maintenance services when full scopes can be clearly defined. Open-end contracts with unit costs and undefined quantities established for a time period, usually one or more years, are commonly used to provide repetitive services when the quantity is unknown. Painting and carpet installation are examples of such procurements. Hourly service contracts for trade workers such as carpentry, electricians, and plumbers are another form of unit cost, open-end contracts. Such contracts can be competitively bid on a unit cost basis or negotiated with selected vendors.

Regardless of the contract form or services provided, an effective quality control and inspection program must be part of the contract administration process in the facilities organization. Timely inspections and feedback to the contractor are perhaps the most important elements of assuring effective contract success. The process must be controlled by facilities management, as it directly affects the quality of services provided to institutional customers. Responsible emergency services can be provided by contractors if properly documented and specified in the contract. Elevator maintenance contracts, for example, must make provisions for emergency services, fire alarms, and emergency generators, and other equipment maintenance contacts must have provisions for emergency responses. It may be appropriate for trained, in-house personnel to respond to specific emergencies, such as passengers trapped in elevators. Such provisions must be clearly documented in the maintenance contract to ensure proper responsibility and accountability of the contractor.

Service contracts such as housekeeping should contain emergency response requirements to such things as water floods in buildings. If the contractor is not required to provide such emergency response, then provisions must be made to provide these responses by in-house personnel under specific conditions detailed in the contract.

In summary, maintenance and service contractors are an effective extension of facilities management capabilities and must be fully managed and controlled by the facilities organization.

30.10   CONSTRUCTION AND RENOVATIONS MANAGEMENT

Many facilities departments provide construction and renovation services to campus customers. This usually involves relatively small projects, often without full scope definition or complete plans and specifications, frequently in occupied spaces and often with relatively short completion time frames. These conditions often preclude fixed-price competitive bidding by construction contractors.

When such work is done in house, the workforce should be completely separate from the maintenance workforce to avoid the migration of maintenance resources to renovation projects. Renovation and construction workload is usually variable over time, and for such work to be efficiently completed, different trade capabilities are required at different times. As most institutions have relatively fixed, stable workforces, it may be difficult to perform this work with the traditional institutional staffing in an effective manner. For these reasons, having contractors perform renovations, with minor construction services managed by the facilities organization, frequently is the most effective way to provide these services.

30.11   SUMMARY

The facilities organization must support the institution's mission and goals. Facilities are not maintained for their own sake or to preserve the facility, but to provide the best possible physical environment for the university community. The preventive maintenance program is the hallmark of an effective facilities maintenance effort. It minimizes system failures, system interruptions, and maintenance service calls over the life of a facility. Unlike services such as housekeeping and grounds care, which are extremely visible, preventive maintenance programs are relatively invisible and therefore require careful management to maintain their effectiveness.

The effective facilities maintenance and service organization must have a strong service orientation, empowering first-line workers and giving them the resources and training to be effective. Most maintenance and facilities service operations are accomplished with little continuous direct supervision in spaces not controlled by facilities management. Therefore, first-line workers and supervisors must institutionalize a customer service philosophy. New technologies and greater sensitivity to customers' needs are leading to a maintenance organization philosophy that is becoming less trade oriented (e.g., carpentry, plumbing, electrical, and HVAC) and more facilities oriented. This facilities orientation is leading to multicraft skilled maintenance people who view their work as maintaining total buildings rather than components or single building systems. These multiskilled maintenance technicians, organized into teams having total maintenance responsibility for several buildings, are the heart of zone maintenance. The zone maintenance concept has been discussed in detail in Section 30.5 of this chapter.

Maintaining the environment in which the university community functions involves a broad range of facility-related services. To effectively provide these services, good communication with customers is essential and must be managed by the facilities leadership. It is the responsibility of facilities managers to educate the campus community to be informed consumers of facilities services.

Communication with customers of the facilities maintenance and service organization is critical to successful facilities operations. Organizational concepts such as facilities representatives or building managers, and customer service representatives have been established at many institutions to enhance this communication. The functions of these entities are to expedite accurate and full communication between the facilities organization and the campus customers. This communication is essential to the successful operation of the facilities maintenance and operations organization.

A system of material management that provides the proper material when needed with a minimum of effort by field maintenance people greatly enhances the effectiveness of the maintenance organization. Although inventory control and accountability are essential ingredients of any material supply system, the final test of the system is how well it serves field workers.

The principles of industrial engineering and work control remain essential to the effective maintenance service organization. These techniques, processes, and systems must enhance, not impede, customer communication with line workers, ensuring proper responsibility for maintenance and service work.

Contractor-provided services are a valuable tool of the facilities manager in providing the most effective services to the institution. Maintenance and services contracting are an effective extension of facilities management internal capabilities and must be fully managed and controlled by facilities managers.

CHAPTER 31

Grounds Maintenance and Operations

Roy Peterman
Brigham Young University

With additional contributions by Don Barr, Norman H. Bedell, Denise M. Candelari, Michael Dale, John P. Harrod Jr., James D. Long, John McCoy, and John Nagy.

31.1 INTRODUCTION

The general atmosphere of a campus influences impressions and attitudes. First impressions of the grounds can affect enrollment of students, employment of faculty and staff, and the attitudes of visitors and benefactors. Modern administrators are realizing the effect of well designed and maintained grounds and interior plantings on the learning environment. Grounds maintenance operations contend with unpredictable variables, such as living plants, pests, and weather. Despite these and other challenges, the purpose of a grounds maintenance organization is to provide a continually safe and invitingly attractive learning environment.

Grounds maintenance organizations operate with varying degrees of success. Real success is achieved through leadership by staff well trained in horticultural cultural practices, task management, equipment maintenance and management, human resources management, and scheduling.

During the past two decades, a wealth of information has been collected in this area that provides a new sophistication that allows the grounds manager to accomplish his or her task. There are many disciplines that a grounds manager can use to advantage. Technological advances as well as governmental regulations often require specialized training and even specialized individuals to accomplish many of the tasks confronted.

31.2   ORGANIZATION

In-house Organizations

Successful grounds maintenance organizations are characterized by strong leadership and well-trained and enthusiastic supervisors, complete with adequate staffing and a functional organization. Grounds managers may face different responsibilities based on institutional needs and the manager's abilities. Grounds sections are often charged with additional responsibilities not directly related to traditional services, but which use labor and machinery more economically. Those responsibilities may include such functions as waste removal, recycling, moving, equipment movements, and special events support.

Grounds maintenance groups generally include a variety of disciplines. A site development group may include a landscape architect, a horticulturist, a surveyor, civil engineers, or other grounds construction specialists. In some cases this group may be aligned under an office of design or a unit called "University Architecture" or "Engineering." Gardening maintenance groups are responsible for general upkeep for the grounds and maintenance of plant materials. Teams are often used in this application to create a uniform and consistently maintained campus with responsibilities for the general tasks that, when done all together, create a finished appearance. These tasks may include mowing, edging, and trimming; sidewalk, curb, and gutter cleanliness; waste container maintenance; general watering; and minor sprinkler adjustments and repairs.

A landscape speciality group is often formed to support the gardening maintenance group. This group may include people with special skills to handle the diverse problems the campus faces and may include chemical application specialists, arborists, turf specialists, plant material specialists, green house operations personnel, and athletic events specialists.

Grounds maintenance groups are effectively organized into zones or areas. Each zone or area gives a quantifiable realm of responsibility, allowing the employee to feel personally responsible for his or her area or zone and gives the manager an ability to monitor performance. Some special tasks that do not fall within this concept, such as chemical application, heavy pruning, or planting, may be handled by roving specialty crews, working in support of the area or zone crews. It is essential that the specialty crews have the same goals of consistently outstanding appearance as the grounds maintenance crews.

A possible grounds maintenance organizational chart has been included (Figure 31-1). Although it is often difficult to establish clear guidelines for all functions, it is important that grounds employees clearly understand their functions and management expectations concerning their performance. Many grounds functions will require close and effective coordination with other functional areas within the facilities organization to ensure that effective results are obtained.

A good example of this is snow removal. The grounds crew typically has a clear responsibility for snow removal on sidewalks, exterior stairs, and parking lots, whereas the custodial staff may be responsible for entryways, building steps, wheelchair ramps, or walks leading to the building. Clear, mutual understanding of these responsibilities, and close coordination to ensure that the snow removal task is fully completed, are essential to success.

Contract Organizations

Contract services have become a popular way to accomplish many grounds maintenance tasks or even the entire function. Services that are most commonly contracted include pest control, waste removal, tree care, and mowing. Careful analysis of the advantages and disadvantages, however, should be a prerequisite to contracting for any services. Chapter 17 of this manual and APPA: The Association of Higher Education Facilities Officers' *Contract Management or Self-Operation*[1] discuss important issues that should be considered and provide good general guidance. Once a decision to contract for services has been made, careful development of the specifications is essential to a successful contract. Effective management of the contract services is key to sucess. Close supervision, inspections, and feedback to the contractor are all essential.

Ensuring that the special needs of an educational setting and the flexibility of campus events can be accommodated can be difficult with contracted services. Many colleges and universities are finding that the use of part-time labor is an alternative to contract services that can provide needed cost reductions while still preserving the greater degree of control that may be lost with contract services.

Special Concerns

The grounds managers face the dilemma of environmental concerns as it pertains to their area. Chemical application specialists need licensing and special training and also need to be kept abreast of new developments. The requirements of the Americans with Disabilities Act should be considered in landscape planning.

Environmental concerns have given rise to vegetation programs in most urban areas of the United States. However, many people feel that vegetated areas facilitate criminal activity, and this feeling has promoted the removal of vegetation from public areas. In recognition of this conflict, researchers have developed a plan for the removal of vegetation to maximize its benefits without encouraging criminal activity and planting suggestions.

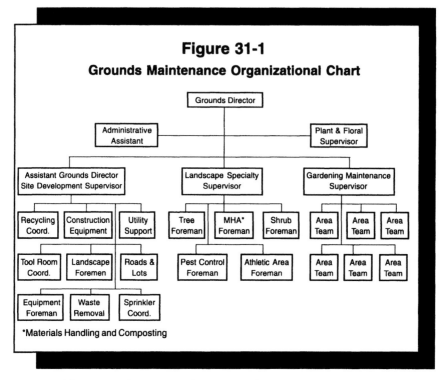

Figure 31-1

Grounds Maintenance Organizational Chart

*Materials Handling and Composting

Specific suggestions regarding vegetation management in specific locations and creation of safe areas that campus patrons can access confidently are as follows:

- Target problem areas; police records should be able to help identify hot spots.
- Maintain sight corridors in targeted areas for easier surveyance in areas of heavy circulation; keep concealing vegetation to a minimum around drinking fountains, parking lots, paths, and other areas where people congregate.
- Convert patches of vegetation into several thin strips. This pattern reduces cover for fleeing criminals yet still offers an attractive visual massing of vegetation.

Vegetation is not the sole cause of crime; it is often no problem at all. Therefore, wholesale removal is unnecessary to create a safe atmosphere.

31.3 STANDARDS AND SCHEDULING

Grounds Inventory and Maintenance Levels

Planning for the performance of grounds maintenance functions should be based on both an inventory of the scope and nature of the grounds to be cared for and a determination of the standards of care. These standards are typically set by the desired level of maintenance established by the administration. Standards should be set for the various zones or areas of the campus and personnel and resources appropriately assigned. Standards may be established on a campus-wide basis or individually according to buildings and functions. The former gives a more uniform appearance to a campus and is much less complicated to carry out.

Scheduling

When standards have been set, priorities should be established according to functional processes, institutional needs, and environmental dictates. After priorities are set, scheduling can begin. Scheduling is a dynamic process, requiring constant updates as the institution changes. Scheduling usually includes time–motion studies to establish time standards. Time reporting systems also often are used to establish time standards. As each individual task is scheduled according to the standard, emergencies can be minimized. Preventive maintenance scheduling that may be less routine is effectively done by a computerized preventive maintenance program. In the area of equipment maintenance, preventive maintenance printouts are essential. Computerized inventories of plant materials and as-built locations of such plant material allows special conditions of maintenance to be integrated into the preventive maintenance program. Repetitive scheduling is most efficiently handled by developing schedules on calendars and sharing them with all affected parties within the facilities organization.

31.4 TRAINING

A well-planned, continuing program of training will benefit any area of the facilities organization. This is particularly true for grounds maintenance activities. In addition to general training applicable to all facilities staff, training should be provided for grounds staff in such specialized areas as the care and operation of grounds equipment; the most effective techniques for the planting, maintenance, and pruning of plant materials; the requirements of governmental regulations; and safety. Chemical application specialists, for example, need special training and licensing and must be kept abreast of new developments in the field. Proper training in the calibration of all types of equipment used for the application of all types of materials is important.

    A range of resources is available to support effective training programs. State extension services can be one of the best sources of training assistance and materials that are both of high quality and specific to local area needs and conditions. State or local nurserymen's associations can be another good source. Community colleges and local technical schools typically offer programs in various elements of grounds care. Videos and other training materials are often available from the suppliers of specialized grounds care equipment.

Training programs will be most effective when they are scheduled regularly and incorporate realistic situations that may be encountered in the work environment. Good documentation of training is essential. Further discussion of staff training and development is included in Chapter 10.

31.5 EQUIPMENT

The increasing costs of labor make proper equipment selection and use more important all the time. The labor required for maintenance procedures can be dramatically reduced when effective and efficient machinery is selected. Proper investigation through the purchasing process includes demonstrations, and collaboration with similar institutions with similar problems can be a valuable part of this process. A proper preventive maintenance program is critical. Down time can waste considerable amounts of money. Not only will down time halt the operator, but it can halt several other crews and disrupt scheduling as well. Factory training should be a requirement of the purchasing agreement. Worn and undependable equipment can disrupt operations and increase costs. A timely replacement program can pay dividends.

Replacement scheduling can allow for budgeting in a systematic fashion without creating peaks and valleys in the needs for replacement machinery. Grounds operations typically require expensive machinery, but the machinery can be cost-effective in diverting labor and conserving resources (such as in compost operations) and diverting landfill and transfer fees.

Leasing equipment can be an attractive alternative to ownership by avoiding the large capital investment costs. Short-term leases or rentals will often be the most cost-effective way of obtaining equipment resources that are needed only on a periodic or temporary basis.

31.6 GRASS AND TURF

Facilities managers should remember three principles of turf management:

1. No lawn is stable. It is either improving or declining in quality. Even when a lawn appears to be in good condition, hidden problems may be beginning.
2. With proper moisture and fertilization a lawn will be established, weeds will be crowded out, and grasses will grow abundantly. However, proper moisture and fertilization does not preclude invasion by fungi and insects.
3. Grass will not continue to grow in dense shade. Beware of miracle advertisements. There are varieties of grasses that tolerate varying degrees of

shade, but none can tolerate the shade caused by dense trees. Some of the problems of dense shade include dry soil caused by the trees absorbing much of the water.

Be wary also of growth regulators—they are not for every lawn. There are a number of restricted-use growth regulators available on the market, and each is effective for different turf situations. Cost factors must be considered to determine the economic feasibility of growth regulator usage. Also, growth regulators turf may cause physiological problems in certain areas.

Nonirrigated lawns usually experience a dormant period from mid to late summer if rainfall falls below levels necessary to sustain growth. This is normal, and the brown-blue grass will turn green with sufficient water.

Seed Selection

Consult a turf specialist or agronomist for seed selection in any given location. As a result of research on turf grass varieties, new selections regularly come on the market. Selection of a combination of grasses suitable for a specific use must be made individually.

Many factors determine seed selection and include use of the lawn; mowing heights; available water and fertility rates; insect and fungus resistance; shade percentage; and intensity, soil factors, and drainage. No common seed mixture is perfect, and no two turf consultants will prescribe the same solution.

Preparation and Planting

Proper preparation for a new lawn ensures the desired end result. In new construction, the site is usually in poor condition for the establishment of turf. All kinds of problems have to be corrected. For example, sheets of plywood may be hidden just below the surface, large concrete chunks may be imbedded in the soil, and buried plaster or lime may cause unknown alkalinity problems. Grounds managers must inspect the construction site before planting a new lawn and correct any abnormalities or unforseen problems will occur for years afterward.

During the grading process, the most important consideration is drainage. No turf area should slope less than a bare minimum of 1 percent. Pockets of undrained turf will result in a multitude of problems.

Areas compacted by heavy equipment should be loosened, or turf establishment will be difficult until the soil loosens naturally, which can take several years. The entire site should be tilled and evenly shaped. It is not necessary to pulverize the soil completely and it is even better if small lumps creating voids of up to 1 in. remain.

A fertilizer chosen for the specific soil condition should be applied and worked into the surface. Soil analysis is helpful, but an experienced turf manager will know the requirements of a given site based on other lawns in the area. Because phosphorus is not mobile, before planting may be the only chance to supplement with this mineral.

Several methods of seeding can be used depending on the site. In small areas, seed might be distributed by hand-carried rotary seeders. For larger spaces, a slicing seeder or drill might be used, and large areas of steep slopes may require hydra seeding. The latter method can incorporate fertilizer, seed, and mulch in one operation.

Mulch is of utmost importance in the seeding process. Mulch provides protection from erosion, retains moisture, and shades the seed bed. Without mulch, the chances of establishing a lawn are greatly reduced.

Irrigation

Irrigation systems, once deemed frivolous and undependable, have become a mainstay for turf and plant maintenance in the current grounds care industry. Such systems not only provide a "cheap insurance policy" for new plant and turf installation, they also serve as a potential source of labor and dollar savings over time. Hand watering with hoses and sprinklers is an inefficient use of water as well as labor. Conversely, an irrigation watering source that controls water usage through specific precipitation rates and precise watering times is more effective.

Water use is typically minimized through the use of an "automatic controller" and a rain sensor. These simple devices allow the system to turn itself on and off at certain times of day and control the length of time each area is watered. In periods of sufficient precipitation, the rain sensor will shut the system down until the moisture level is low enough to warrant additional watering. In expansive commercial jobs, irrigation is often managed through a computer. Such management allows for the control of a system from a central location. Manufacturer-specific programs compute precise watering times and schedules based on regional climate, geography, soil compaction, slope conditions, and specific landscape features to optimize the use of water. In addition, with a change in software, the system controller can also automatically operate fountains, gates, lighting, security systems, and even word processing.

A well or sewer charge meter monitors water used by the irrigation system. This water is not subject to sewage charges. As water management in some communities has become an important issue (sometimes requiring water restrictions), a well will provide a usable source during drought conditions and is acceptable in most communities during restriction periods. It too is exempt from sewage fees.

System selection will depend on the following factors:

- Mechanism desired
- Mode/method of operation

- Specific application
- Size of project
- Water source/water pressure
- Climate

Properly designed and installed irrigation systems require little maintenance. Seasonal watering times may be needed to react to severe deviations from regional weather norms, and winterization and restart will be required in cold climate areas. Proper operation and an appropriate schedule of watering times will control problems from fungus and thatch when combined with a good spraying and aeration program.

Irrigation systems will benefit any landscape setting and will represent an asset to the facilities organization. Once a system is installed, it only requires occasional maintenance and, in colder climates, winterization. Watering amounts are critical. How much water depends on soil type and turf variety and the seasonal changes of the evapotranspiration rate (defined as moisture taken from the soil by plants and air). Potential problems with irrigation include increased fungus growth, thatch accumulation, and soil compaction. A good spraying and aeration program will keep these problems in check. Fungus diseases can also be reduced by timing waterings to minimize the time, during a 24-hour period, that the turf is wet.

Maintenance Problems

A profusion of disease and insect problems exist in lawns. As the density of a lawn increases, so do the problems. The selection of grass for disease resistance is important, and a program of seasonal treatment should be established to cope with standard problems.

One of the common problems of good lawns is the accumulation of thatch. Thatch causes a roof effect and reduces the chances of air and water reaching the roots of the grass plants. It also creates an environment conducive to insect and fungus disease growth.

If this layer of thatch gets too thick, it must be removed or a poor turf will result. Chemical dethatchers have been introduced, but none have proven successful. Mechanical cultivation is the only answer. Core aeration will modify thatch. Annual aeration of heavy-use areas will reduce compaction and allow moisture to permeate the surface.

Mowing schedules and heights must be regulated. In campus situations, particularly where budgets and human resources are limited, mowing often falls behind schedule. Heavy windrows of clippings can create fungus problems. Delayed mowing will also injure the grass plant by allowing the leaf to be cut severely into the "white" area. The optimum is to cut more often and remove less grass per cutting. As summer passes, the cutting height should be raised to allow more leaf surface to remain. Short cutting heights are harmful to turf health and sometimes, depending on the grass variety, cause the plants to die. Some varieties withstand short cutting better than others.

If properly graded and drained, most lawns can become acceptable if management establishes a systematic program of chemical treatment, fertilization, and irrigation. Each area of the country requires specific programs. If they are followed, the lawn quality will improve.

Athletic Fields

The quality turf that is required on some athletic fields will require more maintenance. Irrigation, good drainage, fertilization, aeration, over seeding, top dressing, and chemical application are all necessary to maintain or increase the turf's durability. Many volumes of literature and university courses detail athletic field installation and maintenance. Of course, no maintenance program exists that ensures safe playable turf on overused fields. If such a demand exists for athletic turf, additional areas should be created so that playing fields are not used to the point where the turf is always in a state of decline.

31.7 PLANT MATERIALS

The maintenance problems of plant materials—trees, shrubs, ground covers, and vines—are directly related to the initial selection of the specific plant. If improperly chosen, a plant will be a problem in various ways throughout its life. Many factors must be considered to ensure that the proper plant is chosen for a specific location.

The one problem commonly overlooked is space. Many times the plant selected will grow considerably larger than the space allows, which results in continual pruning and, finally, replacement. A common error of designers is to use several plants—causing overcrowding—where one will do the job.

Poor selection may require continuous treatment to condition the soil acidity, or a plant may be chosen that will not withstand wet conditions or areas of poor drainage. Selecting plants that are not hardy enough for a geographical zone or exposure to sun, snow, or strong winds can be disastrous to certain species.

Soil conditions determine many future maintenance problems. Selection of plants to withstand the various conditions is critical. In urban settings, some plants are harmed by air pollutants; this fairly new problem must be considered.

The person selecting plants for use on campus should be aware of these common problems. Occasionally, a landscape architect or other designer who practices in a different environment and does not have firsthand knowledge of special conditions affecting plants on a specific project is selected to create a planting for a given site. The designer will probably use plant lists for the zone intended, but local conditions may be entirely different. Therefore, consultants must be completely familiar with the area and its problems.

Planting

There are several steps in the planting procedure that frequently cause problems. Preparation of the hole is important. Trees planted in heavy soils with poor drainage die from too much water rather than from a lack of it. When the hole is dug, it should be no deeper than the depth of the ball to plant at the original level. This keeps the plant from settling out of plumb or to a level that will not drain. The hole should be wider than the ball and backfilled with top soil.

The person handling the plant should not lift the plant by the top, which would break or loosen the ball. If the tree will not stand on its own or is loose in the ball, it should be staked. Staking reduces losses caused by wind, people, and other physical problems. If staked and mulched properly, the base of the tree will also be protected from mower damage. The twine or cord used in securing the ball must be completely loosened and cut away from the trunk. If not, the tree will be girdled and die. This problem is particularly evident when nondegradable twine is used.

Methods of planting should follow the standards set by the American Association of Nurserymen in its manual, *American Standards for Nursery Stock*.[2] Those standards also provide guidance for quality and size when purchasing plant material.

Plant Materials Maintenance

All plants require maintenance in varying degrees. There are no plants entirely free of problems. Problems must be dealt with individually because there are thousands of cultivated plants, each with its own requirements. Maintenance common to plants includes watering, fertilization, chemical treatment for proliferation of pests and diseases, pruning, and eventual removal.

Determining proper moisture levels for the numerous varieties of plants is difficult. Most plants should be thoroughly soaked and allowed to drain to the point of being dry before being watered again. Most plants will drown if over watered, but drought conditions can also be damaging. Proper balance can only be reached by learning the peculiarities of individual plants.

Many plants used in cultivated conditions require fertilization. Intensity of fertilization varies according to individual plant requirements. Turf grass plants that are irrigated on well-drained soils require fertilization several times each year, but a good shade tree in natural conditions may require no fertilization. Related to fertilization is the introduction of various trace elements lacking or trapped in the soil. This again relates to improper plant selection. For example, chlorosis (abnormally yellow color) in several varieties of trees is common and must be treated. Chlorosis results from a lack of iron in a plant. Iron is either not present in the soil or is not available to the plant because of some other soil condition. One of the most common causes is an alkalinity and iron-deficiency combination. The same plants located properly will not develop chlorosis.

The various chemical treatment procedures are so numerous that only descriptions of common operations are presented.

Target Pest   The insect or other organism must be identified before proper treatment can be selected. Identification includes timing, resistance, and life cycles.

Toxicity   Pesticide application will become a subject of public concern if not properly monitored. Levels of toxicity are indicated by an LD(50) value. This is the amount of pesticide that is lethal to 50 percent of a test population in a single dose. It is registered in milligrams per kilogram of body weight and is shown both as dermal and oral indicators. This rating does not indicate hazard, rather the killing ability of a chemical. A chemical can be highly toxic but have little hazard potential because of the way it is used (or misused) and the way it is formulated. The lower the number, the higher the risk factor because less volume of material is required to be lethal. LD(50) is not the only way of determining toxicity, but it is a good indicator.

Phytotoxicity   This describes the effects a selected chemical will have on various, nontargeted plants. Incorrect selection and application methods will result in damage to other plants. An example is 2,4-D injury to shrubs and trees when applied improperly to lawns for broadleaf weeds.

Compatibility   Certain chemicals can be combined to resolve combinations of problems. There are many chemicals that should not be mixed or will not mix for many reasons. Awareness of chemical compatibility will avoid many problems.

Other factors to consider when selecting a specific chemical are its effectiveness, residual behavior, method of application, and employees' capabilities to handle application problems.

Shop rules for safety and procedural methods must be established. Reference publications are available that suggest methods of transportation and storage of chemicals, types of protective devices and clothing, and calibration. Information is also available about other use factors of equipment, personal hygiene, and—with the current emphasis on environmental protection—methods of disposal. College and university agencies, such as agriculture schools and health services, state extension services, and manufacturers, are all sources of safety procedures.

The problems inherent in exposure to agricultural chemicals, and the subsequent hazards of absorption into the body, cannot be overemphasized. Health testing procedures are available for use in protecting employees and, in turn, the institution. Chemical absorption can result in absenteeism and serious health problems. A good system of detection can benefit employees and possibly avoid future legal entanglements if hazards are discovered and treated properly.

Chemical Application

A chemical applicator's license or registration is required by most states for at least one person responsible for chemical application. Current federal regulations require employers who transport, store, and apply hazardous chemicals to have a Hazard Communication Standard Program involving obtaining Material Safety Data Sheets (MSDS) from manufacturers for each hazardous chemical on hand. The MSDS must be made available to the employee, and he or she must be trained on how to safely handle each hazardous chemical. For more information regarding the Hazard Communication Standard Prorgam, contact your local or regional Occupational Safety and Health Administration Office.

Pruning

One of the most misunderstood maintenance procedures is pruning, which should not be confused with shearing. The quickest way to ruin plants and the original intent of a plant selection is to allow them to be sheared into round balls or flat-topped cones, unless a formal garden is intended. The use of the hedge clippers can injure plants and ruin the plants' intended purpose.

Pruning should be done to remove dead or damaged branches, retain original plant shape, control the size of the plant, or for renewal purposes. Trained personnel should perform this function at specific times of the year as needed by the type of plant. On larger campuses, or where labor availability is a problem, pruning can be done almost any time. If pruning is done at the wrong time of year, some sacrifices must be made, such as loss of flower or fruit the following season.

Disease control is an important reason for pruning. Some infected plants or groups of plants can be spared the spread of disease if infected branches are removed and destroyed. In some cases, the pruning equipment must be sterilized after each cut to keep from carrying the infection from plant to plant. Dipping tools in alcohol or spraying with a professional strength aerosol disinfectant is effective.

Ivy

The maintenance problems associated with ivy-covered walls involve potential damage to masonry and wood moldings. The two most common ivy plants covering the walls of many buildings are either the deciduous plant, *Parthenocissus tricuspidata,* commonly called Boston ivy, or the vine that does not lose its leaves, *Hedera helix,* commonly called English ivy. Each has its advantages and disadvantages.

Boston ivy grows rapidly, sometimes as much as 10 ft. per year. When it loses its leaves each fall, like most deciduous plants it displays fall colors of bright orange, red, and crimson. Boston ivy climbs readily on nearly any surface by using adhesive discs at the tips of its tendrils. These tendrils do not seek cracks or crevices but adhere to the surface. The vine should not be allowed to grow unrestrained behind and between moldings, joints, or on wood surfaces.

English ivy, a dark green-leafed vine, is a much slower growing plant and, in some conditions, not nearly as hardy as Boston ivy. In certain locations, it will freeze back to the ground and must be removed. In the proper location it is an excellent plant as a ground or wall cover. Some smoother surfaces do not have enough texture to support English ivy.

Both vines provide shading of walls and insulation against bright sunshine. Ivy's value as insulation has not been determined but is significant. The added aesthetic value probably outweighs the negative attitudes, costs to control growth, and potential damage.

31.8   INDOOR AND TROPICAL PLANTS

Interest in indoor plants has risen dramatically in recent years. Although the idea of growing plants inside is not new, it has never been so popular. Now tropical foliaged plants in building interiors have become as necessary as they are fashionable. Plants not only beautify surroundings, but they create an atmosphere of warmth and life. Most important, office plants provide a comfortable environment that improves worker morale and efficiency.

There are various architectural uses for plants, and architects and designers include interior foliage plants in their plans as a standard practice for several reasons. Plants can be used as a visual screen and to diffuse sound in open-plan space designs; to help soften hard lines or surfaces and add texture to nondescript surfaces; to give people directional clues; and to complement furnishings with color, texture, size, and shape.

Plant Selection

When choosing plants for interior landscaping, or "plantscaping," there are three primary considerations: location, type of plant, and size of plant. Location determines which types of plants can be used most satisfactorily and the size suitable for the particular surroundings. Because indoor conditions are usually not conducive to vigorous plant growth, plants that are already the desired size, or slightly smaller, should be chosen. Indoor plants rarely outgrow their allotted space.

Location

Foliaged plants most often found in campus buildings are placed in public areas such as entrances, hallways, lobbies, and lounges. Budgetary limitations often do not allow for plant decoration in private offices. Foliage can be used to personalize a desk, fill the space on top of file cabinets, or create height variation among single-level office furniture.

Light intensity and duration are crucial to plant health. Therefore, it is important to place plants near the light source. The brighter the light provided, the shorter the period of illumination required. Most plants require 12 to 16 hours of artificial light a day, if that light is between 200 and 500 foot-candles (A foot-candle is one lumen of light projected on 1 sq. ft. of area).

Sunlight is the most desirable kind of light but may be unavailable. If artificial light is the sole source, fluorescent is better than incandescent. Fluorescent light produces darker green foliage and more compact growth. Plants grown in incandescent light have a tendency to become weak and spindly.

Insufficient light results in loss of lower leaves, weak stems, pale coloring, and eventual death of the plant. Too much direct sunlight scorches foliage if adequate ventilation is not provided to reduce heat buildup next to windows.

Temperature and Humidity Control

Foliaged plants require moderation in both temperature and humidity. Generally, what is comfortable for people is appropriate for plants. The specific temperature range in which plants thrive is 55°F to 80°F. Avoid placing plants near drafty doors or windows because these spots may be as much as 20°F colder than the rest of the area.

If the heating, ventilation, and air conditioning system removes too much humidity from the air, there are several ways to increase it around the plants. Set plants on trays of moist pebbles, mist leaves frequently, or install a humidifier. Insufficient humidity creates symptoms similar to those of under watering. Leaf tips turn brown, margins yellow, and growth is stunted. Too much humidity over a long period makes plants susceptible to bacterial and fungal invasions. For this reason, misting leaves is not recommended but can be done if plants are regularly monitored for disease.

Care and Maintenance

To ensure the success of interior planting, qualified personnel must be provided. Many grounds maintenance departments have employees with education or work experience in ornamental horticulture or floriculture. Depending on specific departmental policy, the horticulture staff may or may not be required to maintain office and other interior plants. For example, the grounds maintenance horticulture staff may purchase and install foliaged plants, but policy many not allow this staff to maintain them. Office workers where the plants are placed are then responsible for maintenance. In this case, plant care should be assigned to an employee who has an interest in plants.

In large institutions, it is not feasible to have employees care for plants. Hiring a plant leasing firm is an excellent, although expensive, way to maintain plants. These firms are experienced at maintaining, as well as selecting and installing, interior foliage. They can provide the skills needed to lengthen a plant's life indoors and thus reduce replacement costs. In addition to maintaining existing plants, these companies offer purchase-only, purchase/maintenance, and lease/maintenance contracts.

Watering

How a plant grows is determined by the controlled environments of building interiors. Light, temperature, humidity, soil moisture, and nutrients are interrelated and affect the strength and health of a plant. Soil moisture is one of the most significant requirements of foliaged plants.

Light, temperature, humidity, and soil type determine moisture needs. Plants use more water when it is hot, and an increase in light or a decrease in humidity increases the need of water.

Watering should not be a routine practice, such as watering the plants every Friday. By testing the soil with a moisture meter or index finger, one can accurately decide whether plants need water. Ninety percent of plant losses are caused by improper watering, usually over watering. Most plants need to be watered when the top 1 or 2 in. of soil are dry. Some species need more frequent watering, some less.

When it is time to water, add water from above. Subirrigating can cause root rot by saturating the soil airspaces. Use lukewarm water, especially in the winter, because cold water may damage foliage.

Fertilizer and Insecticides

Plants grown in artificial light require only 10 to 30 percent as much fertilization as those grown in natural light. The less light received, the slower the growth rate and a diminished need for nutrients.

A complete fertilizer containing nitrogen, phosphorus, and potassium should be used—preferably one with calcium, magnesium, and iron. Liquid, water soluble, or time-released pellets can be used if they are applied at half strength. A light feeding every three to six months is appropriate. Too much fertilizer applied too often can result in a salt root deterioration.

Insect problems account for only 5 percent of plant losses, but their control is the most difficult aspect of plant maintenance. Before installation, check the plants closely for signs of insects. If insects are found, treat the plant immediately. (One mealy bug can lay up to 600 eggs.) Use recommended pesticides, and follow label instructions carefully. A highly concentrated solution can severely injure foliage, and a weak solution will be ineffective. If a spray is used, treat three times at weekly intervals to kill all of the eggs.

Spraying a plant for bugs in public buildings may be a difficult chore. The Environmental Protection Agency only allows the use of petroleum oils and natural pyrethrins in public areas. Plants should be moved outside, if possible, or to a storage room for spraying.

Cleaning and Pruning

Foliage must be cleaned regularly to keep plants healthy and attractive. A heavy coat of dust clogs the stomata, the cell openings that allow for the exchange of gases and release of moisture. In a greenhouse, plants can be sprayed with one tablespoon of liquid detergent (Dove or Ivory is best) mixed in one gallon of water to rinse off dust. To remove chemical residues or heavy dust accumulations, use a sponge or soft cloth and carefully clean each leaf. This method is time-consuming but does the best job. A feather duster can be used on plants already installed.

Pruning topical foliaged plants can be done to encourage new growth, enhance plant shape, or remove dead or diseased parts. There are two ways to prune: pinching or cutting back.

Pinching of the terminal (end) growing tip is usually done to produce dense, bushy growth. When the growing tip is removed, dormant leaf-producing buds lower down on the stem produce side shoots. This type of pruning is usually done with the thumb nail or scissors. Only $1/4$ to $1/2$ in. is removed, just above the leaf node.

Cutting back is done to thicker or harder stems and requires pruning shears. The cut should not be made flush. Use the natural target pruning method. Dust large, fleshy wounds, such as those on *Dieffenbachia,* with a fungicide to prevent stem rot.

Composting

With increasing environmental regulations aimed at reducing the use of landfills, composting operations are becoming more common on college and university campuses. Wet garbage, yard waste, and other humus material on campus can be composted. Although a composting facility will require grounds space, special equipment, and staffing, it will prove useful in reducing the solid waste stream and in providing soil maintenance fertilizer needs. The composted material can be used in amending soils and as top dressing and to reduce water consumption.

31.9 MAINTENANCE OF OTHER GROUND FACILITIES

Maintenance of paved surfaces, storm drainage systems, campus furniture, fountains, signage, fences, retaining walls, and other external campus improvements is often an important element of the grounds maintenance function. Each of these areas requires special skills and knowledge for effective care and maintenance.

Paved Surfaces

Paved surfaces can range from loose gravel to hand-laid tile or exposed aggregate surfaces. Each material requires its own maintenance methods, but there are some common practices for all. Before any material is selected, consideration should be given to its intended use, type of traffic, cost of installation, cost of maintenance, life expectancy, aesthetic conditions, material availability, and weather and seasonal factors. In regions subject to ice and snow, consideration should be given to the suitability and durability of pavements that will be subject to mechanical snow removal or the use of chemicals for ice control.

The most common problem of paved surfaces is the lack of drainage. This problem often occurs in climates that expose surfaces to continued freeze–thaw cycles. Common to all pavement types is the need for a proper base. A soft sub-base, caused by wet, poorly drained conditions, results in pavement movement, "pumping," and other problems. Good drainage is essential to effective pavement maintenance.

Concrete pavements are widely used and can provide a durable, easily maintained surface. However, a number of problems can result from poor quality in the mixing or placement of concrete. Improper proportioning of materials in the mix, or too much or too little water, can result in a poor quality or low strength material. Overworking of surfaces, retempering, or excessive water can produce surfaces that will crack or spall.

Storm Drainage

Good storm drainage is essential to a well-functioning campus environment. Undersized lines are a frequent cause of storm drainage failure. Ensuring an ample margin of excess capacity is the best policy when dealing with the design of drainage systems, because future growth is likely to add hard-surfaced areas that increase runoff. Adequate manholes are important for cleaning operations. Cleaning of storm sewer inlets should be scheduled as part of the preventive maintenance program to ensure consistent flow capability. Cross connections into sanitary sewers should be sought out and corrected as a matter of resource conservation and current code compliance.

Campus Furniture

A variety of exterior campus furniture is needed for a well ordered and functional campus environment. Furniture can include benches, light standards, refuse receptacles, planters, newspaper distribution boxes, signage kiosks, and fixed or moveable athletic equipment. Design standards and color choices should be established as part of a comprehensive landscape plan for the campus to ensure that campus furniture is in harmony with the architectural character and general design of the campus. This equipment receives hard use and is subject to all the effects of weather, and a regular program for cleaning, repainting, refurbishment, and replacement is important to maintaining the quality of exterior space.

Outdoor Art

Outdoor art and displays represent a peculiar challenge. Special maintenance is often required, such as regular sealing and cleaning of sensitive and unusual materials. At the point of installation of such art, a maintenance plan for these items should be developed in conjunction with the artist and the administration.

Fountains

Fountains and other water attractions are appealing and can add much to the quality of landscaped areas. However, they can represent a significant investment and can be difficult and costly to maintain. Maintenance can require the participation of mechanical and electrical shops for care of pumps, nozzles, and other mechanical features, or of lighting, as well as grounds maintenance staff. Regular water treatment is necessary in warm climates or during the summer to control algae growth, and occasional emptying is required for cleaning or the removal of foreign objects. Regular refurbishment and resealing of fountains and other water features will be necessary.

Signage and Exterior Graphics

A proliferation of non-standard signage and exterior graphics can be the single most significant item creating visual clutter on a campus. Careful planning, and coordinated and compatible design and lettering are essential to a successful signage and exterior graphics program. Strong institutional guidelines and standards should govern all types of signage, visual graphics, posters, traffic control signage, and informational and directional signage. These standards should specifically establish responsibility for management and enforcement of the program. Posters should be limited to strategically placed bulletin boards and kiosks on campus, and these should be regularly cleaned and policed to remove old, outdated material. Building identification and directional information signage should follow set standards using appropriate and consistent colors, locations, sign sizes and shapes, and lettering styles. A consistent, uniform exterior signage program can play an important role in creating a visual continuity and distinctive institutional image throughout a campus.

In many areas, "graffiti" on building surfaces, signage, or other exterior surfaces can represent a major maintenance problem. Institutions faced with this problem have learned that allowing it to accumulate only encourages more; prompt and complete removal soon discourages it. A number of effective cleaning materials are now available to seal surfaces, or to remove all types of paint and other marking materials. Similarly, improperly placed or unauthorized posters or other signage should be promptly removed to discourage their proliferation.

Fencing

Fences are often required to direct traffic, restrict use to designated programs and to secure areas. In some cases fences can be replaced by plants or pavements can be redesigned to provide proper traffic flow. Extensive use of fencing should be avoided. Too much fencing can obstruct view or diminish the aesthetic continuity of space.

Fencing materials should be pleasing and unobtrusive, but still do their intended job. Darker colors help fencing recede visually. Fencing that is compatible with site furniture and existing architecture could represent the character of the surrounding area. Temporary fencing should be avoided. Concerted effort with student organizations to encourage people to refrain from cutting corners or creating paths in turf areas and shrubs can reduce the need for obtrusive temporary fencing.

31.10 SNOW REMOVAL

The removal of snow and ice during the winter months can represent one of the toughest of problems for grounds maintenance crews at institutions in northern climates. During recent years increases in litigation resulting from slips and falls have only put additional pressure on this difficult assignment. Snow and ice removal that is not well handled can have serious public relations consequences for the facilities organization. A successful effort that serves the institution well will be based upon clearly established and recognized priorities, and carefully developed planning and procedures.

Establishing Priorities Although every person on campus would like a high priority for his or her department, parking space, or building, snow and ice removal can be successfully accomplished only by following clearly established priorities that meet recognized needs. At institutions with medical centers that provide 24-hour emergency care, access to this service from major arterial roads and sidewalks is critical. Student health centers that provide similar emergency care would also have a high priority. Second priority is usually assigned to power plant service areas, residence halls, and food services areas. Then the remaining main roads and walks serving classrooms and research buildings are cleared of snow. Handicap ramps and curb cuts, a high priority for mobility of a limited number of people, are of no value until the surrounding streets and walks are in passable condition. Therefore, ramps and curb cuts should have the same priority as adjacent surfaces. Widespread publication of the basic priorities and principles of the snow and ice removal plan can help to forestall much of the criticism and controversy that usually comes with a major snow removal effort.

Employee Call-Back   A workable system of mobilization is of prime importance. The responsibility of initial notification rests with a department and specific individuals that are on campus during nonworking hours. If snow or ice conditions occur during working hours, the removal process can begin as needed. In the evenings, on weekends, and during nonbusiness hours, the responsibility for call-back must be clearly established. The job is often assigned to campus police, as they are on duty 24 hours a day, every day.

The decision to call out employees or contractors is difficult and must be based on experience as well as other readily available information. National and local weather forecast information and contacts with local agencies are important. Managers should consider communicating with the institution's top administrators if conditions warrant restricted movement or school closing. A good rule always is to err on the side of calling people back sooner than proves necessary, or calling back more staff than prove necessary. Once a major storm has gotten the upper hand, it can be almost impossible to recover.

Staffing Requirements   Personnel to form crews for snow emergencies can come from several sources, and duties should be assigned to relate as closely as possible to normal assignments. Custodial employees can remove snow and ice from entrances, steps, and landings. Grounds maintenance employees, usually experienced with heavier equipment, should be responsible for snow removal from sidewalks and curb cuts. Street and parking lot snow removal may be assigned to grounds maintenance crews or to a campus service group that operates heavier equipment and trucks. Equipment maintenance personnel should be on duty to repair breakdowns and install plows. Some institutions supplement their regular maintenance crews with part-time student workers during periods of heavy snowfall, and many institutions have standing contracts to use contractors and their equipment for snow removal.

A strong agreement on participation in emergency call-out procedures should be made with employees and their labor organizations. Some employees may delay their response to emergency calls or not respond at all. The established policy should include method of contact, response time expected, payment for overtime or compensatory time granted, and job responsibilities.

In the case of an emergency condition that could last for many hours or days, it is important to phase the working hours of available labor. Repetitive shifts in cold weather should last a maximum of about 12 hours. Beyond this, efficiency and safety capabilities decrease rapidly. Just before the first snow or ice conditions are expected, hold a brief meeting of all employees and contractors to explain or review the mission and expectations of the operation and delineate responsibilities of the various crews. It is helpful to ask various agencies on campus to send representatives. Interested groups include security, traffic, custodial, someone to represent accessibility issues, and general administration. Problem areas can be discussed and special requests or expectations presented. Instruction and training in the use of equipment and distribution of salt is invaluable. Damage to the campus can be extreme if care is not taken. Plow damage becomes quite obvious as winter snows melt and grass does not grow.

Attention to details can make snow removal a success. Placing stakes at walk intersections to guide operators when the area is entirely covered, or providing coffee and hot chocolate for employees during long hours of overtime improves the operations efforts.

Snow Removal Equipment   Equipment funding, staffing, policy, procedure, local snowfall, and the physical characteristics of the snow removal areas all directly influence the types of snow control equipment best suited for each campus setting. It is important that all of these factors be considered when evaluating equipment, as many varieties of equipment and materials are manufactured to clear, remove, and melt snow and ice.

Once policies, procedures, and staffing patterns are established as mentioned in previous sections, other factors can be considered, such as the physical characteristics of the snow removal areas. Characteristics such as street, sidewalk, and parking lot sizes; vehicle weight restrictions; snow emergency zones; street and sidewalk bollards; and bridges all are factors that need to be considered when purchasing or evaluating equipment. Because funding for equipment is generally limited, it is important to select equipment that will be cost-effective. For instance, will tractors, commercial lawn mowers, and dump trucks be equipped for snow and ice control? Utilizing a piece of equipment that might otherwise sit idle in storage will greatly increase the equipment's return on investment. Will the equipment handle the workload during severe weather conditions? Consideration may be given to contracts with state, county, or local public works to share equipment and routes.

These questions must be asked, in addition to studying manufacturers' recommendations, observing equipment demonstrations, and visiting nearby institutions to observe other snow control programs. Once a program is established, fine-tuning begins through a trial-and-error process. What works in one area may not work in another. As the snow control program is carried out, meetings should be held to discuss and review the effectiveness of the snow control program. The supervisor should implement worthy suggestions from the staff and keep track of which techniques work well.

Be prepared. Organize the snow control program several months in advance. Ensure that supplies are ordered in a timely fashion. Have vehicle maintenance employees winterize equipment to be pulled from storage, and equip the tractor mowers and dump trucks with plows, tire chains, and salt spreaders. Order snow-melting compounds suitable for the local winter conditions, and have them ready in their storage areas. The *Snow Fighter's Handbook,* available from the Salt Institute,[3] contains information pertaining to all phases of snow control. The American Public Works Association schedules an annual snow conference that can be a valuable training resource.

Snow and Ice Control Plan As with any emergency procedure, careful advance planning is an essential element of a successful snow and ice control program. If this planning is to be effective, it must be written in a well-organized and comprehensive snow and ice control plan. Among the basic elements of a good plan are the following:

- Summary of basic policies and procedures for snow and ice control
- Establishment of priorities
- Organization for snow and ice control
- Control and communications
- Assignment of responsibilities
- Personnel assignments
- Delineation of equipment routes

To remain effective, a snow and ice removal plan should be reviewed and updated each fall, well in advance of winter weather, and the updated plan should form the basis for refamiliarization and training for all essential snow and ice control personnel.

31.11 CONCLUSION

A well-maintained campus can be of great importance in attracting students and faculty. To achieve its goals, an institution needs adequate staffing and funding for grounds maintenance and care. The most important factor of all is a leader who is knowledgeable in this subject and has a sincere interest in the work and the people who will carry it out.

NOTES

1. APPA: The Association of Higher Education Facilities Officers. *Contracting Management or Self-Operation.* Alexandria, Virginia: APPA, 1993.
2. American Association of Nurserymen. *American Standard for Nursery Stock.* Washington, D.C.: American Association of Nurserymen, 1986.
3. The Salt Institute. *The Snow Fighter's Handbook: A Practical Guide for Snow and Ice.* Alexandria, Virginia: The Salt Institute, 1982.

ADDITIONAL RESOURCES

Dirr, Michael A. *Manual of Woody Landscape Plants: Their Identification, Ornamental Characteristics, Culture, Propagation and Uses,* third edition. Champaign, Illinois: Stipes Publishing Company, 1983.
Pirone, Pascal P. *Disease and Pests of Ornamental Plants,* fifth edition. New York: Wiley-Interscience, 1978.
Turgen, Alfred J. *Turfgrass Management.* Reston, Virginia: Reston Publishing Company, 1985.

CHAPTER 32

Custodial Services

J. Kirk Campbell
Carleton College

32.1 INTRODUCTION

Custodial service operations can vary significantly from institution to institution regarding the type of services that are provided and how often certain tasks are performed. Many operations have undergone significant downsizing, either by actual reduction of personnel or by not increasing staff while building growth has occurred.

The primary function of a custodial service operation is the internal appearance of the campus, which may or may not include residential facilities. This occurs through various cleaning duties and removal of trash and recyclables to a pick-up area. An aesthetically pleasing and clean environment is vital to the image of the institution and is noticed by all who use or visit the campus.

In addition, other functions may exist within the operation, such as performing setups (e.g., preparing for events and banquets); providing laundry services, limited snow removal, lamp replacement, waste management, and minor maintenance repairs; taking responsibility for various furniture items; contracting other functions; and perhaps performing other services. The multitude of service possibilities adds to the complexity of proper staffing to meet various needs while being at the forefront of customer service and customer contact.

Good leadership of people is critical for the diversity that exists among custodial employees themselves, and the challenge of "doing more, with less, but better" that is usually the expectation of custodial operations. Various methods of staffing are required, along with a definitive quality service approach so that all employees can fully understand the variety of expectations.

The operation is extremely vulnerable to cutbacks and to the assumption of additional duties that do not "fit" into other operations by virtue of usually having the single largest work force within the facilities organization. Contracting out of custodial services, which is not considered a highly technical profession, is often proposed or considered.

32.2   CUSTODIAL AUDIT

A custodial audit is critical not only for documenting current service levels (i.e., what is done and how often), but also for determining whether the programmed quality of service is actually being accomplished.

The first step in an audit is to develop and document the service levels that current labor is providing. This is done by defining each task (activity) and how often each task is to be done (frequency). There may be tasks that are not cleaning oriented, but these will be important later in determining staffing levels or needs.

An audit, once completed, can be used for individual custodial performance inspections, which should measure what tasks are being performed, whether appropriate intervals are being followed, and if the level of performance is meeting expectations.

A clear means of measurement must be determined, and the defined expectations and frequency of service provided must be realistic. A weighting system can be used that accounts for the level of effort that is required in one area versus another.

For example, in Figure 32-1, 80 percent of the effort in public corridors is being spent on floors, while in restrooms only 13 percent of the effort is being spent on floors and 67 percent is being spent on the fixtures themselves.

For further development of audits and scoring mechanisms, refer to *Custodial Staffing Guidelines for Educational Facilities,* published by APPA: The Association of Higher Education Facilities Officers.[1] Figure 32-1 is an example audit instrument from that guideline. Expectations must be clear, with a description for each level so that consistent scoring can be done throughout the organization and by anyone doing the audit. Figure 32-1 will work for any type of space and any task, with the appearance levels and weighting factors chosen by the institution.

32.3   STAFFING CUSTODIAL WORK

There are different methods for determining custodial staffing and balancing workloads. The square footage–per-custodian method is used to determine how staffing compares with other institutions. Cost per square foot is translated from this method and is also used as a comparison.

Major fallacies arise from this thought process. No two institutions are exactly alike, and all institutions vary greatly in the service provided,

Figure 32-1
Weighting Factors for Appearance Items

Appearance Items	Class-rooms	Entryways and Lobbies	Locker/Changing Rooms	Offices	Public Area, Corridor, Stairwell	Research Laboratories with Hazardous Waste	Research Laboratories Without Hazardous Waste	Store-rooms	Wash-rooms
Chalkboards and erasers	15								
Floors	67	52	42	55	80	29	68	89	13
Horizontal surfaces and ashtrays, telephones, furniture, and drinking fountains	8			12	9	3	6		
Lighting and light fixtures	2	1	1	3	3	1	1		1
Lockers and benches			9					6	
Outside walks and ramps (10 ft.)		9							
Showers and drains, including shower floor			35						
Toilets, urinals, washbowls, soap, and drying fixtures						21			67
Trash containers and pencil sharpeners	4		3	23	5	46	22		3
Vertical surfaces: walls, doors, windows, vents, blinds, partitions	4	26	10	7	3	2	3	5	13
Walk-off mats		12							

how often the service is done, and the level of wages provided. To make comparisons based on square footage and cost will paint a completely false picture, either good or bad.

Various Methods Used for Staffing   The *square foot-per-day method* uses the average square footage cleaned during a day. Part of the problem with this method is that some use gross square footage, others use net square footage, and still others use space actually cleaned. The disparities that could exist for comparison are obvious. This is a quick method to use that is easily understood, but it is not based on any recognized time standards.

The *fixture method* involves counting all items (e.g., toilet fixtures, chairs, wastebaskets) that have established times to clean. However, these must be combined with items that have standard times according to square footage or linear feet (e.g., floor scrubbing, mirror cleaning) to develop workloading and labor requirements.

The *type-of-space method* involves developing tasks and frequencies for similar types of space and an average time required for a cleanable square-foot block of that space. This method combines fixture and time standards per square foot of activity without going through the process of counting every fixture. The assumption is that an average time can be established for a 1,000-sq. ft. office space or any type of space and that these times will balance out over a workload assignment.

The *detailed method* involves taking established time standards or standards created by the organization, based on square feet, linear feet, or individual items. This is a detailed approach and requires a highly detailed inventory that must be developed and continually updated.

Variable Factors   The basic methods for staffing are subject to many variables. Thus it is difficult to provide a precise measurement of impact. The following is a list of some variables that can affect work production or time required to do a particular task (or added task):

- Weather (especially snow)
- Training
- Supervisor/management ability
- Age of facility
- Traffic flow
- Equipment/products
- Type of usage
- Surface types

Snow removal can throw all schedules off (especially in a climate that has frequent snow) and constantly does so with the added time required for floor surface cleaning. Having the right types and sizes of equipment to maximize labor utilization is critical. All other functions may require adjustments to workloads in various areas, which can substantially alter the methods rationale and the ability to be accurate.

The Staffing Process   The staffing process is not a precise science. There are many sources available that provide actual times for a task or activity and how long it takes to perform those duties based on the types of equipment, size of equipment, building design, time of day cleaning is done, and traffic. Many other variables can greatly influence the actual time taken to accomplish a job. The reality is that adjustments are made by each institution to properly staff the facilities.

There are guidelines available that provide times and directions on how to compile this information into something relevant, but the mistake of tying the staff completely to these statistics must not be made. This information is a starting point or guideline for process development at best.

The major obstacle to staffing is usually the change that is required to rethink or re-engineer how things are to be done, such as developing new routines, setting service levels, and understanding expectations.

One example is the custodial staffing guidelines for educational facilities (Figure 32-2). Tasks are identified, a frequency is established for how often the task is to be performed, and an average of how long it takes to do a particular task for a given type of space is shown in the general guideline. From this matrix a tailor-made matrix can be created by adding or subtracting tasks and choosing a frequency that better fits the institution's situation. The time required for each task is added up for how long it takes to perform those tasks over 1,200 cleanable square feet of office space. A total of how much cleanable square feet can be done per day by a custodian in this type of space can also be determined.

The guideline provides information on ten different types of space, individual problem samples, and a detailed background, which makes it a functional guideline for determining staffing needs.

32.4   CUSTODIAL ORGANIZATION

Determining full-time equivalents of labor and the shifts on which work is to be performed is vital to the overall structure of custodial operations. Other organizational staffing needs revolve around the major core of staffing required to perform the day-to-day services.

The majority of institutions have generous benefit packages that allow for extensive paid time off in vacation, personal days, and sick leave. Consideration should be given to a "project-and-relief" crew to help balance out day-to-day operations and ensure a measure of consistency in service levels. An ideal percentage would be 15 percent; however, with many programs operating "lean," a 5 to 7 percent level is a "must" guideline.

Figure 32-2

Staffing Service Levels

	Frequency-Adjusted Time					Base
	Level 1	Level 2	Level 3	Level 4	Level 5	Time
Vacuum carpet and	D	A/D	W	W	W	
straighten furniture	21.07	10.54	4.21	4.21	4.21	21.07
Spot-clean carpets	D	W	M	Q	Q	
	10.59	2.12	0.51	0.17	0.17	10.59
Change light bulbs	D/A	D/A	D/A	D/A	D/A	
	0.09	0.09	0.09	0.09	0.09	3.91
Empty waste containers	D	D	D	A/D	W	
	6.60	6.60	6.60	3.30	1.32	6.60
Spot-clean walls, partitions,	A/D	W	M	Q		
and doors	2.97	1.19	0.29	0.10		5.94
Clean telephones	D	W	M	Q		
	4.17	0.83	0.20	0.07		4.17
Empty and clean ashtrays	D	D	D	A/D		
	0.73	0.73	0.73	0.37		0.73
Dust furniture and flat surfaces	A/D	W	M	Q		
	1.44	0.58	0.14	0.05		2.88
Clean trash containers	M	M	Q	S/A		
	0.64	0.64	0.21	0.11		13.35
Dust vents	M	M	Q	S/A		
	0.09	0.09	0.03	0.02		1.93
Dust blinds	Q	Q	S/A	S/A		
	0.06	0.06	0.03	0.03		3.60
Perform interim carpet care	Q	Q	Q			
	0.69	0.69	0.69			43.25
Project-clean upholstered	Q	Q	S/A	A		
furniture	2.12	2.12	1.06	0.53		132.30
Clean windows	Q	S/A	S/A	A		
	0.19	0.10	0.10	0.05		12.06
Project-clean light fixtures	S/A	S/A	A			
	0.91	0.91	0.45			113.50
Perform restorative carpet care	A	A	A	A		
	0.39	0.39	0.39	0.39		98.73
Adjusted minutes*	52.75	27.68	15.73	9.49	5.79	
CSF per custodian†	9,555	18,208	32,041	53,109	87,047	

Frequency codes/daily adjustment factors:

D= Daily/1.00; Q = Quarterly/0.016; A/D =Alternate days/0.50; S/A = Semiannually/ 0.008; W = Weekly/0.20; A = Annually/0.004; M = Monthly/0.048; D/A = Daily adjusted (see notes).

*Adjusted minutes are for 1,200 cleanable square feet.

†CSF = cleanable square feet. Square feet per custodian based on 420 productive minutes per day.

Program modification (reduction of tasks and/or frequencies) should be considered to make this minimum crew possible.

Because of the wide variety of activities custodial operations can assume, considerations may be given for movers, setup people, recycling activities, and other assorted duties that justify regular labor without tapping into regularly staffed, day-to-day activities.

Direct Supervision Supervision is critical for ensuring that standards are met, customers are satisfied, follow-up is pursued, and the myriad of planning, scheduling and paperwork gets accomplished.

The quality of supervision will drive the operation in terms of what can be accomplished. The higher the quality, the higher the results. There is no true ratio of supervisor to employees that can be defined, but supervisors with marginal ability will be able to supervise nine to twelve full-time-equivalent employees, and those with high ability, twenty to twenty-five full-time-equivalent employees.

Training and networking outside the institution are necessary to develop new ideas and ways of doing things for supervisors. Gaining knowledge and finding creative ways to change are important to avoid stagnation and to cope with the continuous complication of laws, new management theory, and financial constraints.

Forepersons, or lead persons, may be beneficial to an organization where shifts or crews do not need a supervisor but do need a designated person to provide limited direction and monitoring. This provides an additional level for individuals to have some responsibility, with minimal added cost. Care should be taken not to split out lead responsibility but to add it to a full-time cleaner who can do both. Keep in mind that lead work requires a minimum of quasi-supervisory duties. This may be in a unionized or non-unionized environment.

Departmental Management Operations vary greatly in size and structure, but there must be one individual who is totally responsible for the custodial operation on a full-time, day-to-day basis. Custodial operations have sufficient involvement in other functions to require an overall manager or director.

Large institutions may develop zones where a manager may supervise 40 to 100 custodians, with several supervisors reporting to each manager, and each manager reporting to a centralized director. This can ensure continuity and overall accountability throughout the institution.

As available resources and accountability are becoming more important, the involvement of employees in decision making has increased. Organizations are moving more and more toward placing traditional supervisory and management responsibilities at the front-line employee level. Self-directed work teams are becoming more prominent and allow the creation of leaner organizations with less structured supervision. Consideration should be given to the development of this type of empowered organization.

32.5 EQUIPMENT, SUPPLIES, AND MATERIALS

Equipment Various sizes and speeds of equipment are available that not only can make the job easier, but also can increase productivity so that more cleaning can be accomplished within the same time frame. Keeping abreast with the newest technology, overcoming old ways of doing things, training, and employee participation in the development of new methods are vital to improved performance. Custodial equipment is a small portion of the overall budget requirement but can produce substantial savings in labor costs.

Repair capabilities must be developed to keep equipment on line, functional, and aesthetically appealing. Frustration can develop within an organization if proper avenues of maintenance are not established. Organizational structures or programs that team the custodial department with other trades may be beneficial in achieving this goal.

Supplies There are many cleaning chemicals available on the market, and manufacturers are constantly improving their products.

The old thought pattern of buying the cheapest product will only frustrate employees attempting to do the work and will build inventories that are excessive. A premeasured system can save up to 60 percent over a nonpremeasured system, cut down inventory (which essentially is "money on the shelf"), and provide consistency in product type and use.

Partnering may be a strong consideration for those institutions in a location that would make such an arrangement beneficial. Committing to one or two vendors for all supplies and equipment can result in the following benefits:

- Training
- Vendor-retained inventory
- Good service and repair
- Quality products

Although this process may be contrary to some purchasing requirements, generally such requirements can be worked through to the benefit of the organization.

Material Safety Data Sheets Material safety data sheet information must be supplied for all products. Many products have this information on the container itself and provide matching labels for spray bottles and other containers in systems that are divided up for use.

An annual training review should begin with basic chemical information and introduction of any new products. The information must be readily available for any employee inquiry or accident that may occur.

32.6 OPERATIONAL PRACTICES

Student Labor The use of students as custodial workers is quite common, ranging from two hours per week per student to up to 20 hours per week. Time of need is the primary determinant of the ability to get students, as early mornings may not a popular time, and evenings might be. Turnover can be an issue, as student's schedules and needs change from semester to semester. Also, it is important that the students receive necessary training so that they can perform their jobs successfully and ensure that all Occupational Safety and Health Administration (OSHA) requirements are met. Students can be a valuable and inexpensive source of labor with proper training, scheduling, and supervision.

Contract Services Contractors can be a valuable resource for the custodial department in functions not frequently performed (e.g., high window washing) or for special projects that are unusual and not normally permitted by time constraints or specialization.

Other possible uses of contractors may be to assume management/supervision of an organization. Contractors are often used in the hope that greater efficiency, better quality, or greater cost savings can be obtained by replacing in-house staff with outside workers making lower wages. This process should be given careful and thorough consideration for short-term and long-term effects. A good standards program will help evaluate contractor versus in-house effectiveness. Chapter 17 provides a more detailed discussion of this issue.

Quality Assurance Inspections can be informal (i.e., walking around and observing) or formal (i.e., where all tasks of the jobs are considered at set intervals and duly commented on). The combination of both processes is the best plan to ensure that quality is obtained consistently.

Employee Participation Labor/management committees, quality assurance committees, total quality groups (or other nomenclature) may be an option. Providing avenues of input for union or nonunion employees can help to shape the operation's future, produce trust, improve morale, develop understanding, and result in better production or re-engineering improvements.

These committees can be used for creating department policies and procedures, choosing equipment and supplies, or developing inspection forms. However, if the employees are unionized, the business agent from the main union hall must be a cooperative participant for these committees and processes to be successful.

Safety   A safety committee should be formed or safety issues made part of a quality committee's responsibilities. Management must be responsive in providing any reasonable safety provisions or developing an equitable solution for any safety concerns.

Training should be provided in blood-borne pathogens, asbestos awareness, electrical awareness, and other areas in which knowledge could prevent harm or ease concern. Safety can adversely affect morale if not properly addressed or if an effective mechanism for resolution is not established.

The Americans with Disabilities Act and Reasonable Accommodation
Active management participation is required to ensure compliance with the Americans with Disabilities Act. Job descriptions are rapidly changing, and there is a need for sufficient detail that essential job duties are identified. It is appropriate, and legally required, that reasonable accommodation be made, yet it is recognized that this must be done within the budget and effectively. This is an area that requires extensive attention, especially in custodial work, where back injuries, carpal tunnel syndrome, and reactions to chemicals are prominent.

Motivation   Motivation is positively affected by developing trust, respect, understanding, and a sense of fairness throughout the organization. Individual awards are rarely an effective tool for an organization unless they are distributed in an equitable and meaningful fashion and are obtainable by all.

If employees have input, feel they are being heard, observe positive action, and have a sense of fair play, morale will be good. Positive recognition and a participative management style are the biggest motivators for employees. Chapter 9 provides a more detailed discussion of this issue.

Absenteeism and Tardiness   Both absenteeism and tardiness need to be addressed at the earliest possible time when a pattern of possible abuse may be occurring.

Tardiness is easier to spot, but a check in the system is needed, with a reasonable grace period provided. However, if an employee has a habit of coming to work a few minutes late, this should be addressed until the problem is corrected. A policy must be clearly defined that sets the expectation that employees will be at work on time.

Absenteeism is more difficult to deal with and usually stretches over a longer period of time. With the extensive benefits given by institutions, there are many days that can be legitimately used, but patterns of abuse and the use of leave without pay are about the only items that can be corrected. Policies that involve a quota, and that may affect everyone, are not the proper way to address the issue. Individual attention is the best way to correct absence problems.

Policy and Procedure Manual   A well-developed policy manual is important for overall understanding by employees. A manual should include current policies, organizational expectations, service levels, institutional information that is pertinent, and safety information.

A "how-to" manual may be created separately that provides instructions on cleaning steps and processes. This should be a stand-alone training and instruction manual.

Customer Service   Providing an organization that meets the needs of the customer is an ever-increasing challenge as budgets get tighter, institutional demands change, and expectations grow for additional and better service. Service levels and services provided should be developed and reviewed continually so that priority tasks are performed at the right frequency and lower priorities are properly placed. Regular and frequent contact with customers to obtain their feedback, input, and understanding is critical.

Often in organizations, stagnation ("we always have done it this way") is the prevailing approach. The challenge is to continuously rethink how things are getting done, or "Is there a better way?" New ideas, rethinking, vision, and creativity will help determine future success and survival as customers insist on changes. Change is inevitable, but the response to change will be a critical factor in a better organization.

There cannot be a "can't do" or "won't do" attitude; instead, a positive approach is needed in trying to meet customer needs—not necessarily wants—with the available resources and labor. This can be done only with dialogue between the organization and the customer. Chapter 5 provides a more detailed discussion of this issue.

32.7 TRAINING

Training is needed not only to improve professional expertise, but also to ensure quality and efficiency within the organizational program. It can be a great morale booster or motivational tool that tells employees at all levels that they are important, their job is important, and they are worth spending money and time on. The confidence developed from gaining knowledge and developing skill proficiency is well worth the effort.

Employees   "How-to" cleaning training should be given to all employees, with refresher or new product training provided periodically. Often vendors will provide this training and have the skill and material to do so. Various safety training programs must be provided for liability and prevention purposes. Specialty training that can provide certification in pool care, carpet care, repair, or other areas for specific people who do these duties should be given, even if off site.

In addition, informational meetings should be held at regular intervals by management. These meetings allow for questions to be asked and answered and information provided on what is happening elsewhere in facilities and/or the general institution and projects that are coming up. Employees want to be informed and want to feel that they are important enough to have money and time spent on their improvement. Informational meetings, where questions can be freely asked, give unlimited potential to help develop the vision and thought processes of the total organization.

Supervision/Management   Training in this area should consist of more than coming up through the ranks and on-the-job training. Education is a vital link to creativity and vision and must be encouraged. Positive idealistic philosophies toward life and working with employees are good, but miss the realistic way to handle the most commonly faced situations. Attendance at organizational seminars and meetings may be the best resource for a reality check.

Networking is important for new ideas. Networking can help determine what has worked and what has not worked well, and what improvement ideas may be useful in the organization. Memberships and networking may need to be a forced issue (some people just do not want to invest the time), and funding should be supplied by the institution. Development of information resources can provide future assets that will help in times of needed change in the future.

32.8   SCHEDULING AND SHIFT ALTERATIONS

There is considerable opinion on when cleaning should be accomplished. Customer expectations should be the determining factor when establishing cleaning schedules. Often, however, historical processes or employee desires set schedules, sometimes not with the customer or maximum efficiency as the primary concern. Each institution must review its operation and what needs to be accomplished and schedule staff accordingly, with the customer and maximum effectiveness in mind.

The "deep night" or "graveyard" shift from 10 pm to 7 am is perhaps the most difficult shift for individuals, and studies have indicated that production can be down by as much as 40 percent over morning or afternoon/evening schedules. This shift should be used only as a last resort.

Consideration should be given to some overlapping of custodians' and customers' schedules. This makes custodians feel that they are a part of the institution (an important morale factor) and adds to accountability for their performance. Shifts should be considered in a quality committee, with employee participation and with the goal of customer satisfaction and maximizing effectiveness.

Institutions that have a workforce that cleans both academic and dormitory space may consider a 4 am start, splitting full-time employees between academic (4–8 am) and dormitory cleaning (8–12:30 pm). Other creative schedules could include using part-time regular help for peak cleaning periods; working 10-hour shifts 4 days per week; using students for peak periods (perhaps job sharing); and working people in groups or teams.

Special Changes School breaks, summer conferences, and dormitory turnovers can result in periods where a different temporary shift may be required or appropriate to accomplish periodic activities that are not part of the normal routine. These may be periods of less institutional activity that allows projects to be accomplished with teams, or they may be peak times (e.g., requiring quick turnover in dormitory rooms) when each employee must be utilized in order to make the deadline. Flexibility in schedules, being reasonable and providing as much notice as possible, is often important in athletic areas or requirements for special events.

Activity on weekends is growing throughout institutions, and weekend duties for setups, special cleaning, and normal cleaning (dormitory areas, heavily used restrooms, student centers, and so forth) are required. A schedule in which employees alternate every other weekend, with Friday or Monday off, works well for full-time employees and results in a more tolerable personal schedule.

32.9 RECYCLING

Recycling has become part of most custodial organizations' responsibilities. What is recycled and how it is recycled vary greatly from institution to institution and are mostly affected by local requirements and location (i.e., rural vs. urban). The custodial department is most often a key part of any recycling effort. A full discussion of recycling is provided in Chapter 33.

32.10 SPECIAL CONSIDERATIONS

Computers The use of computers in custodial work is increasing, with many software packages available that cover custodial operations. With increasing expectations for flexibility and accountability, it will be imperative that usable information be available for quick answers to proposed changes or questions. Information such as type of space; square footage; task/frequency, with realistic times for workloading; budgetary impact; and other pertinent information should be part of the database.

It should be cautioned that many programs offer material that may not be of value; contain times that are not realistic; are not flexible enough to produce varied reports, charts, and graphs for management; or may not be user friendly. Time and effort need to be taken to obtain the program that is appropriate for the user and the institution. Once a program is purchased, gathering and inputting data should be done by the institution, which can greatly reduce the cost of getting the program set up.

Consultants　Custodial consultants have value in areas where an organization needs new ideas for efficiency and production, a review because of senior administrative changes, an increase in expertise in the custodial area, help in designing good contracts, training support in specific areas, or advice regarding reorganization issues. It is not effective, and it is expensive to have consultants come in and count square footage or numbers of items, or bring "canned" ideas to the institution. A consultant should have substantial background and understanding of higher education and its unique needs and criteria.

Beware of consultants who do not have the proper background and expertise and cannot properly identify the needs or concerns of the institution. A former salesperson of custodial products who becomes a consultant probably does not have the proper background. Pricing should be reasonable and not in the multiple-tens-of-thousands range for a campus visit and a report of ideas and recommendations. An idea of the desired result must be developed, and only a consultant with the background and experience that can fulfill that need should be hired.

Uniforms　Uniforms can be great identifiers and add a sense of security for customers, especially during off-peak hours. The cost is not great, but getting proper attire that is acceptable to the existing workforce can be a major obstacle. Tasteful, school-oriented uniforms can make people feel part of the institution and create loyalty. This is a good issue for an employee involvement committee to address. Uniforms are not an absolute need but may be a worthwhile consideration.

Customer Surveys　An annual survey of the customers served is a good way to find out if service levels are adequate and provides an opportunity for customers to express positive or negative concerns.

Surveys can be great marketing tools to get information out to customers and to solicit ideas on how to improve services. A good survey will ask not only the usual questions about the cleanliness achieved, but will also address such issues as responsiveness, flexibility, and employee demeanor. It is important that those customers who provide a comment receive personal follow-up with a response. There is nothing that will impress customers more than personal contact and concern.

Medical Waste Special training and precautions—not only in handling medical waste by custodians, but also in proper waste stream disposal—must be ensured. There are county and state variations in regulations for handling this material, but without strict controls and handling, large fines may be assessed for improper disposal.

In some cases inoculations may be required.

Hazardous Waste The area of hazardous waste has greatly expanded in recent times and will continue to do so. Again, this is governed by county and state regulations, but its impact can be reduced if proper purchasing of materials can be assured.

Proper training defining hazardous materials, precautions that should be taken, and proper methods of handling and/or packaging for disposal must be given. Purchased materials should be biodegradable (e.g., water-based wood [gym] finish), eliminating as much as possible hazardous products and petroleum products that were previously used. The best control is at the entry point.

32.11 SUMMARY

Custodial work has become and will continue to become a more complicated operation, with additional involvement in other activities. "Doing more, with less, but better" is the trend of the present and the future. Professionalism and expanding skills and knowledge will be required more and more as the custodial operation evolves into a stronger role within the facilities organization.

NOTE

1. APPA: The Association of Higher Education Facilities Officers. *Custodial Staffing Guidelines for Educational Facilities.* Alexandria, Virginia: APPA, 1992; revised 1997.

CHAPTER 33

Solid Waste and Recycling

Phillip R. Melnick
Pennsylvania State University

33.1 INTRODUCTION

For many people the phrase "out of sight, out of mind" would be most appropriate for describing municipal solid waste management. Their responsibility ends once an item is put in the trash. In these times of increasing regulatory oversight and societal pressures, this is the last sentiment facilities managers can afford about the waste generated at their institutions. Instead, we must be concerned with the eventual disposition of an item before it is even purchased.

This chapter will attempt to define solid waste and outline various approaches to solid waste disposal. Just as there is no one plan for solid waste disposal that can be used by all institutions, there is no single strategy that will work within an institution. Instead an integrated approach that relies on the hierarchy of reduction, reuse, recycling, and disposal is needed to effectively manage each institution's municipal solid waste.

Solid waste management in colleges and universities is taking on increasing importance as most states move to regulate disposal practices and as the university community, especially students, demands greater environmental stewardship. The complexity of this task may seem insurmountable. This does not need to be the case. A systematic approach to the disposal of institutional wastes can be undertaken to ensure compliance with all regulatory bodies and the demands of varied constituencies, and it starts with a better understanding of the entire disposal process.

33.2   SOLID WASTE DEFINITION

Municipal Solid Waste

Municipal solid waste is the most general category of waste as defined in Subtitle D of the Resource Conservation and Recovery Act of 1976. The Environmental Protection Agency (EPA) defines municipal solid waste as durable and nondurable goods, containers and packaging, food waste, and yard waste. Other Subtitle D wastes include municipal sewage sludge, industrial nonhazardous waste, small-quantity generator waste (hazardous waste), construction and demolition waste, agricultural waste, oil and gas waste, and mining waste.[1]

Non-Subtitle D Waste

Other wastes that are not Subtitle D wastes but may be part of the waste stream at a college or university include infectious waste, chemical waste, hazardous waste, low-level radioactive waste, and special handling wastes. Often these wastes are the responsibility of the safety officer or health physicist on the campus. Chapter 36 covers these issues in more detail, but a general description of these wastes follows.

Infectious wastes are typically generated as a result of biological, medical, or pathological activities that take place on campus. Waste is classified as hazardous by the EPA for one of two reasons: 1) it is listed in the Resource Conservation and Recovery Act as hazardous because it can cause injury or death or damage or pollute air, land, or water; or 2) it is ignitable, corrosive, reactive, or toxic.[2] Examples of such wastes include heavy metals, solvents, asbestos, fluorescent light bulbs, laboratory animal carcasses, and contaminated oils. Low-level radioactive wastes are typically generated through experiments and laboratory research. Wastes requiring special handling include coal ash and septage and sludges.

Although the EPA classifies all wastes, many states reclassify them more stringently or differently. It is important for facilities managers to check with local or state authorities to determine how wastes are classified in that state. The state, not the federal classification, will most likely be what managers will have to know to determine proper disposal methods for any waste generated on campus.

33.3   LEGAL ASPECTS OF WASTE DISPOSAL

Legislation

Environmental concerns have been prominent among issues addressed at the federal level. More than fifteen major federal laws that pertain to the environment in some manner have been passed since the 1970s. States have followed suit, with similar activity seen in state environmental legislation.

Like most federal legislation, Subtitle D of the Resource Conservation and Recovery Act, the section that pertains to municipal solid waste, does not directly influence practices at the institutional level. Instead, the mandates of the legislation, and the regulations promulgated as a result, dictate or delegate to the states the responsibility to enforce new environmental standards. The Resource Conservation and Recovery Act espouses a national philosophy and strategy for disposal of municipal solid waste, especially landfill requirements. It also empowers the EPA to require states to submit solid waste plans. State legislatures, in reaction to these requirements, pass laws that result in regulations at the state and local level.

Regulatory Agencies

Although the Resource Conservation and Recovery Act may not specifically declare how wastes are to be disposed of, through the philosophy set forth and the state solid waste plan requirement, the Act has been the driving force for the many state regulations concerning municipal solid waste disposal. The EPA also is empowered to control municipal solid waste disposal through rule making. Each state has its own regulatory agency that is responsible for environmental policy and enforcement and acts much like the EPA. These bodies will most directly affect solid waste disposal practices at institutions of higher education. Most of the detailed requirements facing solid waste professionals come about through EPA and state agency rule making. A good source for EPA rule-making decisions is the *Environmental Reporter,* which is available through the Bureau of National Affairs. Many states have bulletins that serve this same purpose.

Flow Control

Flow control is a growing component of municipal solid waste management. In short, flow control is the designation of approved disposal sites for various categories of municipal solid waste, including recyclables, by local or state authorities. Flow control has become popular among state and local government officials because it allows them to control where waste is taken, thus ensuring financial viability of government-funded or privately contracted waste facilities. It also provides a way for local authorities to determine the extent to which waste can be imported from outside the immediate service area of the disposal facility. The ability to limit importation is viewed by local authorities as critical to guarantee future disposal capacity and also to reduce any liability that may result from improper disposal of materials contained in the imported waste.

In a Federal Supreme Court decision, *Carbone v. Town of Clarkstown,* the Court ruled that flow control is a violation of commerce laws and is unconstitutional. However, the future of flow control is still in question. Legislation has been introduced in the U.S. House of Representatives that will allow flow control at the state and local level.

Flow control is important to facilities managers because it can greatly affect disposal options. Local flow control may restrict competition, thus reducing disposal options and possibly increasing costs. On the other hand, lack of flow control may discourage the development of additional disposal capacity, especially in rural areas. This could mean long hauls and increased transportation costs for disposal. It would be wise for facilities managers to follow this issue closely in the future, especially at the local level. If they don't, they may find out too late that their waste management plan is no longer valid.

33.4   INTEGRATED APPROACH TO DISPOSAL

Different Solutions for Different Wastes

The current accepted philosophy in municipal solid waste management is to follow an integrated approach. This entails considering the type, location, and quantity of wastes generated and the available disposal options. At the core of integrated solid waste management is the notion that not all wastes are created equally, nor should they be disposed of equally. Managers must weigh factors such as collection, storage, logistics, and disposal cost to determine the best alternative for a particular category of waste.

This integrated approach gained increasing favor in the early 1990s when the "recycle at any cost" philosophy was found to be economically unsustainable. Although managers might recognize the moral imperative to recycle, the cost or ability to do so may force a different decision to put waste in a landfill or incinerate. The hierarchy of steps for integrated municipal solid waste management is first to reduce or reuse to the greatest extent possible. The second step is to recycle what cannot be reused. The third option is incineration, preferably with energy recovery. Finally, landfilling is the least preferred alternative.[1]

This strategy can be implemented as a true hierarchy or as a menu approach where managers pick and choose what works best for their situation. Many would argue that it should be implemented as a hierarchy, but that may not be a realistic proposition, given the amount and type of resources devoted to municipal solid waste management at the institution. For example, incineration may not be a viable alternative because of the substantial capital investment required. Most incinerators require several hundred tons of waste per day to offset the extensive cost of construction and operation. This need to "feed the incinerator" may run counter to reduction and recycling goals.

Source reduction, also referred to as *waste reduction,* can be the most effective way to handle solid waste. The idea of not having to handle the waste in the first place has merit in terms of cost and operational considerations. Unfortunately, waste reduction is limited, while recycling has made great strides as the primary strategy in waste disposal on college campuses. That can be attributed to the difficulty instituting waste reduction practices. In *Making Less Garbage: a Planning Guide for Communities,* Bette K. Fishbein and Caroline Geld state, "Local solid waste planners, engineers, and managers know the steps involved in designing and operating landfills, incinerators, and collection systems. They know that recycling involves collection, separation, and marketing of materials. However, while acknowledging source reduction as the most important priority, most have only a vague idea of how to get businesses and citizens to produce less waste."[3]

This sums up why recycling programs, and not source reduction, have taken the strongest foothold on campus. Later in this chapter there is more in-depth coverage on source reduction. Because recycling receives more attention from both administrators and students, establishing and conducting a recycling program will be the focus of this chapter.

33.5 MUNICIPAL SOLID WASTE DISPOSAL

Refuse Disposal and Collection

It is impossible to talk about recycling without discussing refuse disposal. Because recycling will change the way the institution disposes of refuse, implementing a recycling program provides an opportunity to reevaluate the way that is done. Resources will have need to be shifted from refuse disposal to recycling. Whether the institution proactively changes its refuse disposal system or does it as a result of the changes that recycling brings, the manager must consider how recycling will impact refuse disposal.

There are many choices available for refuse disposal. One choice everyone is faced with is contracting the service or providing it with in-house forces. The following discussion is based on providing the service in house. If collection services are contracted, the vendor usually provides the storage devices and vehicles. However, it is still important for managers to carefully consider what type of system will best suit conditions on their campus. Later in this chapter the contracting option will also be discussed.

Two of the most important aspects of handling refuse are that it be disposed of properly and it must be disposed of in a timely and efficient manner. All states have strict regulations on how refuse must be collected and disposed of, and it is critical that the institution's system complies with these requirements. Garbage will always be generated, and it will be generated on a regular basis, with some allowances for seasonal variations and special events that take place on campus. Custodial staffs are usually responsible for removing refuse from buildings and can often be valuable in efforts to assess the volume and type of trash generated in a building. The design and operation of refuse disposal systems must ensure that the storage capacity and collection frequency are adequate. The unsightly and unsanitary conditions that can result from inadequate storage capacity and collection frequency must be avoided.

The basic components of solid waste disposal are 1) removing the refuse from the buildings, 2) storing it outside of the building, and 3) collecting and transporting it from these storage locations to its disposal destination. The choices for each of the steps in the refuse disposal process are similar to those for the recycling process, so much of what is covered here will apply to recycling as well.

Waste Removal   Removal from the building can be accomplished in a number of ways. Some buildings can be equipped with trash chutes and a trash room. Usually these chutes empty into large containers or wheeled carts. Trash chutes can be a tremendous time savings for your custodians, but they are becoming rare because of maintenance concerns and space constraints. Trash chutes can be effectively incorporated into a recycling program by modifying the chute system or alternating collection days for refuse and recyclables.

Removing waste from individual offices, food service areas, rest rooms, and common areas with trash carts is a more common practice. Waste is collected from the source, transferred to a larger barrel or cart, and then moved to another storage device for pickup.

Waste Storage   Storage devices can be carts, dumpsters, or roll-offs. The use of plastic liners or bags can help improve sanitation and prevent litter from blowing away when the waste is transferred to an outside storage device.

Carts. Carts range in size from 55 gallons to 120 gallons and above. Popular sizes are 60 and 90 gallons. Prices also vary according to size and model, but managers should expect to pay between $60 for smaller carts and $250 for larger carts. Carts are suitable for areas where space is inadequate for a larger storage device or where vehicular access is restricted.

Carts can be manually emptied or automatically dumped into the refuse collection truck. Most carts have wheels and can be moved to a location that is easily accessible by the collection vehicle. They are much less expensive than dumpsters or roll-offs. The disadvantages of carts are 1) they are subject to theft because of ease of mobility, 2) they are less durable than other collection devices, and 3) they have a much smaller capacity than a dumpster. Theft of carts can be overcome by chaining and locking the carts. Drivers of the collection vehicle or custodial employees will then have to unchain the carts before they can be dumped. Carts also must be cleaned thoroughly and frequently if they are placed indoors.

Carts can be used with rear- or side-loading trucks. If the carts are manually emptied, no modifications need to be made to the collection vehicle. Both side- and rear-loading trucks can be purchased that automatically tip the carts. These trucks range in price from $50,000 to $80,000. The increased capital investment can be justified by reducing the labor needed to transfer the waste from the cart to the truck.

Dumpsters. Dumpsters also range in size from 2 to 12 cubic yards. Most dumpsters are steel and have lids. Stationary compacting dumpsters can be used in areas where there is a high volume of refuse. These compacting dumpsters operate much the same as compacting roll-offs, which are discussed in detail later, except they are serviced by front-loading vehicles. Compacting dumpsters can increase storage capacity by two or three times. They require an electrical power supply and are more costly to purchase, operate, and maintain than noncompacting units.

Usually dumpsters are placed outside of loading docks or receiving areas. Often they must be placed away from buildings so that the collection vehicle can reach them. Lids are important for aesthetics and sanitation and to contain costs. Haulers prefer that dumpsters have lids to keep the refuse protected from precipitation. Haulers are customarily charged based on weight, and wet garbage weighs more and costs more to dispose of. Dumpsters are even available with locking lids, which can be used if illegal dumping is a problem, although any lid can make it more difficult to transfer waste into the dumpster. A lid can be a maintenance problem, but it is not only a good idea to use a lid, lids are required in some areas. Plastic lids have become the industry standard, and these lids are easier to use than their heavy steel predecessors.

Dumpsters range in price from $450 to $900. They must be well painted because they are highly likely to rust. Maintenance, however, is minimal. A dumpster will have to be repainted every two to six years depending on the type of dumpster and the local weather conditions. Every 15 to 20 years the floors in the dumpster will have to be repaired if repair is preferred over buying new dumpsters. Lids and hinges will also have to be replaced as they become damaged or worn. Some dumpsters come with wheels, which will also be a regular maintenance concern.

Dumpster collection trucks can be rear loading or front loading. Both versions operate much the same, but dumpsters and collection vehicles must be correctly matched. Front-loading dumpsters are easier to service because the driver has an unobstructed view to align the truck and dumpster. However, there is a greater risk to tree limbs and building overhangs when tipping a front-loading dumpster than when tipping a rear-loading dumpster, because front-loading dumpsters must be lifted in the air above the truck.

Rear-loading dumpsters are more difficult to align because the truck must back up. This must be considered with regard to efficiency and safety. Backing up great distances to approach a rear-loading dumpster can be time consuming and dangerous, given the high volume of pedestrian traffic on campus. Cameras can be installed at the rear of the vehicle, with monitors in the cab, to improve driver visibility and safety. Driver training and safety must be primary concerns for the operation of any in-house collection vehicle. All drivers must have a commercial driver's license to operate these vehicles. All activities related to refuse collection must comply with applicable regulations, such as the Occupational Safety and Health Administration's requirements regarding blood-borne pathogens.

Front-loading trucks range in price from $85,000 to more than $140,000; rear-loading trucks cost 30 percent to 50 percent less. Front-loading trucks also have 10 to 20 percent greater capacity than rear-loading trucks, ranging in size from 10 to 13 tons or 35 cubic yards. This can be important when the disposal point is a great distance from campus. Another advantage to front-loading trucks is that the driver rarely needs to leave the cab, which results in greater efficiency and less potential for on-the-job injury resulting from getting in and out of the truck, as is often the case with rear-loading vehicles.

Maintenance for a refuse collection vehicle is expensive, but proper maintenance is essential to ensure limited downtime. A good preventive maintenance program will reduce the need for unplanned maintenance, resulting in less downtime and better service. Tires are probably the single greatest maintenance expense incurred, and choosing tires wisely can save thousands of dollars over the life of the vehicle. Retreaded tires can prove to be an economically and environmentally beneficial choice.

Roll-offs. Roll-offs are another way to store and transport refuse. A roll-off is nothing more than an oversized dumpster that, rather than being emptied into a truck, is rolled on and off the truck. Roll-offs range in size from 15 to 30 cubic yards. These storage devices are appropriate for areas where a large volume of trash is generated on a daily basis but require a large, easily accessible area for placement. Roll-off containers cost several thousand dollars, depending on their size. The maintenance required is similar to that required for a dumpster.

Roll-offs are serviced by a lift-hook truck, which has a hoist and a rail chassis. Prices for lift-hook trucks begin at $45,000. A lift-hook truck backs up to the roll-off container and pulls it on or off the chassis of the truck. The problem encountered with a rear-loading dumpster—backing up to the container—also applies to a roll-off. Another problem with roll-offs is the loss of storage that results while the truck takes the container to the disposal site. This can be overcome by having the truck bring an empty container when picking up the full one. However, this necessitates even more space because of the maneuvering required to exchange the empty container for the full one.

Roll-offs also come in a compacting version. This requires an electrical hookup, and installation is usually integrated with the building itself. Compacting roll-offs extend the time needed between pickups, but capital and maintenance costs for compacting roll-offs are much greater than for noncompacting roll-offs. The same lift-hook truck can service both types of roll-off containers.

Disposal Destinations There are three primary disposal destinations for waste transported from campus. None of these should affect how waste is collected, but it is important to know what options are available. Most will be regulated by the local municipality or other local authority. The three facilities for disposal are a transfer station, a landfill, or an incinerator. Each type of facility may have regulations concerning what can and cannot be taken there. Incinerators especially may ban highly combustible substances. In suburban or rural areas, institutions may not have a choice where waste is taken, but larger cities may provide several options for disposal. Operationally, any of these collection and disposal systems can work. It is up to the manager responsible for solid waste to determine what system or combination of systems is best suited for his or her campus environment.

33.6 ESTABLISHING A RECYCLING PROGRAM

Local or State Requirements

The first step in developing a recycling program is to determine the requirements of state and local authorities. Most states, counties, and local municipalities have some laws or regulations concerning recycling. Often states pass legislation that establishes broad guidelines regarding the types of material to be collected and the communities to be served, based on population. Also, goals for waste reduction and recycling and time frames to accomplish those goals are set at the state level. Many of the details of how to do these things are delegated to counties or local municipalities.

Although some areas may place restrictions on where refuse may be taken, recyclables are usually not similarly restricted. Managers should be free to decide where to take recyclable material based on convenience, economic return, or services offered. Since the late 1980s, when recycling became popular again, many municipalities, cities, and counties have hired recycling coordinators to manage the recycling efforts of local government. These coordinators are the first people to contact for information about what needs to be done and how it can be accomplished. The EPA and the environmental compliance department in many states have hotlines that provide information about recycling. The EPA hotline number is 800-424-9346.

Administrative Support

Administrative support is vital to the successful implementation and operation of a recycling program. Lack of support results in an ineffectual program. A recycling program does not need to be mandatory to be successful, but it is much easier to implement if it is mandatory and the president or chancellor of the institution goes on record supporting the recycling program. Because recycling activities cross all lines within a university, support is needed from both the business and academic arenas. Support provided by the chief business officer but not by the provost is a recipe for a failed program. Faculty must understand that their leadership supports the recycling effort; otherwise their participation will not be sufficient.

Recycling programs will likely require additional funding beyond what is already budgeted for solid waste disposal, if not for the long term, then certainly in the short term as the program is begun. Upper administrative support in the form of funding to develop and operate the program is essential. The recycling ethic seems to always be to do as much as possible with as little funding as necessary. This attitude, although admirable, is limiting. Adequate funding can go a long way in designing an effective program that will have maximum impact on the environment. Budgetary support from upper administration can make or break a recycling program.

Policy Statement

One measure of administrative support is the willingness of institutional leaders to adopt a recycling policy. This policy is needed not to force people to recycle, but as a symbolic affirmation of support. The policy statement should cover the goals of the program and why recycling is important to the institution. It should also cover waste reduction and purchase of recycled products, both of which will be covered later.

Recycling Coordinator

Ideally, a recycling coordinator should be appointed to oversee the recycling program. This position should be dedicated to this task only, if possible. This is especially important in the start-up phases of any new recycling program.

The responsibilities of the recycling coordinator should include primary development of the program, supervision of any employees dedicated to the recycling efforts, training and education, equipment specification and selection, market development, quality control, governmental relations, and liaison with students and administration. Currently, interest in recycling is high among students, so it is important for the recycling coordinator to have a good rapport with students.

There is some logic to establishing the recycling coordinator's office in the area of the department responsible for custodial or refuse services. These areas of facilities operations will be most involved in performing the recycling activities. By establishing the office nearby, communication with these areas is enhanced, increasing the involvement and cooperation of the employees who will be most affected. This organizational closeness to the workers who are doing the recycling is helpful to the coordinator in determining what is or is not working.

As recycling becomes more commonplace and established throughout North America, the value of having a recycling coordinator will become more apparent. Hiring a full-time recycling coordinator demonstrates a strong commitment and allows the program to mature as the recycling infrastructure develops and changes in the industry take place. Just as an engineer would be hired for a designing position, appointment of a professional to fill the recycling coordinator position should be considered as the industry develops and regulations become more complex.

Recycling Committee

The recycling program on campus has the potential to be highly visible. Students especially may show a great interest in being involved in the formulation of recycling policy and strategy. As mentioned earlier, recycling crosses all areas of the university. For these reasons, it is helpful if a committee is formed to involve the interested parties. Representatives should be recruited from the general student population, student government, campus environmental coalitions, facilities management, housing and food services, residence halls, and faculty, to name a few.

A committee structure can provide feedback on the mood of the campus. The recycling coordinator can gain valuable information on whether it is realistic to ask the university community to perform certain functions in an effort to implement recycling. For example, suppose the design of the program calls for faculty to separate recyclables in their offices and take them to a centrally located collection point in the hallway; a committee may provide feedback as to whether this added responsibility will be accepted or rejected by faculty members.

The recycling committee can be a standing body that continues to meet once the recycling program is implemented, or an ad hoc task force that disbands once the program is established. The university community is transient, so a standing committee can assist with the constant reeducation that is necessary with an ever-changing student body. However, committee continuity will suffer because of this transience, and the committee process can be made even more frustrating by the continual loss of momentum when committee members leave and are replaced.

The Waste Stream and Disposal Process

A key ingredient to successful handling of municipal solid waste is conducting a waste stream analysis, also called a *waste stream characterization* or *assessment*. This process involves getting to know how much is being disposed, what is being disposed, what percentage of the waste stream is represented by different types of waste, and how this entire process occurs. The waste stream analysis is the cornerstone for all decisions regarding disposal options.

Waste Stream Analysis

The waste stream analysis is a systematic study of the disposal process. It looks at the life cycle of waste from the time it is created on campus until its eventual disposition. "Assessment of the campus waste stream and of individual buildings is important for properly implementing recycling."[4] It is important to know what types of wastes are in the waste stream, where they are generated, and how much there is; this will enable the manager to design a recycling program that targets the materials that are most prevalent. It also indicates where the department needs to concentrate its efforts and how it should go about waste collection. Ultimately, it will also provide the basis for the manager's economic cost-benefit analysis. This will be discussed in detail later in the chapter.

The starting point for the waste stream analysis is a survey of campus buildings. Buildings should be classified so that managers can use the data to tailor their collection efforts. Common classifications are office/administrative buildings, classroom buildings, laboratory/research buildings, residence halls, dining halls, special-use buildings, and other buildings. Special-use buildings may be gymnasiums, union buildings, or auditoriums; the "other buildings" category can be used for mixed-use buildings. Buildings can be categorized in any way that suits the institution's particular situation. The classification allows managers to make specific decisions based on trends that become apparent within building classes.

For example, the manager may choose not to collect a certain material, such as plastic, in laboratory buildings, because the results of the waste stream analysis show that the amount of plastic disposed of in those buildings does not justify the effort. On the other hand, the manager may find that a considerable quantity of plastic is being disposed of in the waste of the dining halls, and therefore, plastics recycling might be initiated in those buildings.

Once buildings are classified, the manager can then examine several buildings within each class to gain insight into how and what waste is generated and disposed of in the building. He or she can make preliminary assumptions on whether the classifications are valid based on the findings. *University and College Solid Waste Reduction and Recycling,*[4] a manual produced by the University of Illinois Center for Solid Waste Management and Research, offers methodology for what the manual refers to as *waste stream assessments* and also waste compositions for campus settings.

The main portion of the waste stream analysis is collecting refuse from a number of buildings in each class. The collected waste is sorted, separated further into categories, and weighed. The categories of waste can be revised to best suit the institution's needs. An example of the results of a waste categorization is shown in Figure 33-1.

This process of categorization will tell the manager how much of each type of waste is generated on campus and where it is coming from. As described in the example of plastics above, it will guide decisions on what materials to collect and where the materials will be found. It will also indicate what will be needed to collect these materials.

Waste should be collected over a period of time to accurately reflect the disposal cycle and should be collected from each building class on different days of the week. For example, waste generated from residence halls over the weekend is likely to be quite different from waste collected during a weekday. There will likely be a greater concentration of glass and cans in the weekend waste stream than in waste collected during the week.

Once the manager knows what types of materials are in the waste stream and where they are generated, he or she can use this information to decide how to recycle. Preliminary assumptions can be made regarding the amount of labor that will be involved in the collection process; the types, numbers, and placement of collection receptacles inside of buildings; and storage and collection equipment that might be used outside the buildings. Ultimately these decisions will be determined by the arrangements made with the haulers, recyclers, and processors who will take the recyclable goods; however, the waste stream analysis is the starting point for all the decisions.

Reviewing Custodial Operations

The custodial workforce will most likely play a critical role in the recycling process. Managers should give considerable attention to the way custodial services are currently provided and what will have to be done to begin recycling. Work tasks, frequency of service, and staffing levels will all be subject to change.

Custodial employees will be on the front lines of the recycling battle. They will be the group most affected by the changes created by recycling,

Figure 33-1
Waste Stream Analysis

Pounds	Residence Halls	Dining	Office	Classroom	Research	Total
OCC	71.00	880.50	411.00	147.50	453.50	1,963.50
Mixed	442.50	211.00	3,034.50	629.00	1,747.50	6,064.50
News	147.50	57.00	175.50	97.50	183.00	660.50
Magazines	110.50	215.00	69.50	6.00	340.50	741.50
Subtotal	**771.50**	**1,363.50**	**3,690.50**	**880.00**	**2,724.50**	**9,430.00**
Aluminum	210.50	46.00	31.00	76.00	35.50	399.00
Bimetal	18.50	98.00	4.00	23.50	130.50	274.50
Subtotal	**229.00**	**144.00**	**35.00**	**99.50**	**166.00**	**673.50**
Glass- clear	175.50	139.00	50.50	41.50	38.50	445.00
Glass- brown	39.00	0.00	0.00	5.00	60.00	104.00
Glass- green	47.00	3.00	11.00	5.00	1.00	67.00
Subtotal	**261.50**	**142.00**	**61.50**	**51.50**	**99.50**	**616.00**
PETE- clear	40.00	14.50	18.50	11.00	28.50	112.50
PETE- green	6.50	3.00	2.00	1.50	0.00	13.00
HDPE	45.50	75.00	2.50	2.50	11.50	137.00
SubTotal	**92.00**	**92.50**	**23.00**	**15.00**	**40.00**	**262.50**
Recycling Totals	**1,354.00**	**1,742.00**	**3,810.00**	**1,046.00**	**3,030.00**	**10,982.00**
Refuse	1,061.50	1,613.50	1,238.50	1,355.50	1,771.50	7,040.50
Total	**2,415.50**	**3,355.50**	**5,048.50**	**2,401.50**	**4,801.50**	**18,022.50**

Percent	Residence Halls	Dining	Office	Classroom	Research	Total
OCC	0.39%	4.89%	2.28%	0.82%	2.52%	10.89%
Mixed	2.46%	1.17%	16.84%	3.49%	9.70%	33.65%
News	0.82%	0.32%	0.97%	0.54%	1.02%	3.66%
Magazines	0.61%	1.19%	0.39%	0.03%	1.89%	4.11%

Subtotal for Building Type	52.32%	56.74%	36.64%	73.10%	40.63%	31.94%
Subtotal for Waste Stream		15.12%	4.88%	20.48%	7.57%	4.28%
Aluminum	2.21%	0.20%	0.42%	0.17%	0.26%	1.17%
Bimetal	1.52%	0.72%	0.13%	0.02%	0.54%	0.10%
Subtotal for Building Type	3.74%	3.46%	4.14%	0.69%	4.29%	9.48%
Subtotal for Waste Stream		0.92%	0.55%	0.19%	0.80%	1.27%
Glass- clear	2.47%	0.21%	0.23%	0.28%	0.77%	0.97%
Glass- brown	0.58%	0.33%	0.03%	0.00%	0.00%	0.22%
Glass- green	0.37%	0.01%	0.03%	0.06%	0.02%	0.26%
Subtotal for Building Type	3.42%	2.07%	2.14%	1.22%	4.23%	10.83%
Subtotal for Waste Stream		0.55%	0.29%	0.34%	0.79%	1.45%
PETE- clear	0.62%	0.16%	0.06%	0.10%	0.08%	0.22%
PETE- green	0.07%	0.00%	0.01%	0.01%	0.02%	0.04%
HDPE	0.76%	0.06%	0.01%	0.01%	0.42%	0.25%
Subtotal for Building Type	1.46%	0.83%	0.62%	0.46%	2.76%	3.81%
Subtotal for Waste Stream		0.22%	0.08%	0.13%	0.51%	0.51%
Recycling Totals	60.93%	16.81%	5.80%	21.14%	9.67%	7.51%
Refuse	39.07%	9.83%	7.52%	6.87%	8.95%	5.89%
Total	100.00%	26.64%	13.33%	28.01%	18.62%	13.40%

and they will also be the ambassadors of the recycling program. It is important that their job duties are considered and their cooperation is cultivated.

The addition of recycling will have great potential to add to or change the responsibilities of custodial employees. Even if managers are able to implement a recycling program without it resulting in a net addition of duties, what custodial employees do will change; the nature of recycling dictates that. For example, student workers are often employed to handle recycling, as opposed to having custodial employees do those tasks. However, custodians will still be affected, even though others will be doing the recycling. Change can be intimidating to custodians, because the methods and basic activities of custodial services have remained relatively unchanged in comparison with other areas of facilities management that have been greatly affected by technology.

Work tasks for custodial employees will change when recycling begins. If a central collection system for recyclables is used, then emptying the recycling receptacles will be a task added to the custodians' responsibilities. If managers are asking faculty, staff, and students to do only minimal separation, then custodians may have to do some sorting. These are just two examples of how the tasks of the custodial workforce may change. These changes need to be considered when the recycling program is designed.

The frequency with which custodial services are provided will also have to be reevaluated. A common task for custodians is the emptying of trash cans in offices. A comprehensive recycling program will result in fewer things being thrown in that trash can. For example, suppose custodians previously emptied the office trash cans twice each week; with the advent of recycling, they may need to do this only weekly. Conversely, emptying of recycling receptacles will be an added responsibility in a new program or an increased responsibility in an existing program that incorporates additional recyclables to the program. The frequency of cleaning services will be affected by what is collected and how it is collected.

Finally, staffing levels will also have to be reexamined. It is easy to say that staffing levels should not change, because recycling is not adding to the amount of garbage produced. However, as stated previously, recycling does add to the tasks that custodians will have to perform. Whereas before recycling the custodian may have had only had one trash can to empty, after institution of a recycling program, he or she may have several to empty, depending on the way the recycling program is set up. It is possible that no employees will have to be added. Differently designed recycling programs may actually reduce workload in certain areas (e.g., the twice-weekly trash can emptying example cited above). However, there is an equal chance that employees *will* have to be added, especially if collection of the recyclables outside the buildings will be handled by an in-house workforce.

Other Recycling Efforts

The next step in designing a recycling program is to become familiar with other recycling efforts and opportunities that are happening locally, regionally, and nationally. Like politics, all recycling is local, so what is happening closest to the campus will be the first place to look. When designing a recycling program, what is happening across town is more important than what is happening across the country.

Much of what can be done will be dictated by the recycling activities occurring locally. The greater the distance recyclables must be transported away from campus, the more expensive it is likely to be. However, managers should not limit themselves to local recyclers or markets. If the infrastructure for recycling is not in place or complete in the local area, then managers should expand the search for opportunities. Even if there is an active recycler locally, managers should check to see if better markets are available at a greater distance from the campus. Many states have recycling organizations that can be contacted to learn how recycling works in the local area or state.

Also, managers should network with other universities and colleges. The Internet is a good source for connecting with fellow recycling coordinators at campuses across the country. Currently there are several active recycling discussion lists managers may choose to subscribe to in order to learn about how recycling issues are being addressed on campuses across the country. Managers can check with computer support personnel to learn how to access these lists. Another good source is the National Recycling Congress. This national group has an active caucus of university and college recycling coordinators that may be of assistance. The Experience Exchange coordinated by APPA: The Association of Higher Education Facilities Officers should also be able to provide assistance matching one institution with another institution of similar size.

Once managers have determined what their options are, they can then begin to develop a recycling program. The details of setting up a recycling program on campus are in large part dependent on how the selected recycler requires the recyclable materials to be sorted and delivered. For example, if an available market allows commingling of high-density polyethylene (HDPE) and polyethylene terephthalate (PETE) plastic, then the program will be established differently from one where the recycler requires that these two types of plastic be kept separate. By taking this approach, managers can design their program "from finish to start."

Designing a Program "From Finish to Start"

Designing a program from finish to start means that what is collected and how it is collected is based on the end market for the recyclable. Whether the institution operates some sort of processing facility or contracts with a recycler, all recycling processors will have requirements on how the recyclable material must be sorted, prepared, and delivered. That it is why it is helpful to think through the process backward, from delivering the materials to the recycling facility to collecting them in the buildings.

The first component in this reverse thought process is the transportation of the recyclable material. The method in which the material is transported will be based primarily on the distance to be traveled, the volume of material generated, the type of processing facility, and how many different materials are being taken to the same facility. Transportation of recyclables goes hand in hand with storage.

As mentioned earlier, many of the methods described for storing and transporting refuse can also be applied to recyclables. Carts, dumpsters, and roll-offs can all be used for recyclables in much the same manner in which they are used for refuse. The collection vehicles that service these, when used for recyclable material, are also identical or only slightly modified. There are other pieces of equipment that are intended primarily for collection of recyclables.

Roll-offs are a good example of a refuse collection system that can be modified to collect recyclables. Roll-offs can be compartmentalized to accept sorted recyclables. Often this approach works well at multifamily housing units. Side-loading vehicles that are also compartmentalized are a good example of a vehicle that is designed primarily for recyclables. The compartmentalized roll-off can be used with the same lift-hook truck that serves other roll-offs, and the compartmentalized side-loading truck can be used with the wheeled carts mentioned earlier in the section titled "Refuse Disposal and Collection."

Compacting roll-offs were discussed earlier. Compaction also comes into play with recyclables, as some recyclables, such as cardboard, #10 steel cans, and plastic bottles, are voluminous. Compaction capabilities in the collection system for these recyclables may offer cost savings and savings in operational efficiency that will have a short payback. Recyclable bottles and cans often still contain liquids, and the collection system must be designed with this in mind.

Once the manager has determined what options are available in terms of recycling end markets and what suits the institution's needs for collection and storage outside of the buildings, he or she can then begin to design the separation and collection systems inside the buildings. There are a myriad of considerations once this stage is reached; a brief discussion of some of the more important factors to be examined follows.

Separation is the basis for recycling. Recyclables have to be separated somewhere and somehow. If the end market can take commingled materials, then the collection system inside the building can be simplified. Commingling, simply defined, is the practice of mixing different types of recyclables together to be separated later through further processing. Not only can collection be simplified, but more materials can be collected in a commingled program. If commingling is not an option, then each recyclable targeted will have to be collected separately. Usually this is accomplished by having the person who generates the waste separate it, a process known as *source separation.*

Source-separated recycling programs are popular, because automated processing systems that can sort commingled recyclables are not widely available or financially feasible. The cost to process commingled recyclables is usually high. However, the tradeoff for the lower processing costs that come with source separation is higher collection costs. Because of the problems and cost associated with commingled processing, the recycling coordinator is faced with the dilemma of making the collection system less convenient or more complex to meet the criteria of the processor that is taking the recyclables. Source-separated programs require greater commitment from participants. They also are more labor intensive than a commingled program, because custodians will have more receptacles to empty. One problem with any recycling program is contamination, or nonrecyclables being mixed in with recyclable materials. Source-separated programs usually have less contamination than commingled programs, because source separation requires more proactive participation. Clear labeling of all collection receptacles is important to help reduce contamination.

Some end markets can take commingled recyclables without further processing. Rough paper manufacturers (e.g., tissue paper, paper plates, etc.) are able to take various mixes of paper. Some of the newer papermaking technology will allow commingled collection of certain types of paper, such as newspaper and magazines. This option may afford the manager the opportunity to simplify the collection system. There has been some criticism of this practice, because mixing higher grades of paper with lower grades devalues the higher-grade paper.

Another consideration is whether custodians will collect material in offices and residences, or building occupants will be expected to take the recyclables to central collection points in the building. This decision is a perfect example of the cost of convenience. It is a known fact that the more convenient and simple a program is, the greater the participation and chances of success will be. The recycling coordinator should strive to achieve convenience and simplicity. However, these have a cost that must be taken into account when designing a program.

Another guiding principle is consistency. The recycling program should be as consistent as possible. There are valid reasons to vary what materials are collected, but how these items are collected should remain relatively constant. If aluminum cans are collected in the main administrative building in a green 20-gallon plastic barrel, then such cans should be collected in a residence hall in the same way. If white ledger paper is taken to a brown dumpster and garbage to a black dumpster at the performing arts center, the procedure should remain the same at the engineering building. If aluminum and steel cans are commingled at the dining hall, the same should be true at the residence hall. This consistency will promote habits, and habits lead to more productive recycling behavior. As seen in the preceding example, color coding of collection receptacles is one way to gain consistency.

As recycling programs strive to achieve greater participation and percentages of waste recycled, many institutions have opted to construct and operate their own material recovery facility. A material recovery facility can provide much greater flexibility and control over the recycling process, but it also brings a high level of risk. Owning a material recovery facility implies the need for marketing materials and taking on additional costs to construct and operate the facility. This decision requires considerable planning and a long-term commitment before it is undertaken.

Pilot Programs

It is always a good idea to test decisions under real-life conditions. Pilot programs allow such testing. There are several things that should be tested with a pilot program. First, time studies should be conducted on how the workload of the people involved in recycling will change. If additional time is needed to recycle or if time savings are possible, a pilot program will provide an idea of what those demands on time will be.

A good pilot program will also test the equipment to be used. Managers should try to create as accurately as possible the conditions they anticipate will be in place. This is a tall order, but striving for this will help managers make the best possible decision. During the process of conducting a pilot program, managers should also be able to test the results of the waste stream analysis. The pilot program will not replicate the results of a full-scale program, but it can provide an idea of what is needed to recycle the amount of material the waste stream analysis has indicated as being available for recycling.

Economic Analysis

An economic analysis can be made using the results of the waste stream analysis and the pilot program. The waste stream analysis will provide the quantity figures needed to determine the value of the recyclables chosen for collection. Those same numbers can be used to assist in sizing the collection receptacles and equipment. This will allow a determination of how many of each type of receptacle, dumpster, truck, etc., must be purchased and how much this will cost. Judgments on staffing can also be made using the time studies conducted in the pilot program and applying them on a broader scale to the entire program based on the amount of recyclables in the waste stream.

Many administrators are looking for recycling to reduce overall costs. In some cases the combination of revenue from the sale of recyclables and avoided disposal fees from not landfilling or incinerating the waste can provide an overall savings. Funding for recycling programs is often granted based on projected future savings. Managers should not be tempted to overlook costs or inappropriately assign them to another cost center, because this will come back to haunt them in the end. Managers should be thorough in their economic analysis and refrain from making promises that cannot be delivered. There is nothing worse than having to reduce the scope of a recycling program after several years because of a lack of financial support.

Recycling will always add layers to an existing waste disposal system, and managers should be careful not to overlook the costs associated with recycling, including opportunity costs. If managers do not add employees but shift them from other areas of the organization, this cost must be assigned to the recycling program. If a custodial task is replaced with recycling responsibilities, thereby sacrificing performance level, this opportunity cost should also be considered.

33.7 PROGRAM IMPLEMENTATION

Implementing a campus-wide recycling program is a major undertaking because of the amount of communication that must take place. Strategies for implementation can vary depending on the scope of the program and the size of the institution. Timing for the startup of recycling can coincide with the beginning of the school year or semester, or it can be tied to another important date, possibly Earth Day. The start of the school year is an especially advantageous time to start a program, because many people on campus are starting a new year, and it is a good time to teach them new habits.

For large campuses, it may be necessary to phase in a recycling program. This can be done according to geographic regions of campus, by college or department, by custodial districts, or by other means that suit the particular situation. Managers may also want to phase in the recyclables that are collected, starting with a few and building as the program grows. Phasing in a program makes the task more manageable. It allows for greater attention to detail and for making adjustments as feedback is received on how the program is working. A disadvantage to phasing in a program is the need to sustain momentum and excitement over an extended period of time. For a smaller institution, phasing in a program implementation may not be necessary.

A large-scale educational effort is necessary for successful implementation. The variety of audiences that must be addressed dictates a diversified approach. Strategies for educating the university community in general will be discussed in the text that follows. An important group that must be educated is the custodial staff. There are a couple of reasons why this is important. Obviously, the custodians must know how to handle the recyclables and how their jobs will be affected. Second, and maybe even more important, is that the custodians will promote the recycling program to the university community in general.

Getting custodians to buy into the program and accept ownership should be a primary goal of the training effort. Because the custodians are in daily contact with the university community, they are in an excellent position to also educate the community. If custodians are energized about their role in the recycling program, then they are likely to convey that enthusiasm to the people with whom they come in contact.

33.8 MEASURING THE PROGRAM'S SUCCESS

It is important for managers to know how the program is doing so that they can fine-tune the program and provide feedback to the university community. There are various ways to measure success. The most common is to measure the quantity of waste recycled as a percentage of total waste. This is commonly referred to as *diversion*. Another measure is to survey buildings (e.g., offices, laboratories, and residence hall rooms) to determine the percentage of people who are participating. Managers may also want to conduct a waste stream analysis to determine how much and what type of potentially recyclable material is not being recycled. Many recycling processors charge customers for excess contamination, so managers will want to measure this aspect of your program so that corrections can be made if necessary.

33.9 EDUCATING THE UNIVERSITY COMMUNITY

The importance of education cannot be overstated, especially at start-up. A recycling program that is perfectly designed will not be successful if the university community is not educated to participate. At the outset of the recycling program, a massive educational effort must be conducted. In addition, education must be continually undertaken because of the transient nature of the college campus. There are many ways to communicate the message of recycling to the varied constituents at the institution. It would be impossible to discuss here all the possible methods for educating the community, but the following suggestions are offered to provoke thought about how educational efforts can be structured.

Managers should try to institutionalize the recycling educational efforts. The more education that happens automatically each year as part of the normal operations of the institution, the better educated people will become. Examples of ways to institutionalize educational efforts are 1) having an article appear in a magazine that is distributed to residence hall students at the beginning of each academic year, 2) putting posters on buses, 3) setting up table tents in the dining commons, 4) issuing reusable mugs to new students and employees, and 5) promoting recycling at welcome activities and in housing information packets. Recycling information can also be included in orientation sessions held for all new employees.

Managers should use the various media at the campus. Student newspapers are typically eager to run stories about recycling and to provide a forum to educate people about recycling. Managers should use the editorial section of newspapers to focus on aspects of the recycling program. A letter-writing campaign can be staged to focus attention on different recycling issues.

Many departments have their own newsletters. Managers can network with the people responsible for putting the newsletters together and gain some publicity by contributing stories to these departmental publications. Managers may also consider advertising in the newspaper if the budget will allow it. If the institution has a faculty/staff newspaper, that can be used as well. All of these publications reach a select market segment and can be effective in reaching important audiences. Managers should give as many interviews as they can tolerate. Radio can be used to continually remind people about recycling, and managers should try to get the campus and local radio stations to provide public service announcements promoting the institution's recycling efforts.

Many facilities offices already have established contact networks, and these contacts can be a valuable communication pipeline. Not only does the network already exist, but in many cases formal methods of communication, such as service bulletins or regularly scheduled meetings, are already in place. Managers should use this contact network whenever they need to convey changes or new initiatives in their program. If there is no established network, managers can develop one by identifying people who are interested in recycling by requesting volunteers through the various means mentioned earlier.

Presentations are also helpful, especially at start-up. Again, this is a perfect opportunity to use the contact network. These presentations can be used to educate those who come and prepare them to educate others in their offices. Presentations should also be made to interested student groups. Managers can work with the institution's residential life office to set up meetings with residence assistants or other persons responsible for student conduct in the dormitories. Again, those attending the meetings should be trained so that they can spread the word to others.

Recycling is a favorite topic for term papers and student projects. It is helpful to have an information packet prepared that can answer questions from students who are writing papers. Special materials can be prepared to distribute to building occupants. Recycling guidelines or information packets that help students live a "greener" life can be distributed on a regular basis.

Electronic education is a growing area that has unlimited potential. Managers can use electronic bulletin boards, electronic mail (e-mail), and voice mail to get the message across. Not only are these methods effective, they are environmentally friendly as well.

Managers may also want to initiate contests that reward students or offices for recycling. Labeling of receptacles was mentioned earlier in the discussion on contamination; it is an important part of the educational effort. Labels should be placed on everything, from a desk-side paper bin to dumpsters or roll-offs outside of buildings. Not only do labels reduce contamination, they are also a constant reminder to recycle.

Managers should be creative in developing their educational programs, looking for new ideas to get the message across and refusing to pass up an opportunity if at all possible. The more institutionalized the educational program becomes, the less taxing and time consuming it will be to promote the program.

33.10   CONTRACTED SERVICES

The institution may not have the capability or the desire to operate an in-house refuse and recycling collection service. If this is the case, there are numerous opportunities to contract this service. Many national and local companies provide a full complement of refuse and recycling services.

Request for Proposal

The starting point for contracting this service is putting together a request for proposal. The request for proposal can be written just as it would be to contract any other aspect of facilities management. Important areas to cover are the frequency of service, type of equipment to be used, redundancy in collection equipment, safety history of the company, bonding capacity, access to material recovery facilities and other disposal facilities, recyclable market contracts, and educational programming available.

Evaluation Process

The evaluation process should consist of several stages. The first step should be a review by purchasing officials or the department responsible for contracts. This step will serve to eliminate those proposals that do not meet the technical and legal requirements of the specifications. Vendors that do not have the bonding capacity, insurance coverage levels, or a client list demonstrating a proven capability to provide a level of service similar to that required can be eliminated from further consideration.

The next step is to conduct an analysis of the service proposed. An evaluation sheet should be developed to objectively score the various proposals. One way to do this is to rank different service aspects according to what is important to the manager and the institution. For example, if frequency of collection is more important than type of collection trucks used, this should be reflected in the number of points assigned to these service categories. The goal of the evaluation sheet is to provide an objective analysis by assigning a numerical ranking to each vendor. Once the list of vendors has been narrowed down to a select few, it may be helpful to have each vendor make a formal presentation. Although written proposals are essential, presentations can be helpful to discern the extra level of service provided that might set one vendor apart from the others. These presentations can also be helpful in clarifying parts of the proposal or for answering additional questions that come up in the course of evaluating the written proposals.

Cost–Benefit Analysis

Cost of service is always going to be an important factor. However, the low bidder should not necessarily be accepted if the services proposed do not meet the stated criteria. Cost can be included as one of the criteria and assigned a point value like other criteria, and the vendor who scores the highest number of points can be chosen. Alternatively, cost can be pulled out of the rankings and applied separately after all other criteria have been evaluated.

Vendors' proposals will also have to be compared with known costs and services if these services are currently being provided in house. If in-house service is contracted out and abolished, this is serious, long-term commitment, because it will be difficult and expensive to reestablish in-house capabilities.

33.11 OTHER RECYCLING OPPORTUNITIES

Composting

Composting, the natural process of breaking down organic material, can be used to recycle organic material such as leaves, grass, brush, and tree trimmings; food waste from food service operations; and even low-grade papers such as paper plates, towels, and other nonrecyclable papers. Composting does not have to be a highly technical operation. To determine what must be done, managers should check with the state in which the institution is located regarding composting regulations.

Composting requires a fairly level area with access to water. Compostable material must be placed in windrows and turned on a regular basis to get oxygen to the microorganisms that cause decomposition; these microorganisms need aerobic conditions to survive. This can be accomplished simply by using a loader, but specialized equipment known as *windrow turners* is available to do this job. Windrow turners can be stand-alone pieces or attachments to a loader. For composting of food waste or municipal solid waste that is not highly sorted, in-vessel composting systems may be necessary to ensure proper composting or to control odor.

Composting is becoming more popular as states pass regulations that ban yard waste from being put in landfills. Composting can significantly add to the amount of material that is recycled. The finished compost can be returned to campus as a soil amendment in planting beds, thus completing the loop.

Construction and Demolition Debris

Renovation of campus buildings will produce construction and demolition waste that can be recycled. Lumber, metal studs, roofing material, glass, furniture, scientific and computer equipment, suspended ceiling grids, fluorescent light diffusers and reflectors, piping, duct, and much more can be recycled. Markets for this material might be difficult to find but are available. Managers should make recycling of construction and demolition debris part of the contract if they can.

Garage Services

Garages generate numerous types of waste that can be recycled. Markets for most of these items are readily available. Tires, lead acid automotive batteries, glycol and antifreeze, auto body parts, oil, and solvents and part washing fluid can all be recycled. Because there are many recyclables concentrated in this one area, it is important to work with the service garage facility to institute recycling programs in this operation.

Other Recyclables

Recycling programs on college campuses include many of the same materials. Common items that are collected include aluminum beverage cans; steel cans; clear, green and brown glass bottles; HDPE and PETE plastic bottles; newspapers, computer printout, and ledger paper; and corrugated cardboard boxes. The acronyms listed are universally used in the recycling industry.

There are other recyclables that are less commonly collected. These include mixed office paper, magazines and catalogs, and books; aerosol spray cans; paint; other plastics such as polyvinyl chloride and polystyrene; furniture; carpet; white goods (i.e., appliances such as refrigerators or stoves); clothing; and fluorescent light bulbs and fluorescent light ballasts. These lesser known materials should be included in the program if possible. Some, such as magazines and office paper, are rather abundant; others are not found in great quantities. Managers can consider adding less common materials based on the incremental cost to collect them. If these can be commingled with other materials already being collected, the cost of adding them to the program will be less than if they have to be collected separately.

Special Events

Special events must have recycling services provided, but they also offer a chance to promote recycling in a positive and creative way. Recycling should be considered at the beginning and end of the school year when students are moving and disposing of large volumes of waste. Clothing, furniture, boxes, carpeting, and books are items to target at this time. Reusing such items should also be considered. Athletic events, especially football games, are examples of events that should be targeted for recycling, along with concerts and festivals.

Buying Recycled Products

Recycling is more than collecting materials on campus and sending them off to be processed. To truly consider an item recycled, it must be returned to the consumer market as a usable product. This means the institution should complete the loop by buying products that are made with recycled materials. The need to purchase products made with recycled materials can be included in the recycling policy statement.

Identification of Recycled Products There are many products that are made with recycled content. Every day more and more products are added to this list. Identifying recycled-content products that are equivalent in price and quality to virgin products is the first step in the process.

Managers should enlist the support of other units in the identification and evaluation of recycled-content products. It may help to form a committee for the purpose of selecting recycled-content products. Purchasing, publications, in-house print shops, copy centers, housing, food service, and the facilities organization all are units within the institution that have the potential to purchase recycled-content products and can help with the identification and evaluation of recycled products.

Manufacturers that the institution currently buys from often make equivalent products with recycled content. Managers should check with them to see which products come with recycled content. Some products the institution is using may already have recycled content. Next, the manager should look to other vendors who sell similar products. Recycling trade magazines and recycling organizations can also be consulted for information on recycled-content products. Many states have departments that actively promote businesses that manufacture and sell recycled products.

Evaluation of Products   Involving the end user is critical to successful integration of recycled-content products into the mainstream. The end user must be comfortable with the product. If the end user has not been involved, chances are greater that they will reject the product, perceiving it to be inferior or as not meeting their requirements. Products must be evaluated for price and quality for a fair comparison with the virgin product.

Barriers to Buying Recycled Products   Some systemic barriers to purchasing recycled products may exist in the institution that will have to be overcome. Many institutions have decentralized purchasing. All purchasing paperwork may go through a central office for administrative approvals, but the decision making about what product to specify and buy is done in the various departments requesting the item. This makes it difficult to educate people on the need and ability to buy recycled-content products. Education is key to advancing the use of recycled-content products.

Specifications can also pose barriers to recycled-content products. Many of the products used on campus have been used for many years, and the specifications probably have not changed in that time. Intentionally or unintentionally, these specifications may prevent the use of recycled material in a product. In the early 1970s, many inferior products containing recycled material hit the market. This made a lasting impression on many people, who still believe that such products are inferior. For this reason, specifications that do not allow recycled-content material may have to be changed before recycled-content products can be bought.

Price can also be a barrier. Some recycled-content products are more expensive than virgin products. Some reasons for this could be the need to recover the investment in expensive recycling technology or higher unit costs owing to low demand. Price may be a real obstacle, especially in times of budgetary retrenching.

Price preferences are a tactic that some institutions have implemented to overcome the problem of higher priced recycled products. A price preference policy allows spending a certain percentage above the price of a nonrecycled product to purchase a recycled product. Many governments have effectively used price preference policies to promote the purchase of recycled products. "Piggyback" purchasing programs are offered by state governments; these programs allow public schools to buy products on state contracts. This can be helpful in overcoming pricing problems because of the purchasing power of large government agencies.

33.12 WASTE REDUCTION

Although it has not received nearly the attention that recycling has, waste reduction is a primary component of an integrated solid waste management plan. As mentioned earlier, reduction is the first level of the waste management hierarchy. Many institutions have not concentrated on reduction as a result of the rush to recycle created by the many recycling laws enacted by state governments. Also, as mentioned earlier, it is hard for administrators to get a handle on how to implement waste reduction strategies. Some ideas for waste reduction follow.

Purchasing Decisions

A good place to start is to change the products that are bought. Packaging has the important purpose of protecting the product. However, some packaging is excessive or not needed at all. Managers can seek to reduce waste by buying products that are packaged responsibly. Try not to purchase products that are packaged in a wasteful way or in material that cannot be recycled.

One area of waste reduction that is gaining attention is the use of new products or equipment that result in less use of the product. Concentrated janitorial products that are packaged in bulk are an example. New painting application equipment drastically reduces the amount of paint that is used and wasted. Products that are naturally based or less harmful to the environment can also be purchased.

Often chemicals are purchased in larger quantities because prices are lower when the chemical is purchased in bulk. Many times these chemicals are never completely used and must be disposed of. Disposal of excess chemicals is expensive, but the high cost of such disposal is rarely calculated in the cost of purchasing in bulk.

Computers and Computer Systems

The advent of computers in the workplace was heralded as the beginning of the paperless office. Those in the recycling business know that nothing could be farther from the truth. Reducing the amount of paper generated by computers can have a dramatic effect on waste reduction efforts.

Printouts from both microcomputers and mainframes create enormous amounts of paper on campuses. Reducing reliance on printouts is a difficult task, as people have developed the habit of printing everything. Changing this mind-set involves both using computer technology more efficiently and educating people on ways to do this.

Computers can be programmed to reduce waste, but many reports are formatted in a wasteful way. Leader sheets and trailer sheets can often be eliminated or reduced. Formatting of reports can also be changed. Users and programmers can be encouraged to put as much on a page as possible.

Computerized administrative systems can replace paper forms in almost every aspect of university life. Student academic transactions, personnel transactions, and departmental transfer of funds or services are examples of activities where electronic processing of transactions can eliminate the need for paperwork. These systems are usually university-wide systems, but the process can be applied across the local area. File servers and local area networks allow access to and sharing of data in ways that can eliminate the need for printed reports. New reproduction equipment can store electronic images, allowing users to request reproductions of all or part of what they need and in any quantity.

Electronic mail (e-mail) can be used to eliminate paper. Interoffice memos can be sent on e-mail, and distributions can be created so that a note can be sent to multiple people with one keystroke. The Internet is an efficient way to exchange information. As long as e-mail is not printed out, using it will reduce waste.

Less high-tech waste reduction strategies are also available. Double-sided photocopying is the most well known way to reduce paper use. Diligent updating of mailing lists, greater use of routing lists for reports, and read files that circulate information are just a few examples of actions that can be taken in the office to reduce waste. An alarming development in relatively recent years is the proliferation of junk mail received in the office. People should be encouraged to take steps to have their names removed from mailing lists. Recycling coordinators can also take the lead in developing programs that reuse items that otherwise would be thrown away.

33.13   SUMMARY

Solid waste management on campus is increasingly more complex and demanding. Environmental regulations require extensive expertise to ensure proper disposal of waste. Students, faculty, and staff demand progressive and responsible handling of solid waste. For these reasons it is important to examine waste disposal practices and look for ways to reduce, reuse, and recycle.

NOTES

1. U.S. Environmental Protection Agency, Office of Solid Waste and Emergency Response. *Characterization of Municipal Solid Waste in the*

United States: 1992 Update. Washington, D.C.: U.S. Government Printing Office, July 1992.

2. U.S. Environmental Protection Agency, Office of Solid Waste and Emergency Response. *Understanding the Small Quantity Generator Hazardous Waste Rules: A Handbook for Small Business.* Washington, D.C.: U.S. Government Printing Office, September 1986.

3. Fishbein, Bette K., and Caroline Geld. *Making Less Garbage: A Planning Guide for Communities.* New York: INFORM, 1992, p. 3.

4. Hegberg, Bruce A., Gary R. Brenniman, and William H. Hallenbeck. *University and College Solid Waste Reduction and Recycling.* Chicago: University of Illinois Center for Solid Waste Management and Research, June 1992.

ADDITIONAL RESOURCES

BioCycle, a journal of composting and recycling. Emmaus, Pennsylvania: The JG Press Inc.

Environmental Industry Association (formerly National Solid Waste Management Association), 4301 Connecticut Ave., N.W., Suite 300, Washington, D.C. 20008. Telephone: 202-244-4700.

MSW Management, the journal for municipal solid waste professionals. Santa Barbara, California: Forester Communications.

National Recycling Coalition, 1101 30th Street, N.W., Suite 305, Washington, D.C. 20007. Telephone: 202-625-6406.

Steel Recycling Institute, 680 Andersen Drive, Pittsburgh, PA 15220. Telephone: 800-876-SCRI.

U.S. Environmental Protection Agency Resource Conservation Recovery Act Hotline: 800-424-9346.

Waste Age's Recycling Times. Washington, D.C.: Environmental Industry Association.

World Wastes. Atlanta: Argus Business. Telephone: 404-955-2500.

SECTION II-D

CAMPUS SERVICES

Editor:
William S. Rose
University of Alaska at Anchorage

INTRODUCTION

Campus Services

N ow more than at any other time, it is critical that facilities administrators form effective partnerships with other campus organizations that provide support to the academic enterprise. The interdependency of various campus services units demands that we make collaborative decisions guided by our well-informed understanding of a variety of functions outside the traditional facilities realm. In some cases, campus functions such as parking, campus mail services, transportation, environmental health and safety, and public safety may be part of the facilities portfolio on a given campus. Most often, they are organizationally somewhat removed. Regardless of the campus structure or setting, issues from master planning to operational decision making require thoughtful consideration of the needs and requirements of these support functions and their effects on the physical environment.

The following chapters discuss parking, campus mail services, transportation, environmental health and safety, and public safety in an effort to provide a foundation for facilities managers and leaders.

—William S. Rose

CHAPTER 34

Parking Systems

Susan A. Kirkpatrick
University of Michigan

34.1 INTRODUCTION

The space required to accommodate parking and traffic circulation on college and university campuses generally absorbs a significant portion of prime real estate. Historically, many campuses have developed physically around interior parking and vehicle circulation patterns. Often the traditional campus layout, reinforced by the perceived need to have parking adjacent to the workplace, represents a conflicting mix of vehicles and pedestrians.

A relatively recent phenomenon, identified through many strategic planning exercises, is the requirement that universities and colleges market themselves to potential students, faculty, staff, and visitors. A successful parking system can present an important and positive first image to visitors and to the campus community. The parking environment can influence a visitor's first impression of the institution.

The concept of campus in-fill, coupled with master planning guidelines that emphasize a pedestrian-oriented core and perimeter parking, has further constrained interior parking and complicated the development of parking policy. Facilities managers are being challenged to develop fair and customer-focused parking strategies, prioritize the use of decreasing interior parking resources, understand the explosion in parking technology, provide cost-effective parking solutions while catching up with deferred maintenance, and address the widely held perception that safe and convenient parking can only be provided next to the front door.

To meet the planning challenge, the parking paradigm must be expanded. In the past, planning focused on the number of vehicles within given parking parameters. However, as the number of available sites decreases and the cost to develop and operate parking systems increases, campus communities are demanding more cost-effective solutions. An expanded parking paradigm must have a more comprehensive view of the entire internodal transportation infrastructure, including alternative solutions to get from home to the work site. For many colleges and universities, the challenge is how to update and improve existing parking systems that no longer meet the needs of a diverse campus community.

This chapter includes a discussion of changing policy and planning issues, as well as a management approach, organizational structures, budget, outsourcing opportunities, parking opportunities, technology, audit control, and image.

34.2   PARADIGM SHIFT: IMPLICATIONS FOR PLANNING

In the past, planning has focused on the number of vehicles within given parking parameters. To improve and match the parking system to the changing needs of college and university customers, facilities managers must first move outside of the current parking parameters.

The Traditional Paradigm

Considerable energy has been focused on the management of vehicles and pedestrians within the boundaries of parking areas. Current parking technology has contributed to this planning focus, with the "pay-on-foot" approach in parking structures, central pay stations in surface lots, "smart cards," debit cards, proximity cards, and so on. Planners are bombarded with issues, concerns, and solutions within the parking space boundaries. Traditional master planning guidelines for many colleges and universities have also contributed to this planning focus.

Many colleges and universities have successfully implemented a pedestrian orientation to the campus environment, resulting in parking located at perimeter or off-site locations. Typically with this approach, the need for transportation has correspondingly increased. The planning focus has remained within the boundaries of the parking areas as planners have attempted to match and manage the vehicle demand to the space available. Over the years, many strategies have been developed to manage the increasing demand for limited parking space.

As parking administrators match various forms and combinations of reserved parking, zoned parking, and open parking to their campus cultures, the planning focus remains directed at single-occupant vehicles within designated parking boundaries.

At many colleges and universities, regardless of the system that is used to manage vehicles within designated parking boundaries at perimeter locations, customer dissatisfaction with parking systems seems to be increasing. Customers lament that convenient parking space is not available, thus hindering their ability to do their jobs, and that the cost of parking is rising—at times, faster than salaries. At the same time, many administrators are being faced with deferred maintenance, escalating costs to operate and maintain a parking system, and increasing customer demand for a decreasing supply of parking.

Shortcomings of the Traditional Paradigm

The following factors contribute to the increasing customer and administration dissatisfaction with parking systems that traditionally were perceived as successful:

1. *New parking structures are costly to build.* These costs typically are shared throughout the system; thus some customers benefit directly from a new structure, whereas others continue to use older facilities and receive no direct benefit from the new one.
2. *Costs to maintain structures are escalating.* As campus communities become increasingly security conscious, image may play an important role. For example, additional lighting may be required, video cameras or security patrols may have to be added, additional windows may have to be added for increased visibility, and emergency telephones may have to be added in elevator cabs or key areas throughout a structure. Because campus communities are increasingly image conscious, aesthetics may play an important role. An effective graphic system may be required to clearly identify levels and areas. Many times, multiple methods of reading the signs are required. For example some people remember colors better than they do numbers or written words. Walls that are whitewashed reflect ceiling light and appear brighter and cleaner. Correct organization of signage will contribute to an image of good organization.
3. *Surface lots are costly to build.* Because planners have become sensitive to the importance of aesthetics and the reduction of future deferred maintenance, initial construction costs have increased. For example, current lots have concrete curbs and gutters rather than bituminous curbing. If buses drive into the lot, a heavy-duty bituminous curbing is used. These costs are typically shared throughout the system; thus some customers benefit directly from a new surface parking lot, whereas others continue to use older facilities and receive no direct benefit from the new one.
4. *Costs to maintain core area surface lots are escalating.* As campus communities become increasingly security conscious, image may play an important role. For example, because pedestrians are walking in the parking area, additional lighting may be required; emergency telephones, video cameras, or security patrols may have to be added; and

landscaping may have to be modified so as not to block lighting or provide hiding places. Risk management issues may have to be addressed through additional snow removal.

5. *Costs to maintain commuter parking lots are escalating.* As parking is expanded to perimeter or off-site locations, transportation costs have to be assessed carefully. Also, the demand for evening and weekend access may increase.

6. *Deferred maintenance is adding up for many older structures and surface lots* where security, aesthetics, and quality construction may not have been a priority in the past. Engineering studies may be needed to 1) identify all deferred maintenance throughout the system and 2) prioritize annual work to minimize the number of spaces taken out of the system and to plan for alternate parking during construction.

Thus it can be seen that costs to build, repair, and maintain structures and surface parking lots are escalating. At the same time, members of campus communities typically are not willing to absorb these additional costs by paying higher parking rates, especially when a perceived value may not be present.

An Expanded Paradigm

Planning that traditionally started once a vehicle reached a parking area now encompasses options for getting from home to the work site. This has become necessary because 1) many colleges and universities cannot cost-effectively operate and maintain the traditional expansion of surface lot and structure parking, 2) the campus community is typically not willing to pay the increasing cost, and 3) the traditional parking system of one vehicle per person no longer meets the diverse needs of all customers.

Consistent with expanding the parking paradigm is transportation demand management. Transportation demand management encompasses planning professions from related fields such as transportation, city/regional planning, campus planning, and parking with the focus on reducing traffic volume through a reduction of single-occupant vehicles. As colleges and universities expand their parking paradigm, the traditional approach using single-occupant vehicles becomes one of many options to choose from on a parking "menu." Ideally, traffic volume and the demand for core area parking are voluntarily reduced as customers choose options that better meet their needs.

What might the menu of parking options to get from home to the work site look like? If the future parking paradigm requires first knowing the customers so as to meet their needs and provide a better service, then the menu will differ slightly from one institution to another. Typically, to meet diverse needs, the parking menu should reflect choices in terms of cost to convenience. For example, the following parking options may be included:

- Reserved space or parking area.
- Core area parking.
- Perimeter area parking.
- Commuter area parking.
- Designated motorcycle parking in the preceding four areas if demand requires.
- Carpool/vanpool parking in the first four areas above if demand requires it.
- Bicycle parking, which might include bicycle storage lockers and the traditional hoop.
- Shared parking resources with the surrounding community, such as with park-and-ride programs.
- Economic incentives provided to promote shared bus services with the surrounding community.
- Ride-home programs through local taxi services may have to be provided to promote perimeter or commuter parking during hours when it is not cost-effective to provide bus transportation.
- Walking may be a parking option that needs only to be promoted. Typically, the implementation of many college and university master plans has resulted in a pedestrian orientation to the campus, where special attention has been focused on providing appropriately placed sidewalks (7 or 8 ft. wide for snow removal, which is also a good width for group walking) with excellent lighting, an effective emergency telephone system, and beautiful grounds.

The traditional parking approach—driving a car to campus and parking in a surface lot or structure—has become one option to choose from for those who choose to pay the price for the convenience. Thus, the menu consists not only of cost options, but also of approaches to addressing diverse lifestyles. Ideally, a reduction in demand becomes a voluntary reduction while at the same time providing the revenue source to maintain existing assets.

The key to implementing a successful menu of parking options, based on the match of cost to convenience, is an understanding of customers' needs and their perceptions of solutions, as well as a willingness by all involved to be receptive to trying new ideas. Another key to success is the flexibility of a parking system to provide multiple options to fit diverse life styles. For example, an employee may desire to drive a motorcycle during favorable weather and have the flexibility to take a bus from his or her home to the work site during inclement weather.

34.3 PLANNING ASSUMPTIONS

As a result of the paradigm shift, the following planning assumptions may differ from those included in the traditional planning approach to provide for a parking system:

- Core area parking space typically will not be expanded and will most likely continue to be reduced.
- As the supply of core area parking space decreases and the demand increases, access will have to be prioritized.
- Interior road systems typically will not be expanded and most likely will continue to be reduced.
- As enrollments increase, pedestrian traffic also will increase.
- Expansion of parking at perimeter or off-site locations may not be a solution for all customers.
- As costs of parking increase and the supply of parking spaces decreases, faculty and staff will look for nontraditional solutions to get from home to work.
- Faculty and staff may be increasingly unwilling to support increased parking costs.
- Costs to operate and maintain a traditional parking system will most likely increase.
- Faculty and staff are increasingly looking for a menu of parking options, in terms of cost and convenience, to meet diverse lifestyles.

34.4   CURRENT AND FUTURE POLICY ISSUES

Most colleges and universities will be addressing, to various degrees, the following policy issues that will affect all members of the campus community and have a financial impact as well.

Current Policy Issues

- Many institutions have rushed to install equipment, such as central pay stations, that provide ease of audit control and 24-hour access. Planners must carefully assess the service provided to determine the extent to which people should be replaced with technology. A smiling, friendly student greeting visitors from an attendant's booth may be a greater asset than the latest technology.
- Parking technology is exploding, and often it is costly to install and operate. As more colleges and universities begin to pilot new programs, questions arise as to how far facilities should go with technology. Do they really need "smart cards," debit cards, and proximity cards? Careful attention should be given to matching technology with the needed service.
- To various degrees, parking administrators are faced with the dilemma of how to handle customers who exit a parking area and cannot pay. Maintaining good public relations usually is more important than the amount of money that would have been collected. Thus, careful attention must be given to 1) monitoring the level of

"no pays," 2) training attendants to handle such situations appropriately without embarrassing the customer, and 3) ensuring that the collection process does not far exceed the amount collected.

Future Policy Issues

- To what extent should current parking resources be held constant and future demand be met through shared resources with the surrounding community? It may be more cost-effective to share bus and parking resources with the surrounding community than to expand the traditional commuter parking and bus service.
- As the supply of core area parking continues to decrease as a result of building expansion and/or the continued implementation of a pedestrian-oriented campus, criteria have to be established to prioritize access.
- As costs to build, repair, maintain, and administer a parking system continue to increase, questions may arise as to whether parking should be subsidized, and an appropriate method for doing so.

34.5 MANAGEMENT APPROACH

Most changes that occur within a college/university setting will affect the parking system. For example, if the psychology department moves and adds a clinic function, this typically will necessitate some form of restricted parking for clients. As changes occur throughout the campus community, parking administrators are required to know about the projected changes, work with customers on creative solutions, and implement these solutions when the change occurs. Strategic planning and a total quality approach are essential for the campus community to perceive that a parking program is successful. Strategic planning provides direction. A total quality approach creatively implements that direction.

Strategic Planning

Numerous strategic planning models and approaches have been found to be successful. The challenge is to determine the model, approach, and timetable that best fit one's culture. The following components are typical for an annual cycle:

- *Statement of direction.* Typically, this involves "futuring"—asking, "If I left the campus and returned three to five years from now, what is the ideal system I would hope to see?"—and is a brief description of what the parking system will look like in three to five years. The driving factor is identification of projected campus changes.

- *Planning assumptions.* Usually five to ten general assumptions are listed. An example might be, "Core area parking will continue to decline."
- *Statement of mission.* A statement of purpose is usually summarized in one to two sentences. Typically, the shorter the statement, the easier it is for staff to internalize it. Some organizations have found it helpful to use a slogan to accomplish a sense of "buy-in."
- *Statement of vision.* The vision describes how the mission will be achieved. Typically, eight to twelve brief statements are included. An example might be, "To be an organization recognized for taking a leadership role in providing innovative services and options."
- *Statement of philosophy.* Typically, this is a brief statement outlining the institution's commitment as to how internal and external customers will be treated.
- *Statement of values.* Values to which the organization is committed are listed and described for a common understanding. For example, initiative, fairness, respect, and integrity might be described.

The second part of the strategic plan will identify specifically how the statement of mission will be accomplished. Often projects and teams are listed and may take the form of or be referenced as goals and objectives, action plans, or key result areas. The third part of the strategic plan generally is the budget. Ideally, the strategic plan has identified the top priorities of the budget.

The key to success for any strategic plan is the degree of buy-in by all staff members. The degree of buy-in from staff will correlate with the degree to which they participate throughout the process.

Total Quality

A total quality approach is essential to creatively implement the strategic plan. A parking system is continually updated to enhance the numerous changes that occur throughout a college and university environment. Any change in the parking system typically requires personnel in all functional areas within parking management to work closely together as a team. The following scenario is an example:

The psychology department is adding a clinic and requires client parking. A successful parking program may require the following functional areas to work as a team:

- Contact person for the customer
- Signage
- Enforcement
- Method of payment; audit control
- Permits
- Access control
- Customer follow-up

Increasingly, successful solutions require multiple functional areas working together as a team, in which team members are knowledgeable about all areas of the parking system and supportive of a total quality approach.

34.6   FUNCTIONS: OVERVIEW

University Organizational Structure

The parking function may be found in a number of organizational structures. Positive arguments may be made for all of the typical structures listed below:

- Parking may be included as part of the facilities organizational structure.
- Parking may be grouped together with transportation.
- Parking may be grouped together with campus enforcement.
- Parking may function as an independent unit.

Regardless of where in the organization the parking function resides, parking, enforcement, transportation, and campus planning must function as a team.

Parking Organizational Structure

The following functional areas typically are found within a parking system.

- *Customer service.* With increased awareness of the importance of providing a quality service, a separate customer service component is essential. Typically included are the permit and revenue collection system; customer contact; modes of communication with the campus community, such as a newsletter or notification of construction schedules or events that may affect parking availability; and planning for special events.
- *Financial.* A recent emphasis has been to eliminate cash handling, to the extent possible, and to process payment electronically. For example, parking for special events, such as football games, could be handled through prepaid permits. Electronic meters provide an electronic reading, with tape, of the amount of money in the meter before collection.
- *Data management.* Use of electronic management and the projected increase in data will require careful data management and constant evaluation to ensure that only pertinent data are tracked and maintained. For example, many software programs monitoring parking structure activity will provide more data than is generally needed to manage a parking program successfully.

- *Facilities management or maintenance.* Tasks may include relamping, sweeping, striping, realigning concrete bumpers, providing regulatory and directional signage, removing snow, using cold/hot patch, regrading gravel lots, and painting.
- *Electronics.* This function may be included with facilities management or may operate as a separate unit. Areas of management may include electronic meters, gates, electronic cash registers, central pay stations, and electronic card readers. The electronics field has expanded rapidly; therefore, trained personnel may be difficult to find. In-house training may be needed.
- *Attendant operation.* Many colleges and universities provide a combination of attended parking areas and electronically managed parking areas. Careful planning is required to decide on an appropriate approach. Many schools are equipped for both methods, offering flexibility for special events and shared use. As attendants increasingly work with computerized cash registers, additional training may be necessary.
- *Special events.* Staffing for special events typically is included as part of attendant operations. A pool of temporary employees is required for events such as concerts, athletic events, and conferences. Staff should be trained in good customer relations, able to give directions and answer questions about the institution, and appropriately dressed so to be identifiable as parking employees.
- *Enforcement.* Flexibility, together with good public relations, may be the key to providing good enforcement. Many schools have relied successfully on student or temporary labor. However, if the parking system requires the use of multiple permits, then permanent staff, with knowledge of and experience with the uniqueness of the system, may be required. Equal enforcement of all areas usually is not required. Potential trouble spots may vary. The supervisor must identify management indicators, such as monthly/yearly revenue and percentage of illegal parkers, to prioritize enforcement resources.
- *Citation appeal process.* An effective appeals process, giving the customer the benefit of the doubt, is critical for good public relations. It also provides parking administration with feedback regarding areas that need improvement. For example, signs may have faded and need replacement. Branches may be obstructing the signs and have to be removed.

34.7 FUNDING SOURCES

Ideally, a parking system should be self-supporting. Typical funding sources are as follows:

- Permits, typically issued to faculty and staff

- Metered parking, typically used by visitors, students, and faculty and staff
- Designated visitor parking
- Parking for special events such as concerts, seminars, and athletic events
- Parking tickets

34.8 BUDGETING

Assuming that the parking system is self-supporting, rates should be set to fund the following components annually.

- Administration
- Maintenance
- Repair and renovation
- Deferred maintenance
- New construction
- A reserve for new construction

Alternate transportation options, such as night-ride-home programs, may have to be considered.

As the annual budget is itemized for projected expenses and revenue, it is helpful to attach an explanation for each item. For example, why is permit revenue expected to increase/decrease, or why are utility costs expected to increase/decrease? This information may be invaluable in the future for projecting trends.

34.9 PRIVATIZATION

Opportunities exist for privatizing a number of functions within a parking operation. The question—Could privatization perform a more cost-effective service and/or provide a better service?—should be applied to all areas of the parking operation. The following five functions may be best to privatize:

1. A local bank or cash-counting service within the institution may be better equipped to count money collected from meters or an attendant operation. Typically, it will have state-of-the-art counting equipment and secured facilities that would not be cost-effective for a parking operation to maintain with limited use.
2. The cost to contract with a local bus service for faculty and staff to ride to work from any bus stop should be compared with the cost to build, maintain, and provide bus transportation from university commuter parking lots.
3. Privatizing parking attendant functions may be more cost-effective if full-time staff are used. This is also an excellent opportunity for student employment.

4. The cost to privatize all or a portion of parking enforcement and the citation appeals process should be compared. Both areas may also provide an excellent opportunity for student employment.
5. The cost to privatize all or a portion of snow removal should be compared.

34.10   PARKING SYSTEMS

Parking administrators have matched various forms and combinations of reserved parking, zoned parking, and open parking to their specific campus cultures. The following are advantages/disadvantages of each form:

- Reserved parking is typically the most expensive, with the lowest occupancy. As space constraints continue to grow with the projected decrease in core area parking, it may be increasingly difficult to provide reserved parking for large numbers of people.
- Zoned parking typically restricts parkers to an area close to their work site. The occupancy rate is generally higher than that for reserved parking. As core area parking space continues to decline, the demand will typically exceed the supply.
- Open parking, commonly referred to as "hunting license" parking, provides a system of parking on a first-come basis and has the highest occupancy rate. However, as the demand for parking increases, the level of frustration grows as customers perceive wasted time in hunting for a parking space.

34.11   TYPICAL PARKING OPPORTUNITIES

The following menu of 11 parking options is listed, in order, from the most expensive to the least expensive:

1. *Reserved space or area.* This option is usually the most expensive and has the lowest occupancy rate. If the demand exceeds the supply, the challenge may be to develop criteria for eligibility that customers perceive to be fair.
2. *Core area parking.* This option typically provides parking within a reasonably short walking distance to most campus functions.
3. *Perimeter parking.* This option generally provides parking that requires a longer walk or a short bus ride to most campus functions.
4. *Motorcycle programs.* Space that typically cannot be used for vehicle parking may be promoted for motorcycle parking. Because less space is required for motorcycles than for cars, a lower rate could be charged, whether in a reserved area, core area, or perimeter area parking location.
5. *Carpool/vanpool* programs. This is a cost-effective approach for those who are willing to contend with the perceived inconvenience of

organizing. Payment choices could be offered, depending on whether reserved, core area, or perimeter parking is used.

6. *Bicycle parking.* A choice of bicycle parking could be offered, from the traditional hoops to bicycle lockers. Because many bicycles are expensive, the provision of lockers may promote the use of bicycles over vehicles. Attended bike corrals are also coming into use.

7. *Park-and-ride parking.* This option may provide an opportunity to share resources with the surrounding community, to reduce operating costs, to take advantage of parking space that may be underused, and to address a unique need of commuting faculty, staff, and students.

8. *University commuter lots.* In general, this type of parking is provided free of charge to the customer. As the supply of core area and perimeter parking has steadily decreased, many parking systems have expanded with off-site commuter parking. Whereas commuter lots once were perceived as an inexpensive approach to expanding the supply of parking, the costs to fund such parking are increasing and should be compared to a park-and-ride approach. Costs of commuter lots may be escalating because of the following:

   a. Security issues are being addressed, typically through extensive lighting, emergency telephones (some colleges and universities are installing video equipment), and additional security patrols.

   b. Transportation costs may be increasing as the demand for extended hours increases.

9. *Ride-home programs.* Ride-home programs are typically provided through local taxi companies during hours when it is not cost-effective to provide bus transportation. Service may be provided to commuter lots or directly to homes within the surrounding community as an incentive to encourage use of a bus, carpooling, or some option other than bringing a vehicle to campus.

10. *Local bus service.* A university may be able to contract with a bus service from the local community to transport employees and students from any bus stop to campus. This may be a cost-effective option when compared with constructing and maintaining university commuter parking lots and providing bus transportation to them.

11. *Walking.* The implementation of many college and university master plans has resulted in a pedestrian orientation to the campus. When this happens, another parking option—walking—has been inaugurated, which may need some marketing for customers to view this as an option during suitable weather. Typically, special attention has been focused on providing appropriately placed sidewalks (7 or 8 ft. wide for snow removal, and also a good width for group walking) with excellent lighting, an effective emergency telephone system, and beautiful grounds.

Flexibility must be built into the parking system to provide customers with easy access to multiple parking options, based on their own unique needs. For example, an employee may be able to participate in a carpool three days a week, be dropped off two days a week, but on occasion need commuter access when a car is the only transportation option.

34.12   PARKING TECHNOLOGY

Like many fields of endeavor, the parking industry is experiencing an explosion in technology. Extensive amounts of data may be tracked and monitored. However, implementation of such technology may be costly. The initial task is to identify the information that is essential to manage a successful parking system. The technology needed may vary greatly from school to school. A successful approach on one campus may not be successful on another campus. The typical challenge is to fund only the hardware and software that are actually needed, but with expansion capabilities for future growth. Effective strategic planning, together with a total quality management approach, will help to identify the likely future direction for use of emerging parking technologies.

Technology Examples

The following are examples of parking technology that is currently on the market:

1. Equipment to monitor parking structure activity. Typically this involves a chip in a card or on a permit, or sensors installed in the pavement. For enforcement purposes, gate equipment may be used. Types of information that may be monitored are as follows:

   - Number and time of entry/exit
   - Occupancy trends
   - Use by permit type, such as student, disabled client, or guest
   - Amount of parking used per parker
   - Identification of maintenance needs, such as a gate remaining open

   The system may include a "Parking Available/Full" sign at the entrance for customers' convenience. If the system operates close to full occupancy on a daily basis, all structures may be networked so that if a particular structure is full, a message sign will direct customers to the next closest structures that have parking available.

2. Central pay stations may be used in structures or surface lots. Individual parking meters are eliminated, and customers are directed to a central location to pay. Advantages for the customer are that a parking receipt is provided and dollar bills may be used, eliminating the need to carry a large number of coins. Advantages operationally are that

1) enforcement will pull a tape at the central pay station to quickly identify vehicles whose time has expired, 2) collection time is saved because collection occurs at one location, and 3) audit control is simplified because a tape is provided that identifies the amount of revenue being collected. If multiple locations are added, central pay stations may be networked to an administrative location for on-time identification of all activity in total or per lot. For example, data included could be occupancy, percentage of illegal parkers, and/or revenue collected.

3. Debit card systems allow the parker to add value to a card from a central location or multiple locations and pay only for actual time parked. Debit card capability may be added to many systems, such as central pay stations or individual parking meters. The key advantage to the customer is convenience (i.e., not having to carry change). Advantages operationally are that revenue is collected up front, collection costs are reduced, and audit control is simplified.

Areas of Caution

Referencing the above examples, astute questions must be asked to determine the minimal level of hardware and software needed to operate an effective parking system, but there must be an understanding of future directions to ensure that the system can be expanded.

For example, in monitoring structure activity, is it necessary to know parking use per customer, or is overall occupancy and identification of peak use enough information to operate a parking system successfully? Networking parking structures with a message sign may be costly. Is the cost worth the customer service? Networking multiple central pay station locations provides extensive information at a glance. Does the need for quick information justify the cost of networking and staff time to track and monitor data? Debit cards are a customer convenience. Careful planning must be done to determine whether one solution is better than other options.

34.13   CASH PROCESSING AUDIT CONTROL

Traditionally parking systems have processed a tremendous amount of cash with audit controls that relied heavily on trust. For example, coins from a traditional parking meter might be collected by a trusted employee without any records at the time of collection. The cash processing audit control system must document a clear audit trail from the point of transaction to verification of deposit and provide protection for employees who are involved in cash processing.

The following questions may provide a quick validity check of the system:

- Is individual accountability maintained at every step in cash processing from the point of transaction to verification of deposit?

- Are different staff members involved in each step of the process?
- Is one staff member accountable for overall supervision to ensure a match or explanation between revenue and deposit?

To the extent possible, special attention should be given to eliminating cash handling from the parking system. Computerization can play a big role, as in the following examples:

- Employees may purchase a permit with a check, by credit card, or through a payroll deduction system.
- Billing internal departments for visitor permits that are used may be done electronically.
- Parking for special events, such as athletic events, may be handled with a prepay system for the season, with renewal done through the mail.
- Debit card systems eliminate the need for daily cash, while providing an electronic record of revenue before collection.

34.14 IMAGE

The parking environment can influence a visitor's first impression of the institution. Many factors will contribute to a positive image. Key factors are the level of maintenance, lighting, signage, and the perception of safety.

Level of Maintenance   An appropriate level of maintenance must be established in the following areas:

- Landscape care
- Striping
- Miscellaneous painting, such as pavement arrows, objects of caution, and structure railing, stairwells, and lobby areas
- Sweeping
- Pavement cleaning, such as of oil spills
- Relamping and cleaning of light fixtures
- Pavement care, such as repair of potholes and cracks
- Replacement of faded, damaged, or missing signs
- Structure window washing
- Trash removal
- Snow removal

Ideally, maintenance should be scheduled during times of low occupancy. However, when this is not possible, prior notification of any closure should be provided for good public relations.

Lighting   Many customers associate safety with lighting level. Incorrect lighting, particularly in parking structures, can create a variety of problems, including shadow zones, sense of insecurity, reduced visibility, loss of direction, and even a sense of claustrophobia. Planners have to be sensitive to the correct illumination, uniformity, color of light, surface colors, and reflectance. A more detailed discussion of exterior lighting is provided in Chapter 26.

Signage Most successful signage systems are those that provide as little overall signage as possible. The following ten guidelines will contribute to a successful system:

1. Letter size and wording should be standardized throughout the system.
2. "Warm fuzzy" wording will contribute to a friendly image. For example, "Please Drive Slowly," rather than "Drive Slowly," may go a long way toward achieving customer cooperation.
3. Sign locations should be standardized as much as possible. Customers typically learn where to look for directional signage.
4. Signage should be coordinated with lighting locations to further enhance signage visibility.
5. High-pressure sodium lighting will distort many colors. If color coding is used as a level indicator, colors should be selected that will not be distorted. For example, red will appear brown, but yellow will not change in appearance.
6. Typically a successful directional system will incorporate multiple approaches. For example, some customers will remember colors better than the printed word. A level or area indicator sign may include the number "2" above the written word "TWO" against a blue background. Such a sign includes numerical and written identifiers, as well as color coding.
7. Traditional colors, such as red, brown, yellow, and blue, will probably have a higher recognition level than trendy colors such as mauve, taupe, coral, and cinnamon. Some customers may not know the name of such colors to use when asking for directions.
8. Signage located around the perimeter of a surface lot rather than within the parking lot will provide ease of snow removal and sweeping.
9. Information panels, campus directories, and "you-are-here" maps should be clustered in pedestrian areas such as structure elevator lobbies and bus pullouts.
10. Standardizing signage, maintaining an inventory, and fabricating and installing in-house typically will provide faster service and a more cost-effective approach.

Perception of Safety Campus communities are becoming more safety conscious. Additional action may be needed to combat a past stereotype that parking structures or areas are not safe. The following strategies may have to be considered:

- An emergency telephone system, highly visible, in standard locations throughout the system
- The addition of glass panels in stairwells to increase visibility
- Video cameras
- Parking attendants during evening hours
- Security personnel walking or driving through the structure
- An increased lighting level
- Well-published procedures to enhance safety awareness
- Well-published safety programs such as escort services

ADDITIONAL RESOURCES

Bhuyan, Sam, Anthony P. Chrest, and Mary S. Smith. *Parking Structures, Planning, Design, Construction, Maintenance, and Repair.* New York: Van Nostrand Reinhold, 1989.

Catalog of Parking Publications. Available from the International Parking Institute, P.O. Box 7167, Fredericksburg, Virginia 22404.

Kirkpatrick, Susan A., and Gary L. Cudney. "Parking Structure Signage and Safety Improvements: A Case Study." *The Parking Professional*, February 1992, pp. 14–22.

1997 Guide to Parking Consultants. Available from the International Parking Institute, P.O. Box 7167, Fredericksburg, Virginia 22404.

The 1997 Parking Buyer's Guide. Available from the International Parking Institute, P.O. Box 7167, Fredericksburg, Virginia 22404.

Parking. Monthly publication of the National Parking Association, 1112 16th Street, N.W., Suite 300, Washington, D.C. 20036.

The Parking Professional. Monthly publication of the International Parking Institute, P.O. Box 7167, Fredericksburg, Virginia 22404.

CHAPTER 35

Transportation

John P. Harrod Jr.
University of Wisconsin, Madison

35.1 INTRODUCTION

Enrollments in institutions of higher education have stabilized during the past years, yet campus size and service areas have expanded. Off-campus teaching and research have proliferated. Students, faculty, staff, supplies, and equipment require increased transportation services. Transportation requirements often exceed the abilities of the campus to meet the diverse demands. The function of a college or university transportation system service is to provide for the transportation needs of the institution with optimal economy. Each request for service must be met in a timely and effective manner. There must be a balance between economy and convenience to users.

Transportation divisions are often required to perform a variety of services; examples include the following:

- Maintenance of automobiles for use by university personnel
- Bus service for faculty, staff, and students
- Taxi pay-by-ride service
- Maintenance and servicing of all the institution's vehicles
- Maintenance of heavy equipment, such as construction equipment
- Maintenance of refuse collection trucks and grounds care motorized equipment
- Two-way radio communication with all passenger vehicles
- Messenger/mail service

35.2 MISSION OF TRANSPORTATION SERVICES

A goals and mission statement should be developed to identify and provide for the transportation needs of the institution. The evaluation of the transportation needs must consider all factors, including the type of campus (urban or rural), single or multiple campuses, public transportation available, size of campus, and extent of extension service. Following is a typical mission statement:

> To provide safe, economical, and appropriate transportation for students, faculty, and staff, and the equipment and supplies necessary to operate the university.

Guiding principles are often established to help direct the activities of the organization. These could include the following:

- To acquire the type and number of vehicles necessary and appropriate to meet the needs of the faculty and staff to perform their duties
- To rent the vehicles to qualified departments on either short-term or long-term arrangements, whichever is most appropriate and economical for their requirements
- To manage the fleet in a manner that ensures the lowest possible operating cost while maintaining a consistently high mechanical reliability and safety
- To organize a staff and maintain a facility that provides superior service, repair, reservation, dispatch, and administrative support to the university customer
- To ensure that the service continues as a financially secure and self-supporting operation as it strives to achieve a nonprofit or break-even status
- To maintain, repair, and dispose of the vehicles in a manner that brings the highest possible return on money spent
- To direct policies and procedures so that they reflect the best way to operate the fleet with the customer in mind
- To think independently and competitively to obtain the best ratio of dollars spent to miles driven
- To operate and maintain an environmentally friendly fleet of vehicles

Transportation managers must foster a climate where there is dedication to central goals while maintaining flexibility for new ideas and participation from faculty and staff.

35.3 FLEET OWNERSHIP OPTIONS

Ownership Ownership of vehicles and supporting equipment requires high initial expenditures for their purchase. Maintenance programs require extensive facilities, equipment, and staff, or maintenance contracts. The large fiscal outlay for such programs can be a major problem for many institutions. An important factor in evaluating outright purchase is that long-term amortization necessitates accurate long-term projections of needs. Another factor is whether there is uniform demand throughout the year. It is difficult to justify ownership of vehicles that may sit idle a substantial portion of the year (i.e., during the summer months and other major academic breaks).

Lease or Rental Options Leasing motor vehicles has become popular in recent years. This trend stems from the growing number of commercial leasing firms in the market and the wide variety of programs they offer. There are several advantages to leasing. A minimal initial investment is required, and the remaining cost is distributed uniformly over the lease period. Maintenance facilities and equipment can be eliminated if full-maintenance lease programs are available. Some lease programs offer an arrangement whereby the customer assumes full ownership of the vehicle at the end of the lease period, whereas others permit buy-out by the customer at the end of the lease.

In general, leasing or renting is often a viable economic alternative to institutional ownership if the vehicles are needed to meet periodic peak requirements or if highly specialized vehicles are needed. The decision to own or lease should be made on the basis of anticipated use, comparison of ownership costs with lease or rental costs, and the availability of vehicles for lease or rent (Figures 35-1 and 35-2).

Closed Lease. Sometimes called a "straight lease," a closed lease terminates with definite conditions; it is not open for further adjustment. A closed lease does not offer the option to purchase the leased asset or to share in the profit or loss on the resale of the vehicle.

Open Lease. The major feature of an open lease is that the lessee shares in the future risk of gain or loss at the time of resale. It can be considered as a time payment plan whereby the lessor and lessee agree to sell the vehicle and share in the gain or loss. The advantage to setting up an open lease is the lower monthly rental fee to the lessee.

Special Manufacturer Lease. This type of lease appears to be a cross between the open and the closed leases, in that neither the dealer nor the lessee have any responsibility for gain or loss in resale value at the end of the lease. The lessee does have the option to purchase the vehicle at market value.

Use of Privately Owned Vehicles A common practice, particularly at small colleges, is to reimburse individuals for use of privately owned cars driven while on official business. Payment is typically based on a mileage rate. Most state institutions have an established rate that applies to all state agencies. The use of privately owned vehicles is frequently authorized when demand exceeds fleet supply or when an employee

Figure 35-1
Ownership Alternatives

Alternatives	Advantages	Disadvantages
Institutional ownership	Flexibility	High initial expenditure
	Standardization	Maintenance in-house
	Economies of scale	Facilities
		Equipment
		Staffing
		With maintenance contract
Lease/rental	Short term/seasonal peak demand	Cost
	Minimal initial investment	Timely availability
	Even distribution of cost	Geographical/logistical problems
	Full maintenance lease	
	Highly specialized equipment	
	Lease with option to buy	
Privately owned vehicles	Demand exceeds fleet supply	Liability
	Flexibility	Documentation
		Accountability

wants to combine official and personal business on the same trip (e.g., taking a vacation trip in conjunction with attending a conference).

35.4   TRANSPORTATION PROGRAM ADMINISTRATION

Vehicle Assignment Policy   Vehicles may be scheduled for use from a central motor pool on the basis of individual requirements, or they may be assigned to units or departments that have major continuous needs. In those cases, equipment may be assigned to the units or departments on a continuing subpool basis. A policy is often needed to define the conditions under which subpool assignments will be made, and such assignments should then be subject to periodic review on the basis of utilization. Emergency vehicles (police cars and ambulances) and specialized equipment (four-wheel drive vehicles) that must be available on demand should not be subject to the same utilization standards in making assignments.

Figure 35-2
Lease/Rental Maintenance Options

	Type	Per Month
No Maintenance	Sedans	$220
	Station Wagons	227
	Pickup	185
	Panel	264
	Carryall	348
Full Maintenance	Sedan	260
	Pickup	227
	Carryall	360

Utilization Standards To ensure that the investment in a transportation fleet is held to a minimum, standards for the utilization of equipment should be established. These standards can be set on the basis of a minimum level of hours of use or miles of operation per month, as appropriate to the vehicle type. Utilization standards should address the vehicle's purpose, distances typically traveled in the area in which it is used, and the length of time each month during which the equipment is available. For example, utilization standards would be much higher for a vehicle that is normally assigned to long-distance trips than for one used only for short trips between campus sites. Similarly, a police vehicle used during multiple shifts seven days a week would be subject to a higher utilization standard than one operated only during a normal five-day work week.

Consideration should also be given to purchasing vehicles that can serve more than one purpose: a mini van can be booked for passenger use for a number of years until mileage builds up and then later can be reassigned as a campus maintenance vehicle.

Replacement Criteria Criteria should be developed for replacing vehicles before maintenance and repair costs become excessive. The major factors in determining replacement are age, mileage, remaining warranties, item repair cost, and cumulative repair cost.

Age and Mileage. A policy for retiring passenger vehicles may be 75,000 miles or a five-year average disposal. This target mileage and age may be tempered based on specific maintenance information about a particular vehicle or group of vehicles.

Item Repair Costs. Normally, no repair should be performed if the cost of repair exceeds 50 percent of the replacement cost.

Cumulative Repair Costs. A limit can be set on how much can be spent on repairs throughout the life of a vehicle. The repair limit can be a percentage of the cost of the vehicle. This can be used as a guide, with the option to retain the vehicle if its condition, availability, and need so justify.

Remanufacturing can be a viable alternative for replacing heavy-duty, high-value equipment, typically construction equipment, cranes, and transit buses. When considering remanufacturing, specifications must be very detailed.

Another alternative for replacing vehicles is to buy used equipment from other agencies. Most such vehicles have high mileage, but the bodies are generally in good condition. On-campus vehicles usually do not get many miles added per year, so this is a good way to obtain usable vehicles at a reasonable price.

35.5   USER POLICIES AND PROCEDURES

State institutions usually are required to comply with regulations applicable to all state agencies. The institution has the responsibility of informing the institution's vehicle users of any regulations. In addition, policies and some regulations unique to the college or university must be promulgated, along with procedures, to employees who use the vehicles. The institution should publish the following information for the college or university community:

- Travel authorization regulations and procedures
- Driver licensing/commercial driver licensing requirements
- Authorized passengers or riders
- Restrictions on personal use
- Request and prioritization procedures
- Usage rates
- Long-term assignment policies
- Special regulations for student use
- Insurance obligations (departmental and individual)

The following policies and regulations are required to be publicized for all users of college and university vehicles:

- Operator care and maintenance
- Availability and use of institutional credit card
- Disposition of traffic violation citations
- Procedure for vehicle pick-up and return
- Instructions governing accidents and breakdowns

Insurance   Insurance can be provided by commercial underwriters or by self-insurance. A variety of property damage, public liability, and personal injury coverage programs are available through underwriters. If an institution chooses self-insurance, premiums can be paid into a vehicle-loss reserve account that is drawn on when needed. Self-insurance can be expected to be more economical over a long period. Regardless of the method adopted, insurance costs should be determined or estimated. These costs must be included when establishing vehicle rates.

35.6   VEHICLE RATE STRUCTURES

Cost Categories and Rate Computation   Rate structures are based on three categories of costs:

1. Amortization of the cost of vehicles.
2. Fixed costs, which include insurance, equipment, salaries, and administration of the transportation system.
3. Variable costs that depend on the amount of use. These include maintenance and repair; fuels, oils, and other lubricants; tires; and repair parts and supplies.

Various kinds of rate structures are used. The one employed depends largely on the accounting requirements of an institution. The following example presents a method for computing rates based on all fixed costs and variable costs. This method can be adapted to virtually any other rate system.

Example Computation of Rate   Before computing rates, certain data need to be determined or estimated. The numbers presented are for example purposes only. The data that should be gathered are as follows:

- Vehicle replacement cost ($7,500)
- Expected life (5 years [60 months])
- Residual value ($1,500)
- Estimated mileage (100,000 miles)
- Scheduling efficiency (70 percent)
- Estimated variable costs ($10,000)
- Fixed costs other than replacement ($150 per month)

Daily rates based on fixed costs are determined as follows:

- Determine net replacement cost by subtracting residual value from actual replacement cost ($7,500 - $1,500 = $6,000).
- Determine monthly reserve accrual required by dividing the net replacement cost by the vehicle's expected life ($6,000 ÷ 60 months = $100). Group all vehicles of the same category together to obtain an average monthly reserve accrual rate per vehicle type.
- Add monthly fixed costs to monthly required accrual to determine total monthly fixed costs ($150 + $100 = $250).
- Determine actual days of use per year by multiplying the number of days per year times scheduling efficiency (365 x 0.70 = 255.5).

- To determine the daily rate, multiply monthly fixed costs by 12 months and then divide that figure by the actual days of use per year (250 x 12 ÷ 255.5 = $11.74).

To determine mileage rate based on variable costs, divide the total estimated variable costs by the vehicle's estimated total mileage ($10,000 ÷ 100,000 miles = $0.10/mile). The vehicle rate in this example would be $11.74 per day plus $0.10 per mile.

Alternative Method of Computing Rate   An alternative method of computing rate is as follows:

1. Determine or estimate data for:

Indirect salaries and budget	$185,320
Quantity of vehicles	246
Total mileage, all vehicles, one year	2,483,073
Anticipated insurance rate per vehicle	$245.20
Original cost	$6,015
Depreciation rate	0.358
Retirement mileage	75,000
Operation cost, average over one year	$0.063
Uncollected accidents, all vehicles, one year	$14,622
Maintenance cost, vehicle type, per year	$1,001.86

2. Determine fixed cost:
   Administrative overhead = salaries and budget ÷ total vehicle miles = 185,320 ÷ 2,483,073 or $0.075 per mile.
   Insurance cost = number of vehicles x insurance cost per vehicle ÷ total miles = 246 x 245.20 ÷ 2,483,073 or $0.024 per mile.
   Depreciation cost = vehicle cost x depreciation rate ÷ retirement mileage = 6,015 x 0.358 ÷ 75,000 or $0.029 per mile.
   Total fixed cost = administrative overhead + insurance cost + depreciation cost = 0.075 + 0.024 + 0.029 or $0.128 per mile.

3. Determine variable cost:
   Operating cost = fuel cost per mile + maintenance cost per mile = $0.063 (averaged by type of vehicle)
   Accident share cost (repairs to transportation service vehicles not collected from other party or departments, data from records): Uncollected cost ÷ total miles = 14,622 ÷ 2,483,073 or $0.006 per mile.
   Total variable costs = operating cost + accident share cost = 0.063 + 0.006 or $0.069 per mile.

4. Total cost per mile = total fixed cost + total variable cost = 0.128 + 0.069 or $0.197 per mile.

5. Minimum daily charge = total cost per mile x average miles per trip = 0.19 x 70 or $13 per day.
   Thus, the minimum daily charge is $13, which includes the first 70 miles. The mileage rate of $0.19 applies to all miles exceeding 70.

Van Pool Rates   Many colleges or universities have either established or supported a van pool operation to conserve energy, reduce pollution, reduce traffic, or provide a service to their employees. A van pool differs from other motor pool transportation in that it is designed solely to provide transportation between home and the work site. Riders are charged a monthly rate for this service, which is scaled to pay all expenses for acquiring, maintaining, and operating the vans. In some cases funds to establish the operation have been provided by the federal and/or state governments. Rates can be derived by using the systems described above for passenger vehicles. Customarily van pool riders are charged a flat rate per month that includes both fixed and variable costs. It is important that the institution review its program with local mass transportation companies to avoid possible conflict.

Truck Rental Rates   Rental rates for trucks can be developed by using the same system as outlined for passenger vehicles. The scheduling efficiency for trucks will usually be lower than for passenger vehicles. Their retention period, however, will be longer. When used primarily for maintenance and construction purposes, trucks are normally charged at an hourly rate. For larger trucks, licensing requirements usually stipulate that a driver be provided. The cost of the driver could be added to the truck rental rate or handled separately.

Bus Rates   Many institutions provide buses for class field trips, athletic team travel, and campus tours. In addition, buses are used to shuttle employees between remote parking lots and campuses or from main campuses to satellite facilities. Because of the high acquisition cost, driver salaries, and operating expenses, bus operations are expensive. Therefore, high usage rates are required to make bus operations self-supporting. However, other factors such as convenience to students or reduction of campus traffic may be of such importance that bus operations are subsidized. In these cases, break-even rates should be calculated so that the required subsidy can be considered in the decision-making process.

Bus rental rates should include a mileage charge and a time rate, such as hourly, day, and daily rates and rates for longer periods. The time rate should incorporate the driver rate, which may include overtime and per diem costs.

Heavy Equipment Rates   Because the amount of travel is a poor indicator of usage for this type of equipment, hourly rates are commonly used. Time-based rates can be determined by using the same factors as passenger vehicles, with the exception that variable costs are calculated based on a unit of time (i.e., hour, day, etc.), rather than the number of miles. Large, expensive pieces of equipment such as cranes, bulldozers, and backhoes are normally retained for many years and therefore require long-term use projections to establish rental rates.

35.7   MAINTENANCE AND REPAIR

Vehicle maintenance and repair can be classified into three categories: driver maintenance, garage preventive maintenance, and repairs. The first two categories are both preventive maintenance, but are separated for description purposes into maintenance performed by drivers and that requiring garage mechanics.

Maintenance by Drivers   A clear policy should be established that directs drivers (operators) of vehicles to periodically check the condition of coolants, lubricating oil, tire pressure, and other routine items that any responsible private owner of a similar vehicle would typically check. The vehicle operator is the first line of defense. Each time a driver signs out a fleet vehicle, he or she should be provided with a checklist of these maintenance items. The checklist should include the following:

- All fluid levels should be checked: engine oil, radiator coolant, automatic transmission fluid, and power steering fluid.
- Tires should be checked for proper inflation, unusual wear, or penetration by foreign objects.
- Other areas to be checked include lights, horn, signals, mirrors, windshield washer/wipers, and seat belts.
- Before moving vehicles, brakes should be tested to ensure adequate operation.
- During operation, the driver should observe operation of instruments, brakes, steering, and engine and power drive components.
- After operation, the driver should inspect the vehicle for fluid leaks or other problems.
- Any problems should be reported to the proper authority.

Preventive Maintenance   Transportation equipment should undergo periodic preventive maintenance inspections and service in accordance with the manufacturer's recommendations. These inspections and servicing programs, similar to a building preventive maintenance program, can be set up on the basis of hours of operation, mileage, or elapsed time. The preventive maintenance program should include oil change and lubrication, timely check and adjustment of the cooling system's antifreeze, engine tune-ups, repacking of wheel bearings, replacement of cooling system hoses, changing of oil and air filters, replacement of drive belts, checking and adjustment of brakes and transmission fluid, and similar measures. The normal schedule is based on regular use under normal conditions. The schedule may be accelerated owing to harsh use, adverse climatic conditions, or excessive stop-and-go operation. It is the transportation system manager's responsibility to ensure that these measures are carried out, including calling in permanently assigned vehicles when required.

Scheduled maintenance should cover the following:

- Oil and filter changes every 3,000 miles for hard service, or every 4,500 to 6,000 miles for normal service
- Air filter change at recommended mileage or time interval
- Coolant service and inspection, with coolant change every two years
- Tune-up depending on severity of vehicle use—some annually, others more often
- Hoses and belts checked for softness or wear, and replaced if necessary
- Transmission fluid changed depending on mileage or time and the fluid's color and level
- Exhaust system visual inspection and leak test
- Brakes and brake fluid inspection
- Other services as required by warranties or local authority ordinances

A service checklist is included in Figure 35-3.

Vehicle Repairs   Necessary repairs, such as replacement of parts, will normally be identified through preventive maintenance inspections or breakdowns. Where repairs are performed will depend largely on the garage facilities at the institution and the current workload. Customers can also assist with reporting problems.

When deciding whether to contract maintenance and repair or to establish a college or university garage, the size of the vehicle fleet will usually be the major consideration. Smaller institutions that own a small fleet of vehicles can seldom operate a garage economically because of the large investment in tools and equipment that it entails. Contract maintenance, full maintenance–lease programs, and servicing agreements may be more cost-effective.

Larger institutions can often realize greater economies by staffing and equipping a garage. However, few institutions find it cost-effective to provide facilities for body work, replacement of major vehicle components, or overhauls. Because staffing and investment in tools and equipment are extensive and the work requirements unpredictable, it is usually more economical to take care of major repairs by contract. A final determination, however, must be made by each institution based on the size of the vehicle fleet, comparative costs, and availability of local facilities for contracting.

Consideration must be given to the environmental issues related to operating a vehicle maintenance facility. Petroleum distillates (i.e., gasoline, oil, and grease) need special handling and disposal. Underground storage tanks must be monitored for leakage. Used oil and grease are probable carcinogens and require added safety precautions. Brakes contain asbestos. Antifreeze is toxic to animals. Batteries contain acid and lead. Air conditioners use freon (chlorofluorocarbons). Recycling and reclamation programs can reduce the overall operational risk and must be incorporated into the standard operating procedures within the maintenance shop. Examples include

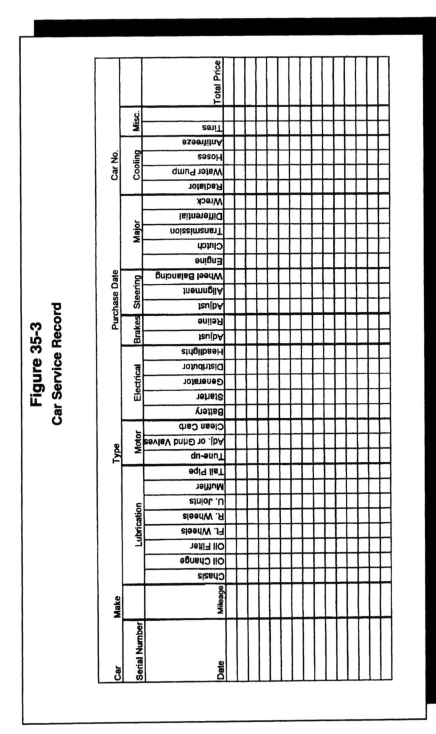

Figure 35-3
Car Service Record

oil filter crushers that capture previously unclaimed oil, antifreeze cleaning equipment, and freon recycling systems.

Vehicle Records A detailed operating and maintenance record should be kept for each motor vehicle. Records should provide a complete history of the maintenance and repair performed, as well as the costs of such maintenance and repair.

35.8 STAFFING

Staff size for transportation service varies with the size of the vehicle fleet, the amount of maintenance and repair handled by the facilities department, and the measure of administrative control that is exercised over use. Small colleges may have only one person to schedule vehicles and maintain requisite records. Larger transportation organizations require mechanics, parts persons, service attendants, fuel attendants, dispatchers, billing and records clerks, and supervisory personnel (Figures 35-4 and 35-5). Note that leasing vehicles does not eliminate the requirements for dispatching, billing, and other administrative functions. Leased vehicles are in the custody of the college or university, but all functions relative to their use and maintenance—except as otherwise provided for in the lease agreement—are similar to those for owned vehicles.

Equipment Management Information Systems Management of transportation services can be greatly assisted by the computer. The management end of transportation is a numbers game. Buying the best vehicles for the lowest cost, meeting customer demand, and recording downtime all influence the success

Figure 35-4
Staffing and Skill Levels
Allows for task assignment and staffing to be used most efficiently.

Level	Description	Percentage of Repair Jobs
A	Top-notch mechanic, all jobs	12
B	Good all around person lacks some skills on complex jobs	38
C	Most jobs within capability; requires close supervision	38
D	Go-for helper type	12

Figure 35-5
Staffing
Vehicle fleet size determines number of staff needed.

Small	Service
	Maintenance
	Administrative (typically very few people)
Large	Mechanics
	Parts personnel
	Service people
	Fuel attendant (or automated pumps)
	Dispatching
	Clerks
	Supervisors
Leased	Administrative
	Fuel
	Dispatching
	Billeting
	Supervision

of the operation. Computers can record data so that it can be organized to generate meaningful reports and enable better decision making.

Computerized systems have been developed to assist managers in the following areas:

- Accounting
- Ownership records and registration
- Equipment repair history records
- Cost reports—maintenance and operations
- Performance standards and targets
- Mileage/utilization
- Downtime/shop backlog
- Preventive maintenance scheduling
- Automated fuel dispensing system

A computer system should provide exception reports as well as scheduled statistical reports, and access should be simple, with safeguards for privileged information. The system should flag performance that does not meet predetermined standards and should be designed so that all individual databases (administrative, maintenance, and operations) are interconnected. A good system will be able to handle a multiplicity of labor-intensive functions according to predetermined schedules and to accommodate updates.

35.9 CONCLUSION

Of all the functions that may be performed by the facilities department, none varies more in scope and methods than meeting the institution's transportation requirements. The rates and availability of vehicles and equipment for lease or rent fluctuate depending on location. Past practices, however, influence current methods. If purchasing a vehicle fleet and establishing the facilities and organization for its maintenance is not presently cost-effective, continuing an operation where one already exists may be the most economical alternative. Administrators need to remember that a careful evaluation of all factors must be made in each situation to determine the best method for meeting the institution's transportation requirements.

ADDITIONAL RESOURCE

Finucane, William S., and John P. Harrod Jr. "Fleet Management." In *Critical Issues in Facilities Management: Work Control.* Alexandria, Virginia: APPA: The Association of Higher Education Facilities Officers, 1988, pp. 70–85.

CHAPTER 36

Environmental Health and Safety

Dr. Ralph O. Allen
University of Virginia

36.1 RISK CONTROL IN FACILITIES MANAGEMENT

Facilities managers have traditionally been concerned with the fundamental operational factors of performance, cost, and schedule. Risk is usually thought of as the opposite of benefit. Most managers desire to be perceived as providing benefit to their organizations and do not want to be thought of as risk managers. Yet it has become increasingly clear that managers who ignore risk, overly delegate its management, or handle it in a haphazard manner will eventually experience problems. There is also increasing recognition that employees' attitudes and perceptions about managers are partially based on their manager's approach to risk management. If employees are unhappy, then a health or safety issue can become a convenient means to gain a sympathetic audience and a platform from which managers and organizations can be criticized. This criticism may lead to investigations by regulatory agencies and eventually litigation. Another important factor that forces managers to foresee and manage risks is that it is considerably more expensive to remedy a problem than to do the job correctly the first time.

It is instructive to consider why managers have ignored or delegated the management of risk rather than placing it on a par with managing benefits. In his book, *Managing Risk: Systematic Loss Prevention for Executives,*[1] Vernon Grose suggested three reasons. Because managers are generally positive thinkers, they may avoid thinking about what could go wrong. A second possibility is that managers may reason that as long as they have enough success, the problems might be ignored, cleaned up after the fact, or perhaps blamed on others. Finally, Grose suggests that managers see their enterprises as an extension of themselves, and they manage risk just as they do in their personal lives. In our personal lives we often trade risks without really considering them. A simple example might be speeding up to get through a yellow traffic light before it turns red. Time is too short to do the physics (time x speed = distance) or to weigh the cost versus benefits (probability of injury and additional insurance costs vs. a few minutes of time).

Although people have the capacity to process, store, and analyze a great deal of information, it is impossible to carefully consider every risk situation in an increasingly complex environment. As the result of this information overload, humans have developed some shortcuts. People learn to categorize things, either by trial and error or through formal instruction, as, for example, "safe" or "unsafe" or "good" or "bad." Once something is categorized as safe, people no longer think about it, but proceed as though it were safe. Even jobs that some might consider unsafe become safe and commonplace for those who have performed the job successfully many times. In addition, people take many risks for recreation and participate in activities such as skiing, hunting, and swimming but are loath to accept risks that are beyond their control, such as toxic wastes in landfills and suspected carcinogens in food. These attitudes are inherent problems that managers face in adjusting their own, as well as their employees', risk behaviors.

Risk Awareness for Better Management

Safety doesn't just happen; it must be cultivated by management. Managers and their employees need to communicate and work together if risks are going to be effectively managed.

If a manager has little regard for an employee's health and well-being, then there is little motivation for the employee to satisfy the manager's desire to accomplish a goal with speed and efficiency. The attitudes of managers and supervisors toward their employees' own personal safety has a great deal to do with the morale of those employees. Management studies and common sense suggest that prevention of serious problems costs less than cleaning up after an accident, particularly one that leads to regulatory enforcement and negative publicity.

Because resources within an organization are limited, managers must have a way to choose among many demands and priorities. A systematic approach to risk management must therefore include some means of ranking risks. Each manager must adopt a rational approach that fits his or her personality; this approach can range from intuitive ranking of risk, utilizing common sense and experience, to the systematic mathematical approach of systems engineering. Managers generally have the most holistic view of an operation and can best consider most of the risks involved. In many cases managers will have historical information such as past loss reports and accident investigations. Some useful information can also be gained by asking "what if" questions to anticipate the possible problems that could occur. Employee perceptions of on-the-job dangers can also be helpful and should always be investigated. Managers should engage in open, collaborative trading of risks. Planning and considering risks in an open manner allows employees to "buy into" a process by which limited resources can be used effectively to control the risks. In some cases merely the open discussion of a risk does a great deal to eliminate the problem by making employees more attentive.

Legal Reasons to Manage Risks

There are many who would argue about the proper or acceptable role of government in protecting an individual's own health and safety at the expense of personal liberty. Supreme Court Justice Stephen Breyer, in his essay on government regulation, "Breaking the Vicious Circle: Toward Effective Risk Regulation,"[2] points out that because anyone might be harmed by anything, it is not possible to regulate all of the risks that fill the world. This is not to say, however, that a lawyer might not attempt litigation over an uncontrolled risk. For managers, the primary issue is negligence. An act may be interpreted as negligent, even if the risk was not understood, if there is some possible causal relationship between the act and an incident that resulted in damage or injury. Because regulations provide a defined standard of conduct, behavior, or expectations, a failure to meet these standards not only will make a manager negligible if an accident occurs, but it will be a violation even if no accident occurs.

It is difficult to understand and comply with regulations. However, the consequences of ignoring regulations can be severe in both real costs (i.e., fines and liability awards) and in terms of the image of the institution.

There are two major regulatory programs pertaining to the safety of employees while on the job. The older of the two programs began with the 1902 Workers' Compensation Act and guarantees employees compensation for on-the-job injuries, including those from chronic exposures to hazardous materials, regardless of fault. The second major body of regulations began with the 1970 Occupational Safety and Health Act, which established minimum standards of safe working conditions for all employees. Initially the emphasis in safety was on compliance with the specific Occupational Safety and Health Administration (OSHA) regulations, but the recent trend has been to look more to the overall effectiveness of the safety program, with an emphasis on educating employees about job safety. There are also numerous environmental regulations aimed at protecting the environment. These are discussed later in this chapter.

Most safety regulations start out as federal initiatives brought about by public pressure or interested lobbying groups such as labor unions and environmental coalitions. After a law is enacted, one or more regulatory agencies must write, publicize, and implement detailed regulations, which are the agencies' interpretations of the law's intent. Although applicable to colleges and universities, new regulations are often written in broad terms for industry rather than for the facilities operations of a college or university. The applicability of some regulations depends on the size of the institution.

Once a federal agency finalizes a set of regulations, individual states may agree to implement federal programs using their own agencies. The state must establish regulations that are "at least as stringent as" those of the federal program. State regulators may, and often do, adopt more demanding regulations.

36.2   ESTABLISHING SAFETY PROGRAMS FOR FACILITIES MANAGEMENT

Understanding Risk and Perception

Managers must develop a plan for their own particular circumstances, but the following observation on risk, perception, and human behavior may help any manager analyze and handle risks that are within his or her control.

There are some activities for which there are risks that are inherent or constant. Working with high-energy sources such as electricity, with dangerous or highly toxic materials, or in tenuous positions (e.g., high above ground or in confined spaces) are all cases where there are inherent and serious risks. The potential consequences in these situations are severe. These are perhaps the easiest risks for safety programs to address, in part because there are often specific OSHA regulations concerning the activities. Other situations carrying inherent risks, such as ice on sidewalks, do not have severe consequences, but a little forethought can identify them even though they may be unavoidable.

A constant factor that managers can anticipate as a cause for increased risk is change. Risk is increased whenever a new procedure, tool, or product is used. Usually the risks are not known when something is new. This becomes an even more complex problem when there is a change from something old and familiar to something new and unknown. Although change increases the chances of problems, a manager should realize that there is always risk when people do not know or are not aware of the potential problems. "Design safety margins" can be incorporated to compensate for ignorance, but it is better to create awareness and provide knowledge about avoiding problems. Managers who try to build a safety system based on strict guidelines are destined to fail. Thus an important part of a complete risk program is to provide solid information and to challenge people to think about their risk activities. In several of the recent key OSHA regulations, the elimination of ignorance by means of specific training has been seen as critical to controlling risk. Incidentally, training records are easy for regulators to inspect, and failure to provide the training is interpreted as an "intentional violation" of the regulation. However, although training can address ignorance in some cases, it cannot eliminate uncertainty. There are often cases where the information is simply not available or clear.

When considering the perceived risks of less familiar materials and the real risks associated with ignorance, it is important to note that familiarity can also affect risks. For most people the complex mathematical estimates and descriptions of risk have a much smaller impact than first-hand experience. It also follows that when someone has been doing a job or working with a particular material for years without any adverse consequences, it is hard for that person to imagine that it is anything but safe.

In addition to those risks and problems that can be easily recognized, there are conditions that happen without a predictable pattern. These apparently unpredictable situations, or those beyond our control, are often called *dynamic risks*. Although managers may not be able to prevent the occurrence of dynamic risks, anticipation can minimize the consequences. If managers develop a work environment and culture where employees are encouraged to think through a job and decide what might go wrong, significant awareness and benefit will occur. Getting input from employees helps focus their creativity in positive ways, and helps everyone succeed in reducing risks.

There are a number of ways that a manager and staff can identify, evaluate, and control risk:

- Learn from your own experiences: Accident investigations should focus on the problems and processes, not who was responsible.
- Learn from the experience of others: Statistics show common combinations of actions or situations that have higher probabilities of leading to serious consequences.
- Use expert advice and become fluent in government regulations as strategies to help identify risks.
- Find effective ways to encourage employees to express concerns and to train supervisors in foreseeing problems.
- Perform work site evaluations and inspections.
- Identify risks prior to an accident by developing scenarios of what might happen. This is sometimes called a *job safety evaluation,* but it need not be complex or difficult.

Effective Risk Management Programs

Many ideas can be discovered by looking at safety programs that have been effective. The first and most important factor is the participation of the people who are being protected. Everyone needs to be involved in risk evaluation and management, but supervisors are the most important employees for establishing and maintaining effective safety programs. First-line supervisors have some management responsibilities but are closest to the workforce and to the problems faced by workers. Upper-level managers are important, because they establish the institutional priority to support the training of first-line supervisors. Managers must provide positive reinforcement for good supervisors and the resources necessary for their employees to carry out their work successfully and safely. Each manager must be clearly charged with carrying out programs in a safe and legal manner. If an individual is designated as the safety director, he or she should serve as a resource or a coach to assist these managers and should not be in charge of the safety program.

Beyond the human resource component, the most important factor in an effective safety program is communication. Safety committees can be used as an effective means of creating and enhancing communication. In many institutions safety committees have not been used effectively and have gained a poor reputation. Rather than dwelling on the negative aspects that often discourage employees from taking an active role in a meaningful safety program, it is more useful to concentrate on the positive communication that can result. If management truly embraces two-way communication with staff, then managers should not strictly control the safety committee. Setting convenient meeting times and places, enabling staff to establish or at least help set the agenda, and encouraging two-way communication sends a positive message to the staff. Safety committee meetings should be short and held frequently, and should have a varied agenda. Committees should comprise first-line supervisors and their staff. In some cases the makeup, agenda, and implementation processes may be governed by union contracts or National Labor Relations Board rulings. When employees raise a safety issue, management should be responsive and explain the reason for its decisions. Rather than giving directives, management can use the committee to establish procedures and programs to deal with employee concerns. A shared vision that accounts for staff concerns and accomplishes organizational goals is easier to implement and benefits the entire organization.

Another important function of safety committees is to establish or suggest training requirements. Employee training should be provided in small doses on a variety of topics. Initially management may want to develop an agenda of some topics and bring in presenters or videotapes. With encouragement, committee members can suggest topics and even be asked to make short presentations about issues in their areas. Accident investigations can be discussed to help develop training materials and identify where training is necessary. Accidents, as well as recommendations from accident reports, should be analyzed and discussed by the safety committee. The committee should also discuss and evaluate possible changes in procedures, policies, or equipment to avoid repetition of accidents. Another area of training focus for committee members might include hints on how to inspect and evaluate risks in their own work areas, how to more thoroughly investigate accidents, and how to transmit what they have learned to their staff. It is clear that safety committee members can benefit from participation and can share their experience and knowledge with others. Rotation of membership is one way to spread participation to more of the staff. Evaluation of supervisors should reward good communication and concern for the safety of their staff. This includes active participation in safety committee activities.

The following lists summarize some of the organizational factors that have had positive and negative impacts on accident frequency and severity in a variety of settings:

Positive Impacts:

- Management commitment to safety
- Proper tools, equipment, and work environment
- Sustained safety training programs
- Continued accident investigation with corrective follow-up
- Monitoring hazardous behavior and accident-prone employees
- Temporary accommodation of injured employees in light duty capacities
- Open communication lines between supervisor/management and employees
- Enforcement of and compliance with State and Federal regulations and standards

Negative Impacts:

- Lack of sincere management commitment
- Broken or inadequate tools and equipment
- Hazardous work environment (e.g., unkept, old buildings)
- Organization restructuring and financial restraints
- No safety training
- No accident investigation
- No light work accommodation
- Negative supervisory attitudes
- Noncompliance with or lack of enforcement of regulations

36.3 HEALTH AND SAFETY PROGRAMS FOR FACILITIES MANAGEMENT ORGANIZATIONS

The demands of regulatory agencies do not always correspond to controlling the risks that are most serious for facilities managers. However, in some cases, when a health and safety program is mandated, the consequences of noncompliance can become greater. Experienced managers make themselves aware of regulatory requirements and allocate the resources necessary to develop programs that meet compliance. It is important for upper-level managers to be aware of regulations and to know how they affect an institution. For example, in the United States, to know how the Environmental Protection Agency's (EPA's) regulations on hazardous waste affect an institution, it is necessary to understand and use the EPA's definitions to determine how much waste the entire institution generates. Institutions should also understand that regulatory agencies will take enforcement actions that single out managers who "should have known" but failed to implement mandated programs. These can be quite costly to an institution and to the individual, through personal civil lawsuits. Citation of violations can also foster negative publicity, even when those who caused the violation are only contractors working for an institution. Finally, it should be noted that most regulatory violations are not discovered through normal agency inspections but through accidents or complaints from employees or the community. When agency inspections do occur, much of the emphasis is on following a "paper trail." Written procedures, documented programs, and training records become important. Although written material is critical during an inspection, a manager's goal should be to have a program that is effective in meeting the spirit of the regulations. If overall programs are effective, the possibility that an incident or complaint will set off the inspection process can be minimized. Again, the most effective method of risk control is to develop a workforce that is risk knowledgeable. Implementing good training programs still remains the most useful tool for enhancing the overall safety of employees.

Regulatory Programs Emphasizing Training

In the United States there are many regulatory programs that emphasize training. In other countries there are similar requirements. The list of regulations for U.S. colleges and universities (in Figure 36-1) is generally derived from the Code of Federal Regulations (CFR), although there are often state and local regulations that implement these federal mandates. Also, there are other parts of the CFR that deal with training; however the section with the most training requirements applicable to facilities operations is CFR 29, which is the basis for OSHA regulations. Figure 36-1 provides a brief description indicating the general content and circumstances or type of people who need training. There are a variety of ways to implement this training, but it should be made relevant to each company's own circumstances and should be documented. Figure 36-2 includes activities in which there are specific types of equipment that may be installed, inspected, or used within the scope of facilities

Figure 36-1

Mandatory Training for Selected Personnel

Title/Standard Number	Description
1. Employee Fire Prevention and Emergency Plans 29 CFR 1910.38 (a)(5) and (b)(4)	A basic course on fire safety is required for all employees. A fire plan must be developed (required elements described by OSHA) and training must conform to these elements.
2. Portable Fire Extinguishers 29 CFR 1910.157G	If portable fire extinguishers are available, the building occupants must be trained in their use. Training should include flight-or-fight decision making and selection and use of extinguishers.
3. Occupational Health–Hearing Protection 29 CFR 1910.95(K)(1)-(3)	OSHA has established a permissible exposure limit of 90 dB and requires a hearing conservation program if noise levels for an eight-hour shift exceed 85 dB. If high noise levels are indicated and engineering controls cannot effectively attenuate the noise, then hearing protectors are required as well as training on the correct use of hearing protectors and participation in the hearing conservation program. Training is recommended for personnel assigned to repetitive work tasks generating high noise levels (i.e., generators, power tool operators, boiler operators).
4. Occupational Health–Ionizing Radiation 29 CFR 1910.96	Anyone working with or near radiation-producing equipment or materials requires, at a minimum, some awareness training. Risks of radiation exposures and regulatory and administrative procedures should be covered.

(Figure continues)

5. Personal Protective Equipment
   29 CFR 1910.132(f)

Training should include an overview of OSHA Personal Protective Equipment (PPE) Standard, including initial work environment hazards assessment with supervisor to determine PPE requirements. Type of PPE required and proper use and care of PPE must be covered for specific work duties.

6. Respiratory Protection
   29 CFR 1910.134 (b)(3)

If respirators are available or used, managers need an overview of the OSHA Respiratory Protection Standard and elements of implementing a respirator program. Initial assessments of the work environment will determine if respirators are needed. Respirators may be required if engineering controls (i.e., fume hoods or other local exhaust ventilation) cannot be used to control potentially hazardous airborne contaminants. Qualitative fit testing and medical approval are required on initial assignment of respirators and annually thereafter for personnel assigned to wear negative-pressure air purifying respirators.

7. Lockout /Tagout Program
   29 CFR 1910.147

Managers need an overview of the OSHA Lockout/Tagout Standard (LOTO). All personnel responsible for machinery and equipment shutdown for maintenance and service must be trained to understand and apply LOTO to control hazardous energy sources during machinery and equipment maintenance and service. In addition, OSHA requires training on standard operating procedures (SOPs) to perform LOTO. This training must be provided by department supervisory management familiar with applicable SOPs for all affected personnel. This includes personnel who must shut equipment down, as well as personnel who may be working on the equipment and will be affected by the shutdown.

8a. Asbestos Awareness
29 CFR 1910.1001

All facilities employees must be given a general awareness of asbestos-containing materials (ACM), including history, health effects, use, recognition of damaged ACM, and description of the asbestos management plan and policy.

8b. Asbestos Task Force
29 CFR 1910.1001

Personnel required to work in asbestos "restricted access" areas should be instructed on proper use of PPE and procedures for safely working in asbestos-contaminated worksites.

8c. Asbestos Operations and
Maintenance
29 CFR 1910.1001

Personnel who will repair/maintain or remove <10 ft. of ACM must take an Environmental Information Association certification course that presents all general awareness information as well as ACM repair/maintenance, small-scale abatement procedures, use of PPE, and respirator fit testing. This is a 16-hour course.

8d. Asbestos Abatement Workers
29 CFR 1926.58

An Environmental Information Association certification course is required for all asbestos abatement workers. Covers all information in the awareness and operations and maintenance courses and presents one day of hands-on training in work site enclosure, decontamination unit, and negative air systems construction. This is a 24-hour course.

8e. Asbestos Abatement in "Your"
Building
29 CFR 1910.1001
29 CFR 1926.58

Building occupants subject to future ACM abatement often develop a concern about activities associated with these projects. Although not required, an awareness course helps

(Figure continues)

describe asbestos abatement procedures in layperson's terms for building occupants. The course discusses daily project inspection and air monitoring procedures.

9. Lead-based Paint and Lead Awareness Training
29 CFR 1910.1025
29 CFR 1926.62

Renovations/painting crew workers, supervisors, and superintendents need a general awareness of lead-containing products including lead-based paint (LBP) history, health effects, uses, university policies, personal hygiene, and removal procedures.

10. Hazard Communication Standard
29 CFR 1910.1200

Under the Hazard Communication Standard, all employees exposed to chemicals and chemical products must be informed about the potential hazards associated with chemical use. Training covers proper labeling, material safety data sheets, emergency procedures, and a description of the written program adopted by the institution.

11. Chemical Hygiene Plan— Hazardous Chemicals in Laboratories
29 CFR 1910.1450

All laboratory employees using hazardous chemicals must receive training designed to reduce occupational exposures to hazardous chemicals in laboratories. Training covers first aid and emergency procedures in the laboratory, safe chemical storage, ways to clean chemical spills, hazardous waste management, and ways to report possible exposure. The written plan must include the way the institution will investigate exposures and provide medical monitoring.

12. Formaldehyde Employee Information and Training
29 CFR 1910.1048

Formaldehyde is frequently encountered in the workplace by laboratory workers, medical students, and research professionals. A comprehensive review of its toxic properties, monitoring requirements, and safe work practices are required by regulation.

13. Confined Space 29 CFR 1910.147	This is mandatory for personnel who enter confined spaces, including supervisors responsible for entries into permit- and non-permit-required confined spaces (e.g., heating plant and utilities personnel).
14. Overview of Excavation Standard 29 CFR 1926.650	Standard defines a competent person who understands these performance-based standards (vs. task-based), which apply when workers must be protected (when excavations are greater than 5 ft. in depth). Factors such as shoring procedures, emergency plans, inspections, etc., can be learned from written material or by training course.
15. Bloodborne Pathogens 29.CFR 1919.1030	Initial and annual training is required for all departmental personnel who may become exposed to blood-borne diseases in their routine work. The institution must devise a procedure to determine which employees may be exposed and offer them hepatitis vaccinations. Training should cover material required by OSHA Occupational Exposure to Bloodborne Pathogens and Draft Guidelines for Preventing the Transmission of Tuberculosis in Healthcare Facilities.

management. In each case the regulation or the American National Standards Institute (ANSI) standard that is recommended also includes an element of training. In these cases the content of the training is not spelled out, as it is for those standards that apply to general programs. In most cases it is reasonable to expect that individuals using such equipment be knowledgeable about the safe operation of the equipment. In addition to mandated training, there are some cases where training is not yet a federal mandate, but there is movement toward developing such programs. Several of these areas are noted in Figure 36-3.

Figure 36-2

Equipment for Which Training is Mandated

Title/Standard Number	Description
Fixed Extinguishing Systems 29 CFR 1910.160(b)(10)	Self-explanatory
Fire Detection Systems 29 CFR 1910.164(c)(4)	Self-explanatory
Operations—Powered Platforms (Type F and Type T for Exterior Building Maintenance) 29 CFR 1910.66	Self-explanatory
Operations—Aerial Devices 29 CFR 1910.67	Self-explanatory
Operations—Fall Arrest Systems 29 CFR 1910.66, 29 CFR 1910.67	Self-explanatory
Servicing Rim Wheels 29 CFR 1910.177	Self-explanatory
Powered Industrial Trucks 29 CFR 1910.178	ANSI B56.1-1988 High Lift Low Lift Trucks
Power Actuated Tools 29 CFR 1910.243(d)	ANSI A10.3-1985 Safety Requirements for Power Actuated Fastening Systems

Training for the following must cover safe operation procedures and equipment inspections for equipment integrity and equipment safety:

Machine Guarding 29 CFR 1910.212	Self-explanatory
Woodworking Machinery 29 CFR 1910.213	ANSI 01.1-1992 Safety Requirements for Woodworking Machinery

Abrasive Wheel Machinery 29 CFR 1910.215	ANSI B7.7-1990 Safety Requirements for Abrading Materials with Coated Abrasive Systems ANSI B7.1-1988 The Use, Care, and Protection of Abrasive Wheels
Electrical Safety-Related Work Practices 29 CFR 1910.331-335	ANSI/NFPA 70-1993 National Electrical Code; ANSI C2-1993 National Electrical Safety Code
Oxygen-Fuel Gas Welding and Cutting 29 CFR 1910.252(a)(4)	American Welding Society A6.1-1966 Recommended Safe Practices for Gas Shielded Arc Welding
Arc Welding and Cutting 29 CFR 1910.252(b)(1)(iii)	ANSI Z49.1-1988 Safety in Welding and Cutting
Resistance Welding 29 CFR 1910.252(c)(1)(iii)	

36.4 REGULATION OF HAZARDOUS MATERIALS

In the operation of an institution of higher education, there are wide ranges in the types and amounts of hazardous materials that are utilized. For smaller institutions, the materials generated by the operation of the facilities, such as oils, excess paints, and small amounts of chemicals used in teaching laboratories, are all that need to be considered. For research institutions and medical facilities, however, the problems are much greater and expensive to solve. Although the following describes some of the materials that are regulated, it is not definitive and does not provide detailed solutions. There is one general piece of advice that is useful for managing all of these regulated materials: Be aware of where these materials are used in your institution, and be sure that they are being *managed*. Hazardous materials management begins with the purchase and continues while the material is being stored, used, and disposed. In some cases responsibility does not end even with disposal. Attempts to solve one problem can create others. For example, in some states the use of fluorescent bulbs to save energy results in hazardous waste disposal costs because the bulbs contain mercury and require collection. All of these factors should be considered and balanced.

There are twenty-nine specific chemicals that are regulated by OSHA, and exposure limits are established for hundreds more in 29 CFR 1910, subpart Z (Toxic and Hazardous Substances).[3] Generally facilities personnel do not work with chemicals or products over periods of time that

Figure 36-3

Nonmandatory Considerations but Recommended for Selected Personnel

Consideration	Description
Employee safety and Workers' Compensation for supervisors and managers	A short overview of the Workers' Compensation law; policies and procedures for claims process should be discussed with supervisors and managers.
Workers' Compensation for all employees	The claims process should be described in detail from the time of the accident to final resolution. All rules and regulations are explained in a step-by-step procedure for all employees. This is a good opportunity to create general awareness.
Accident investigation for supervisors and managers	Motivational overview of concepts and applications in other industries can aid supervisors and managers in conducting accident investigations.
Heat stress	Employees who work in hot environments (and their supervisors) can be provided an overview of the signs, symptoms, and measurement techniques required to prevent detrimental effects of exposure to a hot and humid work environment.
Ventilation for contaminant control	New facilities management employees (project managers and designers) can be provided a course designed to familiarize the architecture and engineering staff with pitfalls and frequently encountered design discrepancies submitted by contract architectural and engineering

	consultants (e.g., recurring flaws and system losses). The eternal philosophical issue of "pay me now or pay me later" should addressed.
Work (back) injury prevention—material handling	Maintenance workers, material handlers, and groundskeepers should be given a presentation that includes descriptions of proper methods and realistic approaches to handling and moving materials correctly to prevent injury to back, shoulders, and wrists.
Office ergonomics—"Avoiding the Painful Desk Job"	For computer users, this course should emphasize low-budget techniques and tips to assist in creating a user-friendly, ergonomically safe workstation.

require occupational monitoring. On the other hand, even short-term exposures to materials that are odorous, cause headaches or respiratory distress, or burn the skin can cause great concern on the part of employees and building occupants. Limiting chemical exposures can be accomplished by ventilation design and operating procedures. For example, many cleaning products, paints, and solvents are likely to irritate building occupants and facilities staff if proper ventilation is not provided. The design of ventilation systems affects energy usage. Efforts to tighten buildings and recirculate air can lead to indoor air quality problems. Many building materials, including glues, carpets, and other finishes, emit gases over time and create problems in well-sealed buildings. Ventilation must also be adequate to avoid mold, mildew, and bacterial contamination, as sensitization to these indoor air problems is becoming diagnosed with increasing frequency.

The use of chemicals and chemical products requires compliance with the Hazard Communication Standard, 29 CFR 1910.1200.[4] Personnel protective equipment such as gloves and splash-proof safety glasses must be provided. Materials must also be stored appropriately. The material safety data sheets required by the Hazard Communication Standard can help identify both storage and personnel protective equipment requirements and can be obtained from the product manufacturer. They can also assist in comparing products to aid in the selection of the safest alternatives. Pesticides are one particular group of chemicals that must be considered carefully. These materials are highly regulated and may require specific licenses for those who are involved in their application. The use of contractors for pesticide application should be carefully controlled and coordinated to avoid problems with building occupants.

Polychlorinated biphenyls (PCBs) are specifically regulated materials that are commonly used as insulating material in transformers and capacitors. Most of these materials should already have been removed from most institutions. However, when they are encountered, PCBs must be monitored and disposed of as hazardous wastes.

Another specific set of regulations requires that all petroleum products stored in underground storage tanks be monitored to prove that they are not leaking. If tanks have been leaking or begin to leak, then cleanup will be necessary. Old, unused tanks must be removed.

Finally, it should be recognized that the presence of some materials can require continuous reporting if the amounts exceed certain threshold limits. Local emergency planning committees have been established under the Superfund Amendments and Reauthorization Act. It may be necessary to report to the local emergency planning committee because of such things as swimming pool chlorination systems or ammonia in large refrigeration units. It is prudent to participate with the such committees to ensure that there are integrated plans in place to deal with fire or chemical spills at the institution.

36.5 HAZARDOUS WASTE REGULATIONS

There are a considerable number of wastes that are heavily regulated, and for which disposal costs are often high. It is false economy to ignore these wastes. Cleanup costs and fines for illegal disposal will far outweigh any normal disposal costs. In some cases disposal costs can be avoided by choosing other materials or by purchasing smaller quantities. The clearest way to minimize costs, when it is necessary to generate hazardous wastes, is to keep wastes segregated. For example, if lead-based paints are removed from wood by dry methods such as scraping, the volume of waste is lower and the disposal is less complicated than when solvents are utilized. Likewise, if wood trim covered with lead-containing paint is removed and mixed with other demolition wastes, an entire dumpster full of mixed building debris may require disposal as hazardous waste based on the composite sampling results. Mixing cleaning solvents containing halogens (e.g., dichloroethylene) with hydrocarbons (e.g., used oil or alcohol) makes the disposal costs much higher.

There are specific requirements for the disposal of any materials contaminated with radioactive nuclides. This should clearly be the responsibility of the institution's radiation safety office. In many states wastes from hospitals, clinics, and some diagnostic or research laboratories must be disposed of as regulated medical wastes (e.g., sharps, infectious waste, and chemotherapy waste). In most cases disposal requires incineration or some means of sterilization.

Chemical waste disposal is highly regulated, and improper disposal can result in large fines and criminal punishment. EPA definitions of hazardous waste are often surprising to many managers. For example, photographic developing solutions are hazardous because they contain silver. In addition, there are several characteristics, such as flammability, corrosiveness, reactivity, and toxicity, that can also delineate wastes as hazardous. Wastes from chemistry laboratories and many other institutional research laboratories must be collected and kept segregated to the greatest extent possible. One of the most expensive sources of hazardous waste is unused and unwanted materials, with disposal costs often exceeding the purchase price. Therefore, it is prudent to avoid purchasing excess quantities. Facilities managers should also understand that any material used to clean up petroleum spills is classified as hazardous waste. When petroleum tanks are overfilled, the contaminated soil and anything used to absorb the ignitable petroleum must be disposed of as hazardous waste. It is best practice to promote spill prevention. It is sometimes possible to write contracts that make suppliers responsible for spill cleanups. Waste oils can be recycled, but great care must be used to avoid mixing other cleaning solutions and solvents with the oils. Efforts are under way to further regulate those who recycle oil, as well as companies who recover silver from used photographic developing solution. Finally, facilities managers need to recognize that fluorescent light bulbs (in which mercury levels exceeds 5.0 mg/L), ballasts (which contain PCBs) and batteries (not dry cell) are hazardous wastes. The EPA is considering making changes in how fluorescent bulbs should be managed to encourage their use for energy savings benefits.

36.6 ENVIRONMENTAL REGULATIONS ON UNIVERSITY ACTIVITIES

In addition to hazardous waste and community right-to-know regulations, the EPA is responsible for clean air and clean water standards, which affect most institutions of higher education. The application of the Clean Water Act varies depending on how much of the water system is operated by the institution. Water supplies must be tested for bacteria and for a number of heavy metals. Even if water is supplied, the plumbing may be a source of lead contamination from solders, of copper contamination from pipes, or of organic contamination from solvents, adhesives, and polyvinyl chloride pipes. Old plumbing systems are particularly susceptible to contamination problems, and it is useful to measure lead concentrations. Waste water must also be monitored in some particular cases. When water is treated by a municipal or regional facility, institutions sometimes must have discharge permits, including some sampling and analysis of the waters going to the treatment facility. When an institution has a radioactive material usage license, the Nuclear Regulatory Commission may require some type of environmental monitoring to prove that radionuclides are not accumulating in water treatment sludge. Underground petroleum storage tanks may pollute underground water and therefore must be evaluated, monitored, and remedied.

Clean air standards apply to a number of university activities. Clearly, operation of a power plant and/or an incinerator for medical wastes (point sources) requires emission permits and perhaps emission monitoring. Currently there are efforts under way to evaluate the emissions from sources that are not point source. In some states there is already a requirement to account for the emission of chemicals from facilities. Laboratories attempt to minimize indoor air contamination by using fume hoods, but the chemicals are then dispersed into the air. The federal National Emission Standard for Hazardous Air Pollutants (NESHAP) standards have already been applied to the emission of radioactive materials. In this case the amounts utilized in a particular facility determine whether airborne concentrations may be estimated, modeled using complex mathematical dispersion models, or require specific sample collection and analysis. In the future it may be necessary to evaluate fume hoods for the emissions of any of a long list of chemicals. New clean air standards will require reporting of emissions from heating facilities (including small furnaces) and refrigeration units (for chlorofluorocarbons).

36.7   PLANNING AND MAINTENANCE FOR FACILITIES

Whenever new facilities are developed or old facilities are remodeled, numerous building and life safety codes must be addressed. Code compliance is a complex problem that is made even more complex by the changes that occur over time. Usually, when the functional or programmatic use of a facility use changes or when minimal remodeling occurs, new building codes will apply. Institutions that allow building use to change may find themselves in violation of the appropriate code unless there is some active evaluation and review process in place. There are several common problems that develop at institutions because of limited space and relatively poor central control of space usage. First, hallways and many other locations are used as storage. Generally this is in violation of fire codes that exist to ensure clear and accessible means of egress in the event of fire. Storage of old paper and other combustible material also creates a major fire risk. A second fire safety issue is the penetration of walls designed to limit the spread of fires. Remodeling, especially installation of telephone and computer network lines around buildings, can compromise fire walls. A program should be in place to train facilities staff, outside contractors, and building occupants about wall penetrations and the need to provide fire stopping (i.e., using proper materials and closing openings with fire resistant caulks). At the same time, those involved in installing these cables and wires should be alerted to any asbestos in the area so that they can avoid releasing asbestos fibers. A third fire safety issue is the need to install and maintain fire alarm and suppression systems. Codes have become more demanding in requiring these systems. Remodeling may require retrofitting of fire alarm and suppression systems to meet the new codes. Such systems will limit fires and save lives, but they are expensive. In addition, newly installed systems must be maintained. In the case of alarm systems, care must be taken to choose the appropriate detectors. Once installed, alarms must be evaluated every time they are activated to identify defective or inappropriately chosen sensors. Procedures should also be in place for temporary deactivation of alarms to avoid having facilities workers or contractors trigger alarms when engaged in such activities as welding, cutting wood, or even some cleaning processes.

In some areas fire safety officials have become concerned about chemical fume hoods when flammable solvents may be present. Generally the approach was to provide some protection of the exhaust ducts and fan motors to prevent explosions and the spread of fires. Avoiding leaks in these systems was critical to safe operation and to avoid contamination of mechanical spaces, where facilities personnel could be exposed. The current trend to equip fume hood exhaust systems with an automatic fire suppression mechanism is not only expensive, but could be counterproductive if water reactive materials are used in the hood. The issue of laboratory fume hoods and biological safety cabinets has become more important as employees have developed a greater concern with chemical exposures. Old airflow standards are usually inadequate for real protection, and increased airflow means greater energy costs. Recirculating air emitted from laboratories is unwise, and care must be taken to avoid doing this accidentally. There are numerous examples of retrofitted fume hoods that have exhausted air close to building or air handling system intakes, causing unplanned recirculation of toxic materials. Leaking systems also create these problems. Proper design of systems includes the appropriate sizing and installation of fans, bending the air plenum in the appropriate direction for the fan, and exhausting the air high enough above a building so that the air is not re-entrained. Another issue is to protect heating, ventilation, and air conditioning (HVAC) workers by having clear mechanisms of communication between the faculty and students using the fume hoods and those workers maintaining them. A lockout/tagout procedure must be in place and must show the users clearly that a system is inoperable. In the case of biological safety cabinets, an annual certification process may be required. A similar annual functional evaluation of all fume hoods is appropriate.

Another aspect of HVAC systems is proper maintenance of systems and cooling towers. Growth of mildew and mold in systems can create problems for those sensitized to things such as mold spores. The entire system should be carefully designed (in new buildings) and evaluated (in older facilities). It is not uncommon to close fire dampers during tests and then fail to reopen them, causing an undetected airflow restriction. At times, remodeling and new uses of spaces can change the demands and performance of a system.

A final issue that should be of concern to all facilities managers is emergency preparedness and response. As indicated earlier, there should be a local emergency preparedness committee with which the manager can work to have good contact with emergency responders. The institution's emergency plans should be dovetailed with those of the local emergency forces, and the local forces should be aware of the institution's resources (e.g., medical, radiation safety, hazardous materials specialists, communications facilities). Such planning and cooperation strengthen communication between an institution and local governing bodies.

Utilization of Limited Resources

There are so many regulatory requirements that it is hard for managers to decide how to address all of the issues and still provide an economical operation. Although it is tempting to ignore issues because the level of regulatory inspection is very low, this is an extremely risky approach. A better approach is to develop an overview in which all potential problems are listed, the magnitude of the impact on an institution's operations is evaluated, the range of costs is predicted, and potential partial solutions are implemented. An example of the latter is the management of asbestos. Removal of all friable and nonfriable asbestos may be desirable, but because of the high costs, it is probably not possible. Establishing an operations and maintenance program allows asbestos-containing materials to be left in place. When an area is remodeled, asbestos must then be removed. Many useful approaches can be found by sharing information between institutions. APPA: The Association of Higher Education Facilities Officers is a good source for information. APPA's Web site (www.appa.org), APPA's newsletter, *The Regulatory Reporter*, with features on government regulations, and the annual meeting and seminars are examples of APPA's efforts to keep its members informed. Communication between institutions can help managers get new ideas and avoid "reinventing" a process. An environmental audit is useful to help managers evaluate the issues that are most important to their operations. This need not be a large-scale, expensive process. By incorporating line staff in the audit process, they become more aware of the issues, develop greater understanding, and become committed to institutional risk programs. It is important to have written material for some regulations; these need not be extensive legal documents, but should represent programs that meet the spirit of the law and are actually used in daily operations. There is no worse message for managers to send to staff than programs that are written to satisfy regulators but are not implemented, supported, or enforced. Regulators will also see through this and hold managers responsible.

One final issue that facilities managers should understand is the movement toward making institutions of higher education examples of and workshops for a sustainable environment. Many university presidents have signed what is called the *Talloires Declaration*. Further information can be obtained from the Talloires Directorate, Tufts University, Boston, Massachusetts. To give this declaration more than "lip service," an institution must try innovative programs to improve the local environment. This means looking beyond proper waste disposal to methods of actually reducing waste. Another example is to look at landscaping issues not only in terms of the beauty that landscaping provides a campus, but also in terms of long-term maintenance. Gas-powered leaf blowers pollute the air, are a source of noise pollution, expose staff to ergonomic and respiratory risks, and remove a natural source of nutrients from the soils. There are clearly conflicting interests, but even small victories help send a message to students and the public that these issues are important.

NOTES

1. Grose, Vernon. *Managing Risk: Systematic Loss Prevention for Executives.* Arlington, Virginia: Omega Systems Group, 1987.
2. Breyer, Stephen. "Breaking the Vicious Circle: Toward Effective Risk Regulation." *KF 3958 B74.* Cambridge, Massachusetts: Howard Press, 1993.
3. 29 CFR 1910, subpart Z. "Toxic and Hazardous Substances." *Occupational Safety and Health Standards for General Industry.* Chicago: Commerce Clearing House, October 1, 1993.
4. 29 CFR 1910.1200. "Hazard Communication Standard." *Occupational Safety and Health Standards for General Industry.* Chicago: Commerce Clearing House, October 1, 1993.

CHAPTER 37

Campus Security

Loras Jaeger
Iowa State University

37.1 INTRODUCTION

The role of campus security has changed dramatically in the past
several decades. Security and law enforcement departments that
were once self-sufficient and went relatively unnoticed have
emerged as highly visible and integral participants in the campus community. As the issue of recruitment and retention continues to become
more entwined with student safety, the role of campus security departments will become more critical.

The purpose of this chapter is to highlight the major elements of campus security in today's higher education environment. The chapter will
focus on eight major topics: campus security role, staffing, interagency cooperation, federal reporting regulations, campus safety issues, crime prevention, fire protection, and professionalism.

The chapter will begin by describing the various roles of campus security personnel on a continuum from security guard to university police officer. The chapter will also focus on staffing requirements and the important
role that students play in providing safety services. It will examine interagency cooperation with other law enforcement entities, including contracting for policing services and the establishment of mutual aid agreements.

Another focus of this chapter will be the federal reporting regulations
currently required of all colleges and universities, covering such areas as
crime statistics, how to report crimes, policies on sexual assault investigations, and open campus law enforcement logs.

Campus safety issues will be reviewed, including a discussion regarding contemporary crime prevention programs, modern equipment,
and safe campus physical design criteria. The chapter will also include a

section on fire protection as it relates to campus security personnel and will conclude with the evolution of campus security departments toward a higher level of professionalism.

37.2   CAMPUS SECURITY ROLES

The year 1994 was a milestone in campus law enforcement, as it marked the one hundredth anniversary of the birth of campus law enforcement in the United States. In 1894 two New Haven police officers were hired by Yale University and given full police powers to handle campus law enforcement problems. The hiring of these two officers is widely recognized as the beginning of the first organized and professional campus police department.

Since that humble beginning, the role of campus security/law enforcement agencies continues to be one of change. In early American higher education, campus administrators were more concerned about heating buildings, providing a fire watch, and locking doors than they were with the physical protection of the individuals inside them. Firm discipline maintained by the faculty or administration controlled much of the student conduct in those times. However, as American life changed and the power of faculty to control student conduct diminished, campus administrators looked for ways both to control student misconduct and to provide facility building protection. This has led to the need for various services, including private contract campus security, watch guards, security personnel, and campus police.

Much as *Miranda v. Arizona*[1] and *Mapp v. Ohio*[2] changed municipal law enforcement, the campus unrest of the 1970s and the Crime Awareness and Campus Security Act of 1990[3] sent a wake-up call across this country that all campuses must properly address the issues of personal safety and campus security. Increasingly, concern for the safety of students, faculty, staff, and visitors has caused many college and university administrators to rethink the role of security on their campuses. These concerns have led to a "new breed" of college/university officer. Persons currently entering the security field are younger and better educated and have made a conscious decision to make it a career.

Generally, a college or university will employ security officers, peace officers holding some type of commission or certification, or both.

Security Function

Security officers' powers range from limited or no peace officer power (not commissioned or certified) to full arrest powers. In law violation situations, noncommissioned security officers have the same powers and limitations of an ordinary citizen. Noncommissioned security officers should not be expected to handle crime problems; instead, they should be the "eyes and ears" for the campus. If they see criminal activity occurring, they should report it immediately to the appropriate policing authority.

On some campuses security officers function in a watch guard capacity and at others they function in a higher level security capacity. What confuses the general public and muddies the definition of security is the issue of police powers. Some security departments have evolved into full-service police operations. They have peace officer status through the authority of a city or state government. They function much like a city police officer, and many times, when comparing campus law enforcement officers with city police officers, it is difficult to distinguish between the two.

It is possible to visit a campus and observe persons in uniform wearing a security patch but having limited or no law enforcement responsibility, and then to visit another campus and observe similar persons in uniform wearing a security patch with full arrest powers.

In a watch guard operation, the primary function is the protection of campus property. Such guards walk the grounds, ensure that buildings are properly locked, and have a primary role in fire detection and prevention. Watch guards generally wear some type of uniform, carry radios, monitor buildings, and patrol a regular beat.

In a higher level security operation, personnel are involved not only in the protection of campus property, but also in the protection of visitors, faculty, staff, and students. They are expected to develop crime prevention programs and to become more integrated into the mainstream campus environment.

Generally, campus security agencies are grounded in the philosophy of maintenance of order. They are expected to emphasize orderly activity rather than strict adherence to rules. This logical philosophy stems from the humanistic values held by most colleges and universities. Persuasion rather than force is the model in maintaining order.

Law Enforcement Function

In an attempt to serve the increasing demands by students, faculty, and staff for a safer campus, many university and college administrators are moving toward campus law enforcement with full peace officer (police) powers, rather than depending on the city police department to handle criminal matters. Although both types of departments have similar responsibilities and authority, there are significant differences.

The age range of campus student clientele, generally 18 to 30 years old, has little resemblance to the age range of those living in a city. The academic community tends to downplay or even deny the need for force as a means of controlling disorder, whereas in a city the need for the use of force by police is considered to be a normal part of the job. Education and debate are the backbone of a campus environment, whereas a city relies on political power, economic growth, the tax base, and quality of life as major issues of identity.

A full-service campus law enforcement agency does have some functions similar to those of a city police department. It preserves the public peace and order, prevents and detects crime, protects persons and property, and enforces the law. However, the functions of a campus law enforcement agency are actually even more complex than these responsibilities indicate. It must assume many building and grounds security issues that are not part of the responsibility of a city police officer. Those in the agency must understand the many rules and regulations of the institution and be able to make decisions on when to arrest a student, when to refer a student internally to the school's judicial process, or when to do both.

Campus peace officers are required to attend an academy and successfully complete a course of instruction before receiving a commission or certification. This training generally includes classes on criminal law, constitutional law, firearms, evidence preservation, patrol procedures, traffic control procedures, investigation and arrest procedures, accident investigation, report writing, procedures with juveniles, physical fitness, and self-defense tactics. Campus law enforcement personnel are expected to understand campus policies and procedures, such as those involving student rights, diversity, sexual harassment, sexual abuse, and domestic violence, in addition to federal and state laws.

Campus peace officers, through their commission or certification, have both arrest powers and investigative powers. Arrest powers give the campus peace officer the right to use force if necessary, to search after arrest, and to exercise seizure and restraint of the person arrested. Investigative powers give the campus peace officer the right, when appropriate and constitutionally acceptable, to stop and detain a person, frisk the person for weapons, question, and interrogate.

The decision by a college or university as to the type of department needed to handle security matters is not an easy one to make. When deciding on the type of security department that would meet the needs of the institution, administrators should consider the following:

- Size of the campus
- Campus setting (urban or rural)
- Number of criminal offenses that occur
- Level of violence on or around the campus
- Type of institution (private or public)
- Number of large social events that occur during the year and the history of problems with them
- Relationship with local law enforcement and the community government ("town–gown relations")
- Number of students living on campus

Many administrators have come to the realization that police departments in the cities where their institutions are located do not provide the special care demanded in a campus setting. This realization is giving rise to a rethinking of the role of the campus security department. This rethinking may take the form of security agencies becoming peace officers and taking over the role of primary law enforcement on campus.

The media has placed colleges and universities on notice that they must show greater accountability for what occurs on a campus. This has been reinforced by a federal mandate to report crime statistics. Parents are asking about the selection, training, and equipping of campus officers.

Whether the decision is to hire peace officers or non-peace officer security guards, one issue is clear: The public expects that personnel assigned to campus security/law enforcement functions be carefully selected and well trained.

37.3   STAFFING

The staffing of a campus security/law enforcement agency is the most important decision an administrator can make in upgrading or maintaining a good security/law enforcement agency.

A number of methods have been suggested for evaluating staffing needs for campus agencies. Studies of these methods show that there is no best way to determine staffing needs. Still, the following pertinent questions should be asked by the administrator when considering a request to add staff to a campus security/law enforcement agency:

- What are the number of officers on duty, especially during the nighttime hours, compared with the number of students living on campus? Compare this with similar institutions.
- Does the institution have a medical or research facility? Medical facilities, especially those with trauma units, generally need a larger security staff.
- Is the institution located in a rural or urban setting?
- How large is the institution, including student population, number of buildings, and acres?
- Is the institution separated from the community by distance or by physical barriers?
- What is the level of serious crime on campus, and how does this compare with other campuses of similar size? Refer to the Federal Bureau of Investigation (FBI) publication, *Crime in the United States,* released annually in October.
- What is the crime rate in the community surrounding the campus?
- What are the demographics of the student body? A graduate school with older students tends to have lower crime problems. Is it a commuter campus?
- How many patrol officers are available to respond to a call for service? Do not include officers who primarily work at a desk or who generally do not patrol.
- How many major events occur throughout the year, and is there a history of violence associated with one or more of them? Examples include football games, outdoor concerts, or other large outdoor events during which alcohol is served.

- Are officers assigned other duties, such as building lockdowns or fire extinguisher checks? The greater the number of auxiliary duties, the more likely the need for an increase in staff. Keep in mind that students are excellent resources for auxiliary duties.
- Is the campus adequately lighted? Does it have emergency telephones?

Another way of determining staffing needs is by examining the number of calls for service handled by personnel. A *call for service* is simply a request for the security/law enforcement agency to do something. Such calls can range from a report of a serious crime in progress to a request to remove a cat from a tree.

At a minimum, the agency should analyze the number of calls for service handled by personnel in a one-year period. Then a determination should be made on the types of calls received. How many are truly emergency calls, and how many are routine calls?

Next, the average time required to handle each incident should be analyzed, along with the response time for both emergency and nonemergency calls. Finally, the agency should factor in time lost through days off, holidays, average sick leave, average vacation time, training, and other leave. By conducting this analysis, the administrator should be able to determine how much time in an average eight-hour shift is spent on emergency and nonemergency calls for service. The remaining time can be broken down into preventive patrol activities, crime prevention, training, and breaks, for example. This analysis should also assist the administrator in identifying problem areas on campus.

This should then lead to directed patrol activities. *Directed patrol* simply means that during those times that an officer is not responding to a call for service, the officer is assigned a specific task. For example, if the institution has received numerous complaints of bicycle riders violating traffic control devices, a directed patrol activity might include assigning an officer to a specific traffic control intersection during those times in which the complaints are received.

Selection of a Manager (Chief)

For many years it was common for college and university administrators to look for retired FBI agents, local law enforcement officials, or military officials to fill management positions in a security/law enforcement agency. Current campus security/law enforcement supervisors possess strong leadership skills that complement their expertise and background in campus life issues. Many of those who now hold management positions are well educated, holding at least a bachelor's degree, and have several years of progressive supervision and management experience.

How does one go about finding a truly outstanding person to administer a campus security/law enforcement agency? First, find a successful program and ask how it became successful. What was done to make the program a leader in its field? What do administrators look for when selecting a manager or chief?

Second, clearly spell out the qualifications for the position. Then make sure that the expectations of the position are articulated to the final candidates. Issues that should be addressed include the degree of freedom in decision making, to whom this person will directly report, and what resources will be available to enable this person to accomplish his or her job.

Third, advertise in publications that will get the best results. Two outstanding law enforcement sources are the International Association of Campus Law Enforcement Administrators and the International Association of Chiefs of Police. Both publish magazines that lists job openings. In addition, *The Chronicle of Higher Education* is an excellent publication for security/law enforcement openings.

Fourth, carefully put together a selection committee to review applications and interview final candidates. Make sure that students, faculty, and staff are included. It is also important that the selection committee be made up of persons of diversity.

Fifth, when a conditional offer of employment has been made, a detailed background investigation *must* be conducted. The manager or chief of the security/law enforcement agency generally has daily contact with the media. An incompetent manager has the potential to cause serious public relations problems for the institution.

Utilization of Students

Students have been found to be a valuable resource for security/law enforcement agencies. Students are used to provide escorts, lock/unlock buildings, write parking tickets, direct traffic, and walk campus grounds.

Some institutions have gone even further by using students as university services officers (community services officers). Students are assigned either to walking beats or to vehicles and provide backup to regular officers. University services officers direct traffic, assist at accident or other emergency scenes, and take some routine reports such as bicycle thefts and vandalism reports. They also attend many of the ongoing training sessions attended by regular officers.

Some institutions have even used students as security officers in place of traditional full-time watch guards. This has been found to be both cost-effective and a way for students to work their way through school.

37.4 INTERAGENCY COOPERATION

It is important that agreements be reached between the campus security/ law enforcement agency and other law enforcement agencies in adjoining jurisdictions to provide assistance or outright law enforcement services.

Law enforcement services for a campus can be provided in three ways. First, campuses may contract with another law enforcement agency to provide all policing functions. Second, assistance (or backup) can be provided by another law enforcement agency though a mutual aid agreement. Third, when concurrent jurisdiction is present, law enforcement services can be provided through a letter of understanding clearly spelling out each agency's responsibilities.

Contract for Law Enforcement Services Some institutions have elected to contract either for security services from an outside company or for law enforcement services from a city or county. Contracting for law enforcement services is not done frequently. There is usually some degree of friction between the local city and the university or college. In addition, there is a fear that the university or college will lose control over criminal matters that occur on campus if they allow the local police to handle all calls. Local police are also reluctant to provide any security functions that are routine for campus law enforcement personnel; this generally means that security guards must be hired by the institution.

When a contract is successful, it is usually because the "town-gown" relationship is good, and there is little crime problem on campus. Success is also more likely on relatively small campuses, where local police limit their activity to cruising the campus in addition to their other regular beats and respond to calls for service when asked.

Mutual Aid Agreements Mutual aid agreements are becoming commonplace between law enforcement agencies. It is not unusual for a campus law enforcement agency to have a mutual aid agreement with several other law enforcement agencies while part of a multijurisdictional drug task force. That same campus agency may have another mutual aid agreement with one or two local agencies that allow for the sharing of law enforcement resources when the need for assistance arises.

The mutual aid agreement should include an estimate of the types of resources, in both personnel and equipment, available. At a minimum the agreement should contain the following:

- The legal status of each law enforcement agency participating and the amount of resources available
- Procedures for providing the invited agency personnel with the proper legal authority to act

- The identity of those persons authorized to request mutual aid
- The identity of the person or position that outside personnel are to report to when mutual aid is requested
- Procedures for maintaining radio communication with an outside agency
- Procedures for the mass arrest, processing, and transporting of prisoners
- Expenditures, if any, that should be borne by the requesting agency to compensate the agency providing the mutual aid

A mutual aid agreement should be reviewed at least annually to ensure that it describes the current legal status of, as well as current information about, the agencies involved in the agreement.

Letter of Understanding   Many institutions have their own law enforcement agency, with the local city or county law enforcement agency having concurrent jurisdiction or equal authority on the campus. In these situations a letter of understanding is important to clearly delineate areas of responsibilities.

The letter of understanding should contain the following:

- Clearly spelled out procedures outlining which agency responds to calls for service on the campus by type of incident. This is of particular importance for those campus law enforcement agencies whose agents have peace officer powers but do not carry sidearms.
- Command responsibility between the campus law enforcement agency and the city/county agency, when the two are involved in joint investigations
- Procedures for calling in backup support from either agency.
- Policies concerning the writing of city/county citations on the campus and the arrest/booking of persons taken into custody on the campus.

37.5   FEDERAL REPORTING REGULATIONS

Administrators of a campus law enforcement agency should be well aware of the liability issues surrounding personal safety issues. The media campaign waged by Howard and Constance Clery, with help from Pennsylvania Congressman William Goodling, after the death of their daughter at Lehigh University, has led to the passage of federally mandated legislation requiring the release of crime information by colleges and universities. The Student Right-to-Know and Campus Security Act was signed into law in November 1990 by President George Bush. In the spring of 1994 the U.S. Department of Education issued final regulations to this act.

The act requires that any institution receiving Title IV student aid assistance must prepare and distribute an annual campus security report to all current students and employees by September 1 of each year. It also requires that the report be provided on request to any applicant for enrollment or employment. Although the cost of preparing and distributing is high, no money was appropriated by the federal government to fund compliance with the act.

The Student Right-to-Know and Campus Security Act for the first time required all postsecondary institutions to establish a record keeping system and then report specific criminal incidents to students and employees. The act also requires an annual report containing statistics on the number of specified crimes (murder; forcible or nonforcible sex offenses, including rape; robbery; aggravated assault; burglary; and motor vehicle theft) committed during the three calendar years preceding the year in which the report is issued. The following definitions of murder, robbery, aggravated assault, burglary, and motor vehicle theft are found in the FBI's *Crime in the United States*[3]:

- *Murder:* The willful killing of one human being by another.
- *Robbery:* Taking or attempting to take anything of value from the care, custody, or control of another person or persons by force or threat of force or violence or by putting the victim in fear.
- *Aggravated assault:* The unlawful attack of one person by another for the purpose of inflicting severe or aggravated bodily injury. This type of assault is usually accompanied by the use of a weapon or by means likely to produce death or great bodily harm. Attempts are included; it is not necessary for an injury to result when a gun, knife, or other weapon is used that could and probably would result in serious personal injury if the crime were successfully completed.
- *Burglary:* The unlawful entry of a structure to commit a felony or theft. The use of force to gain entry is not required to classify an offense as burglary. Burglary is categorized into three subclassifications: forcible entry; unlawful entry, where no force is used; and attempted forcible entry.
- *Motor vehicle theft:* The theft or attempted theft of a motor vehicle; this includes the stealing of automobiles, trucks, buses, motorcycles, motorscooters, snowmobiles, etc. The definition excludes the taking of a motor vehicle for temporary use by those persons having lawful access.

The definitions of forcible and nonforcible sex offenses are found in the National Incident-Based Reporting System edition of the *Uniform Crime Reporting* handbook.[4]

- *Forcible sex offense:* Any sexual act directed against another person, forcibly and/or against that person's will; or not forcibly or against the person's will where the victim is incapable of giving consent.

Forcible sex offenses include forcible rape, forcible sodomy, sexual assault with an object, and forcible fondling.

- *Nonforcible sex offense:* Acts of unlawful, nonforcible sexual intercourse. Nonforcible sex offenses include incest and statutory rape.

The act also requires disclosure on the number of arrests for liquor law violations, drug abuse violations, and weapons possessions that occurred on campus.

- *Liquor law violations:* The violation of laws or ordinances prohibiting the manufacture, sale, purchase, transportation, possession, or use of alcoholic beverages. (The 1994 Department of Education final regulations specifically exclude driving under the influence and drunkenness.)
- *Drug abuse violations:* The violation of laws prohibiting the production, distribution, and/or use of certain controlled substances and the equipment or devices utilized in their preparation and/or use.
- *Weapon law violations:* The violation of laws or ordinances prohibiting the manufacture, sale, purchase, transportation, possession, concealment, or use of firearms, cutting instruments, explosives, incendiary devices, or other deadly weapons.[4]

The institution must set forth its policies on how to report criminal activity, the role and authority of the campus security/law enforcement agency, how the institution responds to incidents of crime, and the relationship of the campus security/law enforcement agency with local and state police. The institution must also outline its policies concerning access to academic buildings, residence halls, fraternity and sorority houses, and other facilities under its control. Crime prevention issues must be described, and students and employees must be encouraged to be responsible for their own and others' safety. The institution must specifically outline its policy covering the possession, use, or sale of alcoholic beverages and illegal drugs and its policy regarding sex offenses.

The act also requires institutions to provide timely warnings to the campus community if any of the six specified crimes (murder; forcible or nonforcible sex offenses, including rape; robbery; aggravated assault; burglary; and motor vehicle theft) are reported to either the local police or campus security/law enforcement and are considered a potential threat to students and employees.[5]

The International Association of Campus Law Enforcement Administrators has worked closely with the Department of Education to both better understand the regulations and notify campus security/law enforcement members.

Open Campus Police (Security) Logs One of the dilemmas previously facing a campus law enforcement agency was the release of crime information obtained by it. State laws typically require police to disclose general information about a crime to the public. However, the Buckley Amendment barred campus law enforcement agencies from releasing daily logs or arrest information. This led to lawsuits by student organizations demanding such information under the state release-of-information laws. This effectively placed campus law enforcement officials in the middle. On the one hand, Department of Education officials were demanding that no release be made on penalty of losing federal student aid, and on the other, campus law enforcement officials were being sued for not releasing the information.

This matter was officially resolved with an exception to the release of information. Campus law enforcement officials are allowed to follow the state administrative rules and/or laws covering the release of information. Administrators should become familiar with their state laws and formulate policies covering campus law enforcement release of crime information.

37.6 CRIME PREVENTION

College campuses, for the most part, were immune from this crime awareness frenzy until the 1980s. Television news programs suddenly focused on colleges and universities and "exposed" a wide range of crimes on campus, from murder and rape to drug dealing and petty theft.

To address safety concerns, many security/law enforcement agencies have at least one person dedicated full time to crime prevention. The Standards for Law Enforcement Agencies established by the Commission on Accreditation for Law Enforcement Agencies, Inc., recommend that the law enforcement agency's crime prevention function provides for the following:

- Targeting programs by crime type and geographic area on the basis of an analysis of local crime data
- Targeting programs to address community perceptions or misperceptions of crime
- Evaluating the effectiveness of crime prevention programs[6]

To do this, the campus security/law enforcement agency must develop close ties with the local police and with students, staff, and faculty. Crime prevention should be perceived not as the responsibility of campus law enforcement, but as the responsibility of all those working or attending classes at a campus. Law enforcement's responsibility should be to coordinate various programs. A safety-conscious campus is one that involves everyone and addresses safety issues in a proactive rather than a reactive way.

Personal Safety

The campus security/law enforcement agency must set the tone for the institution. The agency must make it known that it will not tolerate sexual abuse or any act of violence on campus. This must be backed up with strong sanctions against those committing acts of violence. Both the criminal justice system and the university judicial system should take a stand of zero tolerance to violence.

The best way for individuals on campus to avoid becoming victimized is to be alert to the surroundings. Other suggestions are:

- Students should study in areas where there are other people and not be isolated.
- In residence halls students should always lock their doors and report any unauthorized persons on the floor.
- Individuals should walk with confidence and, when possible, not alone.
- Individuals should walk in the middle of the sidewalk, facing traffic and away from buildings, parked cars, and bushes.
- When driving, individuals should always lock their doors and avoid parking in deserted areas.
- Drivers should look inside the car before entering it.
- If followed, drivers should go to the nearest police station or firehouse or to an open gas station.

Many sexual assault incidents are acquaintance or date rapes. The following advice will help students in dealing with social situations:

- Arrange to meet a first-time date in a public place, or go out with a group.
- When first meeting someone new, do not leave with him or her, but go home with people you know, especially if either of you have been drinking alcohol. A date can be arranged later.
- Abstain from or limit alcohol consumption.
- Trust your instincts. If the person or the situation makes you uncomfortable or scared, get away.
- Set sexual and touching limits and stick to them. Be assertive in communicating your feelings.

In sexual assault incidents, the assaulted student fears losing control. The student must be given the option to pursue the matter criminally or to simply file an anonymous report. Sexual assault advocates should be made available to the student if the student desires those services. The student should also have the option of seeking remedy through the college or university's student conduct system.

The student should be encouraged to seek medical attention as soon as possible (i.e., before bathing and/or changing clothes). If there is no provision for payment of the medical examination, the examination should be paid for by the law enforcement agency, not the student.

The campus law enforcement agency is required to make timely advisories to the campus and community of sexual assaults. This can be done through crime-alert bulletins, notification of the media with a press release or telephone call, advisory notices on various information bulletins, and use of the electronic mail system.

What advice should be given to a student who asks, "What can I do to increase my chances of getting away or attracting help if I'm assaulted while walking after dark?"

One standard answer to this question is to never walk alone at night and to utilize the campus escort service. However, many students are not interested in that answer, because it restricts their ability to move around campus. To reduce vulnerability to assaults by strangers, many people have purchased personal security devices. Students should know the limitations of each device and then decide which will best serve their needs.

Before purchasing any personal security device, the student should consider the following:

- Is the device easily accessible? For example, if the device is in the bottom of a purse, it will do little good when suddenly attacked.
- Can the attacker use the device against me?
- Will the device attract enough attention from others in the area to get me help?
- Will the device frighten or disable the attacker?
- Will it provide a "window of opportunity" to escape?
- Once activated, will the device continue to operate without continued activation by me?

There are a variety of personal security devices marketed to students. These include chemical deterrents, pressure-activated alarms, whistles, and noise alarms.

Chemical Deterrents   Chemical deterrents can provide some level of protection. When considering the purchase of a chemical deterrent, it is important to answer the above questions. Especially consider whether the device is easily accessible and whether it will disable or frighten the attacker. Some chemical deterrents have little effect on emotionally upset individuals. Chemical deterrents can also be taken away from the student and then used against him or her.

Pressure-Activated Alarms   Pressure-activated alarms are canisters that emit an ear-piercing noise designed to frighten off the attacker and/or attract attention. Because the device is under pressure, it has a time limit on use.

Whistles   Whistles are used to alert others in the area that an individual is having trouble and to summon help. The success of the whistle depends on the student's ability to blow into it. The advantage of utilizing a whistle is the low cost and not having to worry about dead batteries.

Noise Alarms Noise alarms are battery-operated devices that emit an alarm, generally over 100 dB. They are designed to continually operate when activated and both scare off the attacker and summon help.

Alarm devices are tools to help in defending students against attack. Of more importance is the student's ability to be aware of his or her surroundings and maintain an avenue of escape.

Property Safety A campus is the student's home away from home. Theft of student property is one of the most frequent crimes on campuses. Programs such as Operation ID, Property Safety Awareness, and Campus Watch have been found to be valuable in addressing theft problems.

Operation ID is a nation-wide crime prevention project designed to protect a person's valuables. The program participant is assigned an identification number, and the number is then recorded at the campus security/law enforcement agency. The student is then encouraged to enter that ID number, usually with an engraver or invisible marking pen, on all valuables.

Once property has been marked and a list of all property recorded, Operation ID decals are placed on all doors and windows where entry can be made. If property is stolen, its ID number can be entered into the National Crime Information System computer.

Property Safety Awareness is a program in which students are taught how to be smart with their property. This includes such advice as keeping valuables out of sight when they are not being used, recording serial numbers and model numbers for all purchased equipment, and reporting suspicious activity.

Campus Watch is similar to the Neighborhood Watch program. Campus Watch is designed to acquaint students living together in a residence complex or building so that they can be the eyes and ears for each other. The program covers property safety, theft prevention, burglary prevention, tips for personal safety, safety patrols, self-defense, sexual assault prevention, education and investigation, and crime alert procedures. Monthly newsletters are part of this program, along with special crime bulletins when necessary. A security survey of the student's residence is performed, and suggestions are given on how to live more safely. Participants are taught how to spot suspicious activity and who to call.

37.7 CAMPUS SAFETY ISSUES

Institutions of higher education own millions of dollars worth of buildings, art work, books, research projects, computers, laboratories, and research animals, to name only a few. Of even more value are the students, faculty, and staff who come to work or live on the campus.

Lighting   The illumination of grounds, parking lots, and buildings is important in discouraging criminal activity. Although statistical data on the effectiveness of lighting are inconclusive, lighting does reduce the possibility that someone will be victimized. There are benefits other than personal safety. Lighting reduces vehicle and pedestrian accidents; illuminated intersections make persons assigned to traffic direction more visible; and illumination provides a more friendly environment, which should encourage increased social interaction.

At the very least, a college or university should notify students, faculty, and staff or supply them with maps showing the lighted walks on campus. At least annually security personnel should walk the grounds of the campus to identify areas that require new lighting or additional lighting. Many campus law enforcement agencies conduct this walking survey twice a year, once when plant material is in full bloom and again when plant foliage has dropped. Security/law enforcement officers should make special note of any burned-out or broken lights to ensure prompt replacement.

With additional lighting will come complaints of "light blight," or too much illumination of the skyline. This can be reduced by installing downwardly directed lighting. A more detailed discussion of exterior lighting can be found in Chapter 53.

Emergency Telephones   The installation of emergency telephones on campus has proven to be useful. With cellular technology or the use of radio frequencies, emergency telephones can be installed anywhere, even in remote parking lot locations. These telephones are used for true emergencies, but in some institutions such use has been expanded to include requests for escorts, direction information, or car trouble.

Emergency telephones are generally installed along busy sidewalks and in parking lots. They should be marked in such a way as to make them clearly visible at night from a distance. Operation should be easy, accomplished either by simply picking up the receiver or by pushing a button. The call should be received on an emergency line (911 or other emergency line) so that immediate law enforcement response can occur.

Emergency telephones should be tested frequently, especially during inclement weather. They must also be accessible by persons with disabilities.

Environmental Design for Safety   Campus security/law enforcement agencies currently are being asked for input when new building construction occurs, plantings are installed, or parking lots are built. They provide input into the selection of locks, lighting, alarms, and the proper location of specific plantings.

It is important that the security/law enforcement agency form a partnership with campus planners and architects in this area. The purpose of this partnership is to improve safety on campus.

Alarm Systems   As the call goes out to provide a safer campus environment, inevitably the subject turns to alarm systems. When addressing this issue, one basic question must be answered: Is the intent to scare away the intruder, or to catch the intruder in the act of a criminal offense?

An audible alarm at the location will certainly get the attention of an intruder, but if someone is not in the area to respond, the possibility of apprehending the intruder is slight. The use of an audible alarm on fire doors and exit-only doors are common at many institutions.

There are a wide variety of silent alarms on the market. Examples include panic buttons, magnetic switches on doors and windows, metallic foil tape on windows, infrared beam devices, ultrasonic alarms, and pressure pads, just to name a few.

As with the emergency telephones, frequent checks of the alarm system are a must to ensure its integrity.

Alarm Response Plan   Most campus security/law enforcement agencies respond to a variety of silent or audible alarms. Administrators must review the alarm response policy to ensure the greatest degree of safety to officers responding to the alarm and to persons on the way to and at the scene of the alarm. The goal, when responding to an alarm, is to demonstrate concern for the safety of all persons; to prevent the taking of hostages, should a robbery be taking place; and to improve the chances of apprehending the offender.

At least two patrol units should be dispatched to an alarm. If the alarm is at night in an unoccupied building, then the first responding unit to arrive should cover the front and one side of the building. The next arriving unit should cover the building corner opposite that covered by the first unit.

It should be the responsibility of the first responding unit to direct the backup units to positions where routes of escape can be effectively blocked. When the exterior is secured, the interior of the building can be searched.

When an alarm occurs during the daylight hours and persons are inside the building, responding officers must use extra caution to ensure the safety of persons at the scene. Officers should get to know key personnel in each building by sight. There is nothing to prevent a robber from posing as a building or department supervisor. When responding to a silent daytime alarm, officers should park their vehicles out of sight. Some agencies have procedures in which they call the location and receive a coded message that indicates whether a robbery is in fact occurring.

If a robbery has occurred, the officers at the scene should secure the area to keep evidence from being destroyed. The area should be locked down to keep people from entering the affected area, and persons in the affected area should be advised to not touch anything. Officers should obtain the necessary physical and vehicle descriptions and direction of travel, if known. This information should be immediately broadcast to other officers and neighboring law enforcement agencies. A list should be made of all witnesses to the robbery and statements obtained from each before they are allowed to leave. Witnesses should be encouraged not to converse with one another about the crime until after they have been interviewed.

Keying There are three components to a keying system; design of the system, implementation of the system, and control of the system. These three components are often handled by the same organization, although this is not necessary.

The design of the system lays the foundation for all future security decisions. The design must start with an understanding of the customer's needs and should include such items as an understanding of the level of security desired, the organizational levels and access needs, the temporal patterns of the occupants, and the physical layout of the facilities. It should also include an understanding of available technology such as the traditional brass key system, mechanical pushbutton technology, card access, and others and could even include the more sophisticated approaches of personal identification.

The implementation process can be fairly straightforward once the design is completed. If a traditional keying approach is used, then appropriate hardware can be installed, with cylinders and keys set up to match occupants' needs. Most often schools settle on one supplier for cylinders and keys so that the system is consistent across the campus. If card readers are chosen, then door and door frame preparation are necessary. If the card reader is to be centrally controlled, then additional accommodation must be made for networking the readers.

Key control, if handled improperly, can quickly become a nightmare. An effective key control system must be a top priority of the institution. Professional campus law enforcement organizations recommend that key control be the responsibility of the campus security or law enforcement agency, not of another area of facilities management.

Keys that allow entry into buildings should be a type that cannot be easily duplicated. High-security key blanks are available that will address this concern.

All keys must be signed for by the person in possession of them. The department head should also sign the key card as acknowledgment that he or she approves. An inventory of all keys should be completed once a year by all departments. Any missing keys should be immediately reported.

Keys should be marked, but not so that they are easily identified. Markings such as "SM" for submaster or "GM" for grand master are giveaways of the power of each.

A number of references may be useful, including the following:

- Eaton Corporation, Lock and Hardware Division. *Yale Master Keying Manual*. Monroe, North Carolina: Yale Security Incorporated, 1975.
- Door and Hardware Institute. "Master Keying—Part 1 & Part 2." *Tech Talk*, 1973.
- O'Shall, Don. *Practical Masterkeying of Pin Tumbler Locks*. Secaucus, New Jersey: Haines Publishing, 1984.

Several organizations also may be useful, including the following:

- National Locksmith Association
  1533 Burgundy Parkway
  Steamwood, Illinois 60107
- Associated Locksmiths of America
  3003 Live Oak Street
  Dallas, Texas 75204
- International Association of Campus Law Enforcement Administrators
  638 Prospect Avenue
  Hartford, Connecticut 06105-4298

37.8 FIRE PROTECTION

Campus security/law enforcement agencies have a pivotal role to play in fire protection. It is important that they develop partnerships with other departments responsible for fire inspection, safety, equipment, and drills. Campus security/law enforcement officials have two major responsibilities in fire protection: fire prevention and fire preparedness, should they have to respond to a working fire. Important aspects of fire prevention are covered in Chapter 28, while the role of the campus security/law enforcement agency is covered in more detail here.

Fire Preparedness

Officers should have a working knowledge of portable fire extinguishers and how they function. They should know the location of all standpipe hose stations, fire alarm panels, and electrical panels. They should also know or have access to a list of all locations on campus that house hazardous materials. Any officer who responds to a fire in which hazardous materials are burning could easily walk into a death trap unless he or she is properly trained.

Officers should work closely with those officials assigned to maintenance of the fire alarm system. All fire alarms should be tested frequently.

Officers should also know the general evacuation plans for each building and then participate in fire drills for each building. This should be coordinated with the building supervisor if the institution has such a system.

It is important to have a written plan outlining the role of various departments during a fire and to review this plan on an annual basis. The plan should cover at least the following:

- A list of available personnel and equipment resources, external resources from other law enforcement agencies, officials who should be contacted immediately, and media relations information
- Procedures for notifying the fire department, which should be advised regarding the type of fire, the extent of fire loss at the time, any occupants believed to be in the building, any hazardous material at the scene, and the color of the flame
- Observation of wind direction and, if possible, staying on the upwind side
- Establishment of a control post to coordinate all communications
- Evacuation plans—if possible, without harm to officers or when ordered by fire personnel—to calmly and in an orderly manner remove persons from the building. Officers should not attempt to enter a building if any hazardous materials are involved.
- Establishment of an outer and inner perimeter to provide security to the scene and crowd control of the affected area. It is important that the outer perimeter be far enough from the fire to prevent injury if the building should collapse or an explosion occur.
- Assignment of someone to guard the fire equipment and to monitor the area to prevent looting and other criminal activity
- Direction of all routine traffic from the fire scene
- Provision of assistance to other emergency services that come to the fire scene
- Photographing and/or video taping of the fire scene, including the crowd, if the fire is of suspicious origin
- Maintenance of the inner perimeter after the fire is put out and until the fire investigation is complete
- Provision of assistance to fire officials investigating the origin of the fire
- Ensuring that all traffic routes have been returned to normal
- Completion of all necessary reports of the incident

37.9 CAMPUS LAW ENFORCEMENT PROFESSIONALISM

Much like their counterparts, the city police, campus law enforcement agencies are embracing the concept of professionalism. This continuing movement toward professionalism includes higher standards of admission; adherence to the Law Enforcement Code of Ethics; academy training; successful completion of a field officer program; mandatory in-service training; and adoption of detailed rules, regulations, policies, and procedures. As part of this movement toward professionalism, law enforcement and security agencies are adhering to established standards of competence.

National Accreditation

Law Enforcement Accreditation began in 1979, when four organizations— the International Association of Chiefs of Police, the National Organization of Black Law Enforcement Executives, the National Sheriff's Association, and the Police Executive Research Forum—worked together to create the Commission on Accreditation for Law Enforcement Agencies (CALEA). In late 1983, CALEA began accepting applications from law enforcement agencies. Currently more than 400 law enforcement agencies, including campus law enforcement agencies, are accredited.

Law enforcement accreditation was created as a voluntary process through which a law enforcement agency could seek to verify its conformance to universally developed and tested standards. The four major goals developed for the accreditation process were as follows:

1. Increase the capability of law enforcement agencies to prevent and control crime.
2. Enhance agency effectiveness and efficiency in the delivery of law enforcement services.
3. Improve cooperation with other law enforcement agencies and with other components of the criminal justice system.
4. Increase citizen and employee confidence in goals, policies, and practices of the agency.

Some standards are mandatory; others are optional. The standards also relate to the size of the agency. In other words, some standards may be mandatory for a large agency but may be optional or not applicable for a small agency. Many of the standards require written policies. An example of a mandatory standard for all agencies that requires a written policy is use of force.

There is a great deal of controversy among law enforcement agencies concerning accreditation. Advocates believe that standardization of police departments across the nation is currently the most effective means for continuing the professionalism process. They believe that a system of self-governance is better than regulations thrust on them by the courts or legislature. Also, advocates point out that the accreditation process clearly demonstrates to an agency what is not acceptable behavior.

Those opposed to accreditation point to the cost and the amount of time required to complete the process. They believe the benefits do not outweigh the cost in personnel to achieve it. Others believe the standards are too general and permit accredited agencies to continue to have incompetent police officers on the street.

37.10 SUMMARY

The field of campus security will continue to undergo changes in the future. Security organizations will be called on to be more involved in issues involving personal safety, environmental design for safety, building security systems, and parking lot/grounds safety.

To meet these requests, security and campus law enforcement officials will need to be better educated and trained. An example of change toward a more professional environment in campus security/law enforcement can be found in the International Association of Campus Law Enforcement Administrators, which has begun a study to develop standards for campus law enforcement/security organizations.

The challenges facing campus security/law enforcement are many, but the field is open to those who want to develop an organization that is highly skilled and professional.

NOTES

1. *Miranda v. Arizona,* 384 U.S. 436 (1966).
2. *Mapp v. Ohio,* 367 U.S. 643 (1961).
3. Federal Bureau of Investigation. *Crime in the United States, 1993.* Washington, D.C.: U.S. Government Printing Office, 1994.
4. Federal Bureau of Investigation. *Uniform Crime Reporting: National Incident-Based Reporting System.* Washington, D.C.: U.S. Government Printing Office, 1988.
5. For more information, refer to the *Federal Register, 34 CFR Part 668, Student Assistance General Provisions; Campus Safety: Final Rule*, U.S. Department of Education,Vol. 59, No. 82, April 29, 1994, pp. 22,314–22,321.
6. Commission on Accreditation for Law Enforcement Agencies, Inc. *Standards for Law Enforcement Agencies.* Fairfax, Virginia: Commission on Accreditation for Law Enforcement Agencies, Inc., April 1994.

ADDITIONAL RESOURCE

Powell, John W., Michael S. Pander, and Robert C. Nielsen. *Campus Security and Law Enforcement,* second edition. Newton, Massachusetts: Butterworth-Heineman, 1994.

CHAPTER 38

Campus Mail Services

James E. Ziebold
Iowa State University

38.1 INTRODUCTION

The purpose of a campus mail system is to deliver incoming mail and to route outgoing mail. How this takes place is dependent on such variables as the volume of mail, the size of the institution, the level of service desired, the size of the workspace, and the funds available. The first step in either starting a campus mail system, or improving an existing system, is a thorough analysis of the present operation. The next step would be to establish capabilities that would provide the required services or improve on what is presently being done. This can be accomplished in many ways. Visiting other campuses and observing their operation will present ideas. The U.S. Postal Service is also an excellent source of information.

38.2 U.S. POSTAL SERVICE

Services provided to colleges and universities by the U.S. Postal Service may vary considerably from one location to the next. In some instances the campus mail department may have to pick up the institution's mail at the post office in bulk and unsorted. In a few rare cases, the U.S. Postal Service delivers the mail directly to the department on the campus. The U.S. Postal Service continually evaluates the level of service provided, and campus mail personnel must be prepared to cope with attempts to decrease levels of service.

The *Postal Operations Manual* serves as the basis for the procedures used by the U.S. Postal Service in the distribution of mail. This manual contains only ten paragraphs directed specifically at colleges and universities and covers the following areas:

- Administrative buildings
- Dormitories or residence halls
- Married student housing
- Fraternity and sorority facilities
- Parcel post
- Special delivery
- Forwarding mail
- Noncity deliveries
- Exceptions

Paragraph 615.2 of the *Postal Operations Manual* details the philosophy of delivering mail to schools, hotels, and similar places, and familiarity with it may be beneficial when negotiating with the U.S. Postal Service over service.[1]

Other U.S. Postal Service manuals are the *Domestic Mail Manual* and the *International Mail Manual.*

The U.S. Postal Service has initiated a Customer Advisory Council (CAC) program in which members of the U.S. Postal Service and local mailers meet at regular intervals to discuss changes in postal regulations and local mailing procedures.

On a larger, regional level, there are Postal Customer Councils (PCCs). These groups are also made up of U.S. Postal Service officials and mailers within the area served by the PCC. Some PCCs become rather large in their geographic area and are generally co-chaired by a U.S. Postal Service representative and a mailing industry representative. The PCC provides its members with the latest information on postal rates and regulations.

The U.S. Postal Service sponsors the Postal Forum twice per year. This national meeting provides the latest technology available to the mail industry. The newest mailing equipment is displayed in addition to the latest computer and software mail processing systems that will help with your needs. In addition a number of sessions provide information on proposed regulations.

Finally, the U.S. Postal Service Inspection Division will provide a free on-site audit of a campus's mail service. This audit can be invaluable in identifying practices and procedures that may lead to efficiency improvements and improved security.

38.3   PROFESSIONAL ORGANIZATIONS

There are a number of professional organizations in the United States from which information can be obtained on the latest mailing industry technology.

There are similar organizations in countries outside the United States. The primary U.S. organizations are as follows:

- The National Association of College and University Mail Services (NACUMS) was formed to provide a forum to forward industry concerns to headquarters U.S. Postal Service officials.
- The Association of College and University Mail Services (ACUMS) serves members in the northeastern United States.
- The College and University Mail Services Association (CUMSA) serves members in the southern and eastern United States.
- The University Mail Managers Association, formerly known as the Big 10, serves members in the midwestern United States.
- CUNIMAIL is an Internet forum that brings together many resources of knowledge about campus mail (e.g., U.S. mail, parcel and express shipments, equipment, and management).

38.4 EXPEDITED SERVICES

At many institutions, expedited services (overnight/next day) are handled by the university's mail service. This consolidation is usually done so that the institution will be eligible for volume discounts. Some expediters will pick up at one location but deliver to the buildings. This approach allows quicker delivery and less burden on the campus infrastructure while still allowing the university to receive the benefits of volume discounts.

Some expedited services make available a computerized terminal that allows direct access to their system for information such as package tracking. Some of the services now have homepages on the Internet that will allow package tracking.

Every package should be weighed and recorded, with all packages checked for on-time performance. Expediters will provide refunds if they fail to deliver on time, and this can be worth several thousands of dollars per year.

38.5 MAIL SERVICE FACILITIES

Ideally the campus mail service facility should be centrally located to provide for easy pickup and delivery of the mail. The use of walking routes obviates the need for expensive delivery vehicles. The size of the facility needed varies with the size of the operation. Systems in small institutions may occupy one small room, whereas those of larger institutions resemble— and in many cases are—branch post offices. Because central campus space is usually at a premium, a great deal of innovation is often required to fit the operation into the available space. Depending on the size of the operation, the space requirements should include the following:

- *Incoming mail holding area.* Whether the mail is picked up from the post office or delivered, an area should be set aside for holding incoming mail until it can be sorted. To prevent pilferage, this area should never be in a hallway or other area open to other than mail room personnel. If the mail comes to the mail room in hampers, a dock will be needed to facilitate unloading.
- *Sorting area.* The sorting racks or bins should be located so as to create a logical flow from the incoming mail holding area to the outgoing mail holding area. Because this is one of the most detailed duties in the mail room, care should be taken to ensure that this area is properly lighted. If the sorters stand, fatigue mats should be provided to ease the burden on the sorters' legs and feet.
- *Outgoing mail hold area.* This area should be adjacent to the exit and provide adequate space to stage the campus mail and the metered or stamped U.S. mail prior to delivery. If a presort or international mail service is provided, separate space should be provided.
- *Metering area.* The metering area should be located so that the mail, after it is metered, will be adjacent to the outgoing mail holding area. The metering area requires good lighting, adequate work surface, and enough electrical outlets to handle the scales and postage machines. Because personnel in this area are usually on their feet and standing in one area, fatigue mats should be provided.
- *Administrative area.* This area should be centrally located, with at least one window to allow the supervisor of the mail room to see all activities. In large operations this area is important to provide a quiet area for planning, accounting, record keeping, and employee counseling duties.
- *Employee lounge area.* The sorting and handling of mail is a tedious process that requires continuous concentration. Employee fatigue can greatly affect the accuracy and productivity of the mail operation. It is important to have a lounge separate from the work area for employees to take a break from the work routine.
- *Service area.* This area includes a mail pickup point; stamp sales window; post office boxes in the units, with full service or contract stations; and a loading dock or ramp. A well-lit and clearly identified area should be provided for customer access. There should be a safe and adequate area for parking and unloading of delivery vehicles separate from customer access.
- *Alternate area.* In the event of a disaster, the mail service should have alternative sites identified. These alternative sites may include other universities, government entities, or private vendors. In addition, the logistics of moving the mail to the new location and delivering it to the campus must be planned.

38.6   HANDLING CAMPUS MAIL

Campus Addresses

Developing a campus mailing address is the first step in handling the campus mail. The assignment of a unique campus zip code will make the job of receiving mail much easier. The U.S. Postal Service will issue unique zip codes to large users. This simplifies their sorting process and also decreases the chance of campus mail being misdirected to local addresses.

Campus addresses often pose a problem, especially on large campuses. Large universities have a large number of departments, with frequent departmental reallocations. It is often difficult for the mail service sorting staff to keep up with the locations of departments. One solution to the problem of changing locations is to issue unique box numbers to the departments. The box number always remains the same, regardless of the department location. This makes sorting much simpler, as it is by box number only. Changes will be required only when a department relocation necessitates use of a different delivery route. The routes can be denoted by color coding the box numbers on the sorting bin.

Some systems sort by building name or number. An additional sorting is often required after the mail arrives in the building and is usually done by clerical employees in the building. A simple system that avoids this is to have locked departmental mailboxes located at a central point in the building. The mail service employee can pick up and deliver mail to these boxes. To work effectively, the mail should be sorted to the building box numbers before it leaves the mail room.

Delivery Routes

After the mail is sorted, there should be an effective method of delivering it throughout the campus. Routes should be developed for speed and efficiency. Pickup of outgoing mail should always accompany the delivery of incoming mail. Over time, the mail service should try to make its pickups and deliveries within a certain time frame. This will help customers know approximately when to have their mail ready.

The method for delivery should be by foot, with the use of a cart whenever possible. Vans, cars, scooters, and even golf carts can be used for delivery. It should be noted that some of the small specialty vehicles are expensive and often require more maintenance than standard vehicles. Slow speeds coupled with stop-and-go driving are hard on vehicles.

Interdepartmental Mail

Interdepartmental mail constitutes a large portion of the mail on most campuses. Departments should be instructed to bundle this mail separately to keep it from being mixed with outgoing U.S. mail. This eliminates an additional sorting step. When the departmental mail arrives in the mail room, it should be sorted with the same priority as incoming posted mail. Many campuses use a campus mail envelope for the transmittal of interdepartmental mail. The form most often used is a reusable envelope with a series of blank address lines on both the front and back of the envelope. The user marks through the last lines used and enters the new address. Some of these envelopes can be used as many as fifty times. This system makes sorting easier and saves money on envelopes. The address used on the envelope should be simple and easy for the sorter to read and should be limited to the addressee name, department name, and campus box number. Speed letters are an effective means of campus communication. Many campuses use an 8.5 in. x 11 in. formatted letter that can be bifolded, addressed, and sent as an interdepartmental letter. Some institutions place boxes for campus mail at strategic points on the campus. This service provides a convenient drop and pickup point for campus mail. These boxes may be located in conjunction with U.S. Postal Service boxes, but they should be clearly identified as being for campus mail only.

Mail sorting is one of the most time-consuming and tedious duties in the mail room. At times the workload created by incoming mail may be greater than the available staffing level can handle. It is important to have established sorting priorities. This ensures timely delivery of the most important mail. A good guide to priorities is the U.S. Postal Service's class system, with mail sorted in the following order:

1. First class and campus mail
2. Second class mail
3. Third class (single-piece) mail
4. Fourth class (parcel) mail
5. Third class (bulk) mail

38.7 PERMITS AND OUTSIDE MAIL SERVICES

Permits can save a great deal of postage expense. They guarantee a reduced rate for catalogs and advertisements that are regularly mailed on behalf of colleges and universities. Efforts should be made to reduce the number of these permits to give better control and to reduce expenses. Incoming reply permits can also save money and ensure a higher percentage of replies on advertisements and surveys by saving recipients the expense of enclosing a stamped return envelope.

In recent years there has been a proliferation of outside mail services. These range from carrier services to presort and special handling services. Courier services are extremely helpful for the overnight delivery of mail between isolated campuses and urban post offices. Private institutions may contract with private couriers, whereas public institutions often have local or state government–operated services available. Presort services will take all of an institution's first class mail and presort it by zip code so that the mail can be posted at a reduced presorted rate. The savings between the regular first class rate and the presort rate is split between the presort service and the institution. Many institutions save thousands of dollars each year using these services.

A more recent service has been provided by international mail service consolidators, which will negotiate to pick up and handle an institution's overseas mail for a reduced rate. The service flies the mail overseas, where it is handled by a service similar to the presort service. The mail is posted at a rate far below U.S. Postal Service rates and mailed from the overseas point. There are several companies that provide this service, and savings can be as much as 20 percent.

38.8 ACCOUNTING SYSTEM

Aside from personnel and equipment costs, the basic cost in a mail system is the cost of postage. It is imperative that the mail service institute an accounting system to monitor this expense. This is especially important on campuses where each department is responsible for its own postage expense. In decentralized operations, the mail is posted in individual departments, which have their own postage machine, and is then picked up and disbursed by the campus mail service.

In centralized operations the mail is separated, bundled, and tagged in the department and then picked up by the campus mail service for posting and disbursal. In centralized operations, the tag identifying the department and its account number is used by the postage machine operator as the bill or record for charges for that particular bundle of mail. The postage machine operator posts the postage cost and verifies the piece count to the tag. The tag, which is usually carbonized, is used to post the charge to the proper account in the ledger. One copy is usually kept in the mail service accounting office files, and the other copy is sent to the department with the postage expense bill.

Mechanized mail room accounting systems are beginning to appear on the market. Most are designed to operate with microcomputers and can improve the speed with which reports are processed. Some of these systems are designed to work directly from the postage scales, which will greatly automate the accounting operation. It would be wise to look at more than one "live" system before a decision is made to purchase any system. The software and hardware can be quite costly. Together with repair service and providing for unforeseen downtime, a poorly selected system can be quite expensive.

38.9   CONTRACT POST OFFICES

Many campuses operate full-service post offices. These operations are convenient for students, especially in isolated areas. The U.S. Postal Service will pay some campuses or individuals, either through a bid process or by agreement, to operate contract post offices. The service is usually offered because it would be too costly for the U.S. Postal Service to operate a station in the area. The contract stations are subject to audit by the U.S. Postal Service, as all of the stamps and money orders are furnished by the U.S. Postal Service. Usually the operations are for convenience rather than for profit. However, a contract post office located adjacent to the campus book store or student center will create traffic and business for these operations.

38.10   STUDENT MAIL

The problem of handling student mail can be approached in many ways. Some institutions provide boxes in residence halls, and the mail is sorted and placed in the boxes by mail service employees or student workers. Other campuses have central campus post offices that provide rental boxes for the students. Still others rely on U.S. post offices located in the community to rent boxes to the students. Although this last system may not be more convenient for the students, it saves the institution additional handling expenses.

38.11   MAIL SERVICE STAFFING

The selection of personnel is one of the most important aspects of mail service, as the success of the operation depends on reliable employees. Small operations will require the most versatile employees, because each person will share the burden of the various jobs in the mail room. Larger operations, on the other hand, tend to be more specialized, with some employees engaged primarily in sorting, others in delivery, and still others in posting outgoing mail. Even in centralized operations, it is important to cross-train employees so that they can handle more than one job, thereby making the operation more flexible. Some operations test applicants for incoming positions to ensure that the prospective employee has the mathematics and reading skills to be a productive member of the team. Because stamps cost money, a person without good mathematics skills could create havoc for mail room accounting in only a few hours.

Uniforms are used by some mail operations to identify their employees. This is especially important on large campuses, where the mail service employees go into areas not generally open to the public. Other systems require their employee to wear name badges identifying them as university employees.

The schedule of the mail service will vary depending on the size of the campus and the volume of mail. Part-time employees can reduce the payroll costs, as they can be brought in during peak periods to sort or deliver the mail. Students represent a good source of available part-time help. If the volume is heavy, skeleton or part-time crews can be used to sort the mail during peak periods or on weekends. This is especially helpful during holidays, when the mail room will fill up with incoming mail.

NOTE

1. *Postal Operations Manual.* Washington, D.C.: U.S. Postal Services Headquarters Document Control Division, Issue 7, August 1, 1996. Paragraph 615.2.